AF581800

Quality of Life and Side Effects Management in Cancer Treatment

Quality of Life and Side Effects Management in Cancer Treatment

Editor

António Araújo

Basel • Beijing • Wuhan • Barcelona • Belgrade • Novi Sad • Cluj • Manchester

Editor
António Araújo
Universidade do Porto
Porto, Portugal

Editorial Office
MDPI
St. Alban-Anlage 66
4052 Basel, Switzerland

This is a reprint of articles from the Special Issue published online in the open access journal *Cancers* (ISSN 2072-6694) (available at: https://www.mdpi.com/journal/cancers/special_issues/Quality_Effects_Cancer_Treatment).

For citation purposes, cite each article independently as indicated on the article page online and as indicated below:

Lastname, A.A.; Lastname, B.B. Article Title. *Journal Name* **Year**, *Volume Number*, Page Range.

ISBN 978-3-0365-9528-3 (Hbk)
ISBN 978-3-0365-9529-0 (PDF)
doi.org/10.3390/books978-3-0365-9529-0

Contents

About the Editor

António Araújo

António Araújo is Head of the Department of Medical Oncology at Centro Hospitalar Universitário do Porto and Full Professor at the School of Medicine and Biomedical Sciences, ICBAS and Dental Medical Faculty, University of Porto.

Since January 2023, he has been Head of the Integrated Master Degree in Medicine at the School of Medicine and Biomedical Sciences, ICBAS, University of Porto. He has been a member of the Portuguese National Council for Oncology since 2013 and a member of the Scientific Committee of the Portuguese National Cancer Registry since 2019. He was a member of the International Association for the Study of Lung Cancer Ethics Committee from 2015 to 2019 and a member of its Educational Committee from 2019 to 2023. He was also a member of the Executive Board of the Portuguese Lung Cancer Study Group from 2006 to 2012. He is a founding member of REDICAP—Red IberoAmericana de Cancer de Pulmón—founded on 12th February 2021, which includes oncologic researchers from Spain, Colombia, Argentina, Peru, and Portugal. He was President of the Executive Board of the Northern Regional Council of the Portuguese Medical Association from 2017–2022. Professor Araújo has authored or contributed to more than 120 scientific papers and book chapters. He is a full member of ESMO, ASCO, IASLC, the Portuguese Society of Oncology, and the Portuguese Society of Pulmonology. In 2022, he was distinguished with the Municipal Medal of Merit—Gold Grade—by the city of Porto.

Preface

Cancer is one of the most important diseases in the world because it is the leading cause of death and a major cause of premature disability, retirement, or job loss.

Advanced cancer is becoming a chronic disease with longer overall survival due to the development of new treatments, such as target therapies and immunotherapy. With this accomplishment, the patient's quality of life, as well as the management of drug-related adverse events and their long-term complications, take on a new significance.

All these aspects contribute to the significance of this Special Issue, "Quality of Life and Side Effects Management in Cancer Treatment."

This Issue, addressed to all health professionals involved in cancer therapy and care, covers several aspects of the impact of drug-related adverse events, cancer patient care, and the relationship between quality of life and outcomes. We would like to thank all of the authors who published their research in this Issue, as well as all of the patients who participated in the papers published. We work every day to improve their chances of survival and quality of life.

António Araújo
Editor

Article

Impact of Immune-Related Adverse Events on Immune Checkpoint Inhibitors Treated Cancer Patients' Survival: Single Center Experience and Literature Review

Raquel Romão [1,*], Ana S. Mendes [1], Ridhi Ranchor [1], Maria João Ramos [1], João Coelho [1], Rita Carrilho Pichel [1], Sérgio Xavier Azevedo [1], Paula Fidalgo [1] and António Araújo [1,2]

1 Medical Oncology Department, Centro Hospitalar Universitário do Porto, 4099-001 Porto, Portugal
2 Oncology Research Unit, UMIB-Unit for Multidisciplinary Research in Biomedicine, ICBAS-School of Medicine and Biomedical Sciences, Universidade do Porto, 4050-346 Porto, Portugal
* Correspondence: r.raquelromao@gmail.com; Tel.: +351-22-207-7500

Simple Summary: The widespread use of immune checkpoint inhibitors (ICI) came along with a new challenge for oncologists, immune-related adverse events (irAE). A positive correlation between irAE onset and ICI efficacy has been suggested. However, it remains unsettled. Whether the association exists and if it is affected by cancer type or ethnicity needs further investigation. This study provides additional evidence to support this association by using a retrospective, single-center cohort design to analyze survival outcomes and the development of irAEs of 155 patients. Overall, the study offers new insights into the potential use of irAEs as biomarkers for response and survival in solid tumor patients receiving ICIs, and highlights the need for further research in this area.

Abstract: Immune-related adverse events have emerged as a new challenge and its correlation with survival remains unclear. The goal of our study was to investigate the effect of irAE on survival outcomes in solid tumor patients receiving ICI treatment. This was a retrospective, single-center study at a university hospital involving patients with malignancy who received immune checkpoint inhibitors. Chart review was performed on each patient, noting any irAE, including new events or worsening of previous autoimmune condition after starting treatment with ICI. A total of 155 patients were included, 118 (76.1%) were male, with median age of 64 years. Median follow up time was 36 months. Seventy patients (45.2%) had at least one irAE. Of all irAE, nine (8.1%) were classified as grade 3 or higher according to the CTCAE version 5.0. There was one death secondary to pneumonitis. Median ICI cycles until first irAE onset was 4 (range: 2–99). The objective response rate was higher for patients who developed irAE (18.7% vs. 9.0%; $p = 0.001$), as was median overall survival (18 months (95% CI, 8.67–27.32) vs. 10 (95% CI, 3.48–16.52) months; $p < 0.016$) and progression free survival (10 months (95% CI, 5.44–14.56) vs. 3 months (95% CI, 1.94–4.05); $p = 0.000$). The risk of death in patients with irAE was 33% lower when compared to patients without such events (hazard ratio (HR): 0.67; 95% CI, 0.46–0.99; $p = 0.043$). Development of irAE predicted better outcomes, including OS in patients with advanced solid tumors treated with ICI. Further prospective studies are needed to explore and validate this prognostic value.

Keywords: immune checkpoint inhibitors; immune-related adverse event; survival; prognosis

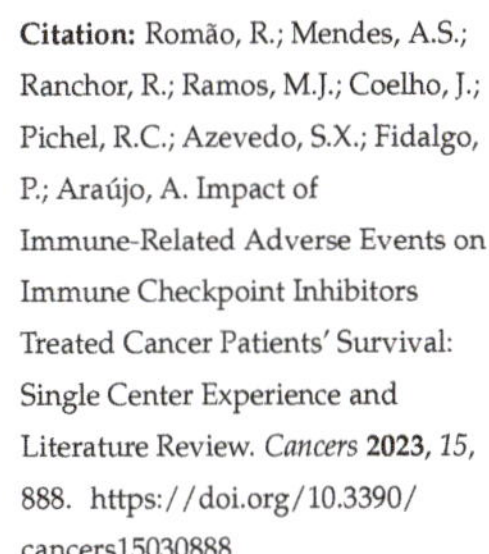

Citation: Romão, R.; Mendes, A.S.; Ranchor, R.; Ramos, M.J.; Coelho, J.; Pichel, R.C.; Azevedo, S.X.; Fidalgo, P.; Araújo, A. Impact of Immune-Related Adverse Events on Immune Checkpoint Inhibitors Treated Cancer Patients' Survival: Single Center Experience and Literature Review. *Cancers* **2023**, *15*, 888. https://doi.org/10.3390/cancers15030888

Academic Editors: Ann Richmond and Constantin N. Baxevanis

Received: 12 December 2022
Revised: 26 January 2023
Accepted: 28 January 2023
Published: 31 January 2023

1. Introduction

Over the last decade, immunotherapy has transformed cancer treatment, improving the prognosis of multiple types of malignancies. Among the different forms of immunotherapy, immune checkpoint inhibitors (ICI) showed remarkable benefits and durable clinical responses in certain patients [1].

Immune checkpoint inhibitors are immunomodulatory antibodies that target inhibitory T cell receptors, enhancing anti-tumor immune response [2]. The most distinguished ICI include cytotoxic T lymphocyte-associated protein 4 (CTLA-4) inhibitor (e.g., ipilimumab), anti-programmed cell death 1 (PD-1) (e.g., nivolumab and pembrolizumab), and anti-programmed cell death-ligand 1 (PD-L1) agents (e.g., atezolizumab, durvalumab and avelumab) [3]. This field is rapidly evolving, with new agents targeting other inhibitory T-cell, such as Lymphocyte-activation gene 3 (LAG-3), T-cell immunoglobulin, and mucin domain-containing 3 (TIM-3), or T-cell immune receptor with immunoglobulin and ITIM domain (TIGIT), being developed to further improve the effectiveness of cancer immunotherapy. These new agents aim to overcome the mechanisms of T-cell exhaustion, which can occur in cancer patients [4].

With the increasing use of these drugs in clinical practice, a new challenge has emerged, immune-related adverse events (irAE) [5]. As this drug class works by increasing host immune response, destabilizing immune homeostasis, and enhancing proinflammatory activity, it can induce inflammatory drug-related side effects. However, the pathophysiology underlying these events is not fully understood [6].

In the context of irAE, any organ or system can be affected. The most frequently described organs are the skin, gut, endocrine glands, liver, and lungs.The type of organ affected, seems to be related with the type and mechanism of ICI. For instance, colitis seems most frequent with anti-CTLA-4 and pneumonitis with anti-P-D1/PD-L1 antibodies. These toxicities are often reported as self-limited and easily manageable, though in a small portion they can be lethal [7]. The management of irAEs is not entirely defined, but there are some guidelines defining strategies on the basis of specific organ toxicity, to improve patient care and prevent life-threatening events [8,9].

Although ICI has been successful in improving response rates, it has been shown that only a small percentage of patients, approximately 20%, have benefited from treatment with CTLA-4 or PD-1 inhibitors [10,11].

The interest in predictive response biomarkers to ICI has grown substantially. PD-L1 expression has been found to show a positive association with response to anti-PD-1/PD-L1 antibodies, but it has also been acknowledged as a flawed indicator. Other biomarkers that have been found to predict response to ICI include the presence of high neoantigen loads and microsatellite instability (MSI) associated with mutations in mismatch repair (MMR) proteins. Additionally, studies have shown that intestinal microbial composition also plays a role in the therapeutic effects of ICI. These biomarkers have proven to be of great importance in predicting the efficacy of ICI, but more studies are needed to determine their full reliability and applicability in the clinical setting. Although ongoing efforts are being made to discover new biomarkers that can enhance the prediction of response to ICI, clinical biomarkers have received less attention in research [12].

Based on the ICI mechanism of action, it has been suggested that irAE onset may exemplify a clinical biomarker for ICI response. There was a suggestion of some association between irAE incidence and better clinical outcomes with ICI, but so far it remains unclear. Meanwhile, some publications have verified a positive correlation across different types of cancer between irAE development and longer survival [13,14]. Other investigations in the area have demonstrated unfauvorable results [15]. If the association exists, whether it will be affected by cancer type, organ-specific irAE, or ethnicity also needs to be explored.

The objective of this study was to evaluate the incidence and potential of irAE as biomarkers for response and survival in solid tumor patients receiving ICI in a palliative setting.

2. Materials and Methods

2.1. Patient Population

This was a retrospective, single-center cohort study of patients aged 18 years or older, with histologically or cytologically confirmed metastatic solid tumors who received at least one dose of ICI (Pembrolizumab, Nivolumab, Atezolizumab and Ipilimumab) as a single therapy, administered intravenously, at Centro Hospitalar Universitário do Porto from July 2012 to January 2020.

Patients with prior ICI therapy and patients with incomplete data were excluded.

2.2. Data Collection

Demographic data, such as patient's age, Eastern Cooperative Oncology Group (ECOG) performance status, diagnosis date, date of widespread disease confirmation, histology type, prior therapeutic regimens, number of doses of ICI, tumor response based on Response Evaluation Criteria in Solid Solid Tumors version 1.1, progression date, and death date or last follow-up visit, were all retrieved from the digital health records. Additionally, irAEs were registered and graded according to Common Terminology Classification Adverse Events (CTCAE) version 5.0, as well the date of the diagnosis and the respective prescribed treatment.

2.3. Statistical Analysis

Standard descriptive statistics were used to summarize all variables. The Fisher exact test was used to determine the association between categorical variables. Objective response rate (ORR) was defined as the percentage of patients who have a confirmed complete response (CR) or partial response (PR) to the treatment.

The overall survival (OS) was defined as the time period from the initial administration of ICI therapy until death. The progression-free survival (PFS) was defined as the time period from the initial administration of ICI therapy until the documentation of tumor progression or death, whichever occurred first. At the end of the follow-up period, patients who were still alive or had not shown progression of their tumor were considered censored in the analysis.

Survival curves were generated by using the Kaplan–Meier product limit method, and differences in OS where analyzed by stratifying for the ocorrence of irAE using the log-rank test. Univariate and multivariate Cox regression were used to identify factors with potential prognostic significance.

IBM SPSS Statistics 27.00.00 was applied for statistical analysis, and $p < 0.05$ was considered significant.

3. Results

3.1. Patient and Tumor Characteristics

The study included a total of 155 Caucasian patients who received ICI therapy. The median age of the patients was 64, and out of the total number of patients, 118 (76.1%) were male. Only two patients in the cohort had a previous history of autoimmune diseases (vitiligo and psoriasis). The study population was composed of patients with lung cancer (n = 76, 49%), melanoma (n = 28, 18.1%), renal cancer (n = 18, 11.6%), head and neck cancer (n = 17, 11%), bladder cancer (n = 11, 7.1%), and other types of cancer (n = 5, 3.2%). A total of 146 (94.2%) patients received anti-PD-1/PD-L1 drugs and only 9 (5.8%) patients received anti-CTLA-4. Out of the total number of patients, 80 (51.6%) were treated with pembrolizumab, 60 (38.7%) with nivolumab, nine (5.8%) with ipilimumab, and six (3.9%) with atezolizumab. The median duration of treatment was six months (range, 0–63 months).

At the initiation of ICI therapy, 101 (65.2%) patients had a performance status of ECOG 1, 38 (24.5%) had a performance status of ECOG 0, and 16 (10.3%) had a performance status of ECOG 2. ICI therapy was used as first-line treatment in 50 (32.3%) patients, second-line in 91 (58.7%), third-line in 11 (7.1%), and fifth or later line in 3 (1.9%) patients. The median progression-free survival (PFS) was 5 months (95% CI, 3.2–6.8) and the median overall survival (OS) was 15 months (95% CI, 11.23–18.77). A total of 43 (27.7%) patients had an objective response and 71 (45.8%) patients had disease control (complete/partial response or stable disease). The objective response rate by cancer type was 36% for lung cancer, 14% for melanoma, 20% for renal cancer, 13% for head and neck cancer, 83% for bladder cancer, and 40% for other cancers.

Patients' characteristics at baseline in the whole cohort and specified by development or not of irAE are summarized in Table 1.

Table 1. Patients and tumor characteristics at baseline in the whole cohort and specified by development or not of irAE.

Variables	Whole Cohort n (%)	irAE n (%)	non-irAE n (%)
Age (years)			
Median (range)	64 (21–86)	64 (36–86)	65 (21–85)
Sex			
Male	118 (76.1)	52 (74.3)	66 (77.6)
Female	37 (23.9)	18 (25.7)	19 (22.4)
Tumor type			
Lung, non-small cell	76 (49.0)	42(60)	34 (40.0)
Melanoma	28 (18.1)	16 (22.8)	12 (14.2)
Renal	18 (11.6)	7(10.0)	11 (12.9)
Head and neck	17 (11.0)	2 (2.9)	15 (17.6)
Bladder	11 (7.1)	3 (4.3)	8 (9.4)
Others	5 (3.2)	-	5 (6.0)
ECOG PS			
0	38 (24.5)	17 (24.3)	21 (24.7)
1	101 (65.2)	46 (65.7)	55 (64.7)
2	16 (10.3)	7 (10.0)	9 (10.6)
Treatment Line			
First	50 (32.3)	28 (40.0)	22 (25.9)
Second	91 (58.7)	37 (52.9)	54 (63.5)
Third	11 (7.1)	5 (7.1)	6 (7.1)
Fifth or later	3 (1.9)	-	3 (3.6)
Type of Immune Checkpoint inhibitor			
Anti-PD-1/PD-L1	146 (94.2)	66 (94.3)	80 (94.1)
Anti-CTLA-4	9 (5.8)	4 (5.7)	5 (5.9)

ECOG PS-Eastern Cooperative Oncology Group Performance Status; PD-1-programmed cell death 1; PD-L1-programmed cell death-ligand 1; CTLA-4-cytotoxic T lymphocyte-associated protein 4.

3.2. Immune-Related Adverse Events

In this study, 70 (45.2%) patients developed irAE, with 25 (16.1%) experiencing more than one event. The median number of ICI cycles before an irAE onset was four (range, 2–99). When examining irAE by organ system, 34 (35.4%) patients had dermatological events, 17 (17.7%) had rheumatologic events, 14 (14.7%) had endocrine events, 13 (13.5%) had neurological or musculoskeletal events, nine (9.4%) had gastrointestinal and hepatic and biliary events, eight (8.3%) had pulmonary events, and one (1.0%) had a renal event (Table 2).

The majority (92.9%) of irAE were grade 1 or 2. The only treatment-related death was caused by pneumonitis.

Most irAE were managed with supportive care, but 27 (38.6%) cases required oral steroids and five (7.1%) required intravenous steroids. In a few cases, treatment escalation with methotrexate or immunoglobulin was used for patients with severe arthritis or necrotizing inflammatory myositis.

ICI was suspended in all patients with irAE grade 3 or more. One patient with preexisting psoriasis experienced a flare of the disease, but ICI treatment was not suspended.

Table 2. Immune-related adverse events in the whole cohort.

Variables	n (%)
Treatment-related irAEs	
yes	70 (45.2)
no	85 (54.8)
Grade of irAE	
<3	91 (92.9)
≥3	9 (8.1)
Frequency of irAEs	
1	33(47.1)
2	26 (37.1)
3	11(15.71)
Type of irAE	
Dermatologic	34 (35.4)
Pruritus	18(18.7)
Rash	14(14.7)
Vitiligo	1(1.0)
Bullous pemphigoid	1(1.0)
Neurologic/Musculoskeletal	13 (13.5)
Myalgias	12 (12.5)
Immune-mediated necrotizing myopathy	1(1.0)
Endocrin	14(14.7)
Hypothyroidism	11 (11.5)
Hypertiroidism	3 (3.2)
Rheumatologic	17 (17.7)
Artralgias	16 (16.7)
Vasculitis	1(1.0)
Pulmonary	8(8.3)
Pneumonitis	8 (8.3)
Gastrointestinal and Hepatic and biliary	9 (9.4)
Diarrhea	2 (2.1)
Colitis	2 (2.1)
Hepatitis	3 (3.2)
Colangitis	1 (1.0)
Colestases	1 (1.0)
Renal	1 (1.0)
Nephritis	1 (1.0)
Treatment of irAE	
Supportive care	66 (94.3)
Oral Corticosteroid	27 (38.6)
Intravenous Corticosteroid	5 (7.1)
Other Immunosuppressor (Methotrexate)	1(1.4)

3.3. Response Rate and Survival Analysis

It was found that the ORR was higher for patients who developed irAE (18.7% vs. 9.0%; $p = 0.001$).

The Kaplan–Meier estimates for overall survival and progression-free survival for patients with irAE were compared to those without irAE, as shown in Figure 1. The results were significant when compared using the log-rank test. The median overall survival for patients with irAE was 18 months (95% CI, 8.67–27.32) compared to 10 months (95% CI, 3.48–16.52) for patients without irAE, with a *p*-value of <0.016. Similarly, the median progression-free survival was 10 months (95% CI, 5.44–14.56) for patients with irAE and three months (95% CI, 1.94–4.05) for patients without irAE, with a *p*-value of 0.000.

(a)

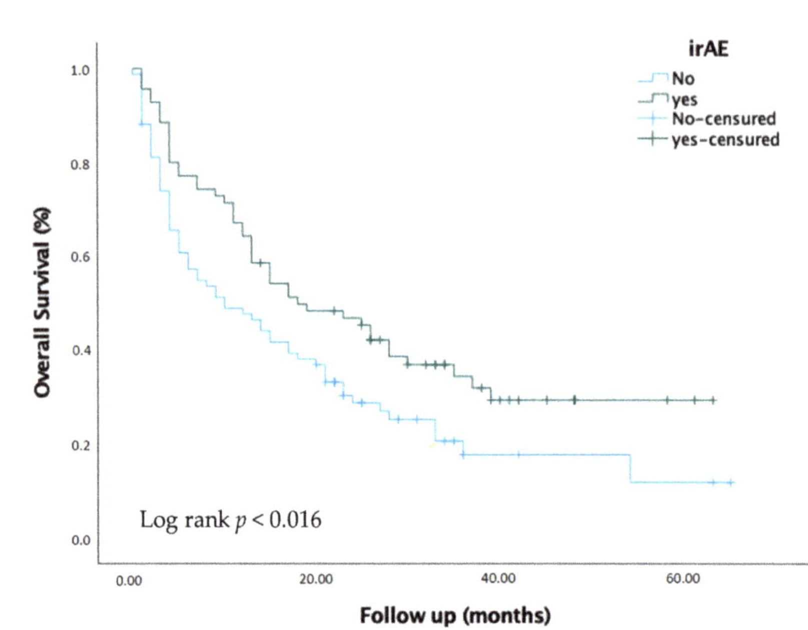

(b)

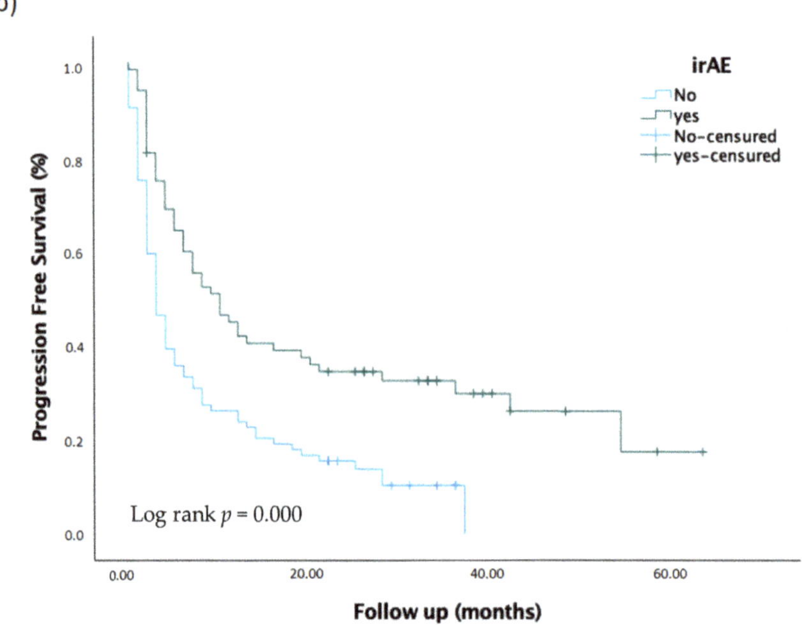

Figure 1. Kaplan–Meier analysis of overall survival (**a**) and progression free survival (**b**) comparing patients irAE and non-irAE.

In univariate analysis, the benefit of irAE persisted. The risk of death in patients with irAE was 33% lower compared to patients without such events (HR: 0.67; 95% CI,

0.46–0.99; $p = 0.043$). A Cox proportional hazards analysis was conducted to identify factors associated with increased mortality, as shown in Table 3.

Table 3. Univariate and Multivariate cox proportional hazards regression analysis of the risk of death.

Predictable Variables	HR Crude (CI 95%)	p Value	HR Adjusted (CI 95%)	p Value
irAE Yes vs. No	0.67 (0.46–0.99)	0.043	0.65 (0.44–0.96)	0.03
Sex Male vs. female	0.73 (0.48–1.12)	0.152		
ECOG PS 1 vs. 0	1.84 (1.12–3.03)	0.017	1.81 (1.10–2.88)	0.020
2 vs. 0	3.50 (1.72–7.11)	0.001	3.73 (1.83–7.62)	0.000
Age <65 vs. ≥65	0.95 (0.65–1.30)	0.771		
Treatment Line 1 vs. ≥2	1.29 (0.85–1.98)	0.230		
Type of tumor NSCLC vs. other	0.75 (0.52–1.11)	0.151		
Grade of toxicity <3 vs. ≥3	0.46 (0.19–1.14)	0.094		

ECOG PS-Eastern Cooperative Oncology Group Performance Status; NSCLC-Non-Small Cell Lung Cancer.

In multivariate analysis, irAE experience and ECOG were included as they were related to the outcome with statistical significance ($p < 0.05$) in univariate analysis. When adjusted for ECOG, irAE experience persisted as an independent prognostic factor associated with better overall survival (HR: 0.65; 95% CI, 0.44 to 0.96; $p = 0.03$).

4. Discussion

Despite the remarkable advantages of ICI witnessed over the last years, significant morbidity due to irAE can be a limiting factor for its widespread use and patients' quality of life. Immune toxicity is unpredictable, diverse, and on several occasions disabling or life-threatening [16]. Overall, in this study, the treatment with ICI was well tolerated. In our sample, 45.2% of patients had an irAE of any grade, with fatal events in less than 1%, which is consistent with the published data [17–21]. The only fatal event was secondary to pneumonitis, which is in concordance with the results presented in a recent network meta-analysis where pneumonitis was also the most common cause of irAE grade 5 [22].

The most frequent adverse events by organ systems are in line with other real-world evidence. Yet, the frequency of neurological events could be considered higher than what was reported before [23,24]. This can be justified by heterogeneity and lack of standardization in irAE documentation/registration. While there is a consensus between the type of irAE and ICI class, as colitis and hypophysitis are more common with anti-CTLA4 and pneumonitis and thyroiditis with anti-PD-(L)1 therapy, this analysis was not performed in our study due to the low percentage of patients treated with anti-CTLA4 [25].

It is uncertain whether ICI toxicities are distinct from standard autoimmune diseases or if the manifestation of irAE is associated with treatment efficacy. However, the potential association between tumor response to ICI and increased risk of developing irAE has been hypothesized since it is believed that irAEs could represent to some level the intensification of the immune response, including anti-tumor immune activity. While an early analysis of patients with melanoma treated with ipilimumab did not show a relationship between PFS and irAE, the following studies have associated a higher response rate with irAE incidence [26,27]. Data regarding outcomes within a single disease entity and irAE are more limited but have been shown in melanoma, NSCLC, and more recently in urothelial and hepatocellular carcinomas [28–31].

In our study, we observed a strong correlation between the development of irAE and patients' response rates as well as PFS and OS. These results add to the robust growing evidence suggesting better oncological outcomes in patients who develop irAE. Although most patients included had a diagnosis of NSCLC, there was no correlation between tumor type and the outcomes.

Therefore, a positive relation between irAE incidence and better clinical outcomes in cancer patients is almost certain, although not completely validated. There are some questions to be further explored, such as the role of ethnicity. Ethnical discrepancies in the overall survival of cancer patients have been commonly observed [32]. Furthermore, it is known that genetic polymorphisms of PD-1 and CTLA-4 are associated with various autoimmune diseases, such as thyroiditis, diabetes mellitus, and rheumatoid arthritis [33]. As comprehensive registration trials might fail to identify racial side effect profile disparities, real-world data could help to overcome this issue. To the best of our knowledge, this is the first study in Portuguese patients and, as enlightened above, the ICI toxicity profile is in line with other populations. However, since it is believed that the Portuguese population has an important incidence of autoimmune diseases, a superior frequency and/or higher grade of irAE could be expected, as reported in a Finnish study [34,35]. Hence, the pathophysiology and genetic background underlying irAE need additional investigation. Recent studies have identified certain genetic variants that may be associated with increased risk for irAEs, as well as variants that may predict a better response to immunotherapy. However, more research is needed to fully understand the role of germline factors in immunotherapy treatment and to develop personalized treatment strategies for patients [36].

This study had some limitations. First, it was retrospective, leading to information bias. Secondly, population heterogeneity was wide. Thus, the weighted average outcomes should be interpreted accordingly. Third, data were analyzed using a log-rank test and a standard Cox model, which introduces a bias owing to the different follow-up times and treatment exposures between patients who did and did not develop irAE. As such, a time-dependent analysis could be an alternative to minimize that bias in future analysis. Fourth, our study is skewed towards patients treated for advanced or metastatic NSCLC and melanoma, which could be explained by the earlier approval of ICI use in these settings. Lastly, a vast majority of the patients were treated predominantly with anti-PD1 drugs, with very few patients receiving anti-PD-L1 or anti-CTLA-4 agents. Thus, it would be difficult to generalize our findings to patients that have received ICI other than anti-PD-1.

5. Conclusions

The development of irAE predicted better outcomes including OS in Portuguese patients with advanced solid tumors treated with ICI. Further prospective studies are needed to explore and validate this prognostic value. The long-term impact of immune checkpoint blockade on quality of life, the detrimental effect of steroid administration on anticancer efficacy, and the biological underlying mechanisms of irAE also need more investigation.

Author Contributions: Conceptualization, R.R. (Raquel Romão), P.F. and A.A.; methodology, R.R. (Raquel Romão), P.F. and S.X.A.; software, R.R. (Raquel Romão); validation, R.R. (Raquel Romão), P.F., and S.X.A.; formal analysis, R.R. (Raquel Romão); investigation, R.R. (Raquel Romão); resources, R.R. (Raquel Romão), A.S.M., R.R. (Ridhi Ranchor), M.J.R., J.C. and R.C.P.; data curation, R.R. (Raquel Romão), A.S.M., R.R. (Ridhi Ranchor), M.J.R., J.C., and R.C.P.; writing—original draft preparation, R.R. (Raquel Romão); writing—review and editing, P.F., S.X.A. and A.A.; visualization, R.R. (Raquel Romão) and A.A.; supervision, A.A.; project administration, R.R. (Raquel Romão). All authors have read and agreed to the published version of the manuscript.

Funding: This research received no external funding.

Institutional Review Board Statement: The study was conducted in accordance with the Declaration of Helsinki and approved by Ethics Committee of the Centro Hospitalar Universitário do Porto (047-DEFI/049-CE), date of approval 7 July 2021.

Informed Consent Statement: Patient consent was waived due to the retrospective and non-interventional nature of the study.

Data Availability Statement: The datasets generated and/or analyzed during the current study are not publicly available. However, they are available from the corresponding author upon reasonable request.

Conflicts of Interest: The authors declare no conflict of interest.

References

1. Ribas, A.; Wolchok, J.D. Cancer immunotherapy using checkpoint blockade. *Science* **2018**, *359*, 1350–1355. [CrossRef] [PubMed]
2. Fritz, J.M.; Lenardo, M.J. Development of immune checkpoint therapy for cancer. *J. Exp. Med.* **2019**, *216*, 1244–1254. [CrossRef]
3. Hargadon, K.M.; Johnson, C.E.; Williams, C.J. Williams. Immune checkpoint blockade therapy for cancer: An overview of FDA-approved immune checkpoint inhibitors. *Int. Immunopharmacol.* **2018**, *62*, 29–39. [CrossRef] [PubMed]
4. Anderson, A.C.; Joller, N.; Kuchroo, V.K.; Hospital, W.; Immunology, E. functions in immune regulation. *Immunity* **2017**, *44*, 989–1004. [CrossRef]
5. Postow, M.A.; Sidlow, R.; Hellmann, M.D. Hellmann. Immune-related adverse events associated with immune checkpoint blockade. *N. Engl. J. Med.* **2018**, *378*, 158–168. [CrossRef] [PubMed]
6. Martins, F.; Sofiya, L.; Sykiotis, G.P.; Lamine, F.; Maillard, M.; Fraga, M.; Shabafrouz, K.; Ribi, C.; Cairoli, A.; Guex-Crosier, Y.; et al. Adverse effects of immune-checkpoint inhibitors: Epidemiology, management and surveillance. *Nat. Rev. Clin. Oncol.* **2019**, *16*, 563–580. [CrossRef]
7. Gangadhar, T.C.; Vonderheide, R.H. Mitigating the toxic effects of anticancer immunotherapy. *Nat. Rev. Clin. Oncol.* **2014**, *11*, 91–99. [CrossRef]
8. Brahmer, J.R.; Lacchetti, C.; Schneider, B.J.; Atkins, M.B.; Brassil, K.J.; Caterino, J.M.; Chau, I.; Ernstoff, M.S.; Gardner, J.M.; Ginex, P.; et al. Management of immune-related adverse events in patients treated with immune checkpoint inhibitor therapy: American society of clinical oncology clinical practice guideline. *J. Clin. Oncol.* **2018**, *36*, 1714–1768. [CrossRef]
9. Haanen, J.; Carbonnel, F.; Robert, C.; Kerr, K.; Peters, S.; Larkin, J.; Jordan, K. Management of toxicities from immunotherapy: ESMO clinical practice guidelines for diagnosis, treatment and follow-up. *Ann. Oncol.* **2018**, *29*, iv264–iv266. [CrossRef]
10. Hodi, F.S.; O'Day, S.J.; McDermott, D.F.; Weber, R.W.; Sosman, J.A.; Haanen, J.B.; Gonzalez, R.; Robert, C.; Schadendorf, D.; Hassel, J.C.; et al. Improved survival with ipilimumab in patients with metastatic melanoma. *N. Engl. J. Med.* **2010**, *363*, 711–723. [CrossRef]
11. Gettinger, S.N.; Horn, L.; Gandhi, L.; Spigel, D.R.; Antonia, S.J.; Rizvi, N.A.; Powderly, J.D.; Heist, R.S.; Carvajal, R.D.; Jackman, D.M.; et al. Overall Survival and Long-Term Safety of Nivolumab (Anti–Programmed Death 1 Antibody, BMS-936558, ONO-4538) in Patients With Previously Treated Advanced Non–Small-Cell Lung Cancer. *J. Clin. Oncol.* **2015**, *33*, 2004–2012. [CrossRef] [PubMed]
12. Park, Y.-J.; Kuen, D.-S.; Chung, Y. Future prospects of immune checkpoint blockade in cancer: From response prediction to overcoming resistance. *Exp. Mol. Med.* **2018**, *50*, 1–13. [CrossRef] [PubMed]
13. Horvat, T.; Adel, N.G.; Dang, T.-O.; Momtaz, P.; Postow, M.A.; Callahan, M.K.; Carvajal, R.D.; Dickson, M.A.; D'Angelo, S.P.; Woo, K.M.; et al. Immune-related adverse events, need for systemic immunosuppression, and effects on survival and time to treatment failure in patients with melanoma treated with ipilimumab at Memorial Sloan Kettering Cancer Center. *J. Clin. Oncol.* **2015**, *33*, 3193–3198. [CrossRef]
14. Petrelli, F.; Grizzi, G.; Ghidini, M.; Ghidini, A.; Ratti, M.; Panni, S.; Cabiddu, M.; Ghilardi, M.; Borgonovo, K.; Parati, M.C.; et al. Immune-related Adverse Events and Survival in Solid Tumors Treated with Immune Checkpoint Inhibitors: A Systematic Review and Meta-Analysis. *J. Immunother.* **2020**, *43*, 1–7. [CrossRef]
15. Suresh, K.; Naidoo, J. Lower Survival in Patients Who Develop Pneumonitis Following Immunotherapy for Lung Cancer. *Clin. Lung Cancer* **2020**, *21*, e169–e170. [CrossRef]
16. Weber, J.S.; Yang, J.C.; Atkins, M.B.; Disis, M.L. Toxicities of immunotherapy for the practitioner. *J. Clin. Oncol.* **2015**, *33*, 2092–2099. [CrossRef] [PubMed]
17. Postow, M.A.; Chesney, J.; Pavlick, A.C.; Robert, C.; Grossmann, K.; McDermott, D.; Linette, G.P.; Meyer, N.; Giguere, J.K.; Agarwala, S.S.; et al. Nivolumab and Ipilimumab versus Ipilimumab in Untreated Melanoma. *N. Engl. J. Med.* **2015**, *372*, 2006–2017. [CrossRef] [PubMed]
18. Weber, J.S.; Hodi, F.S.; Wolchok, J.D.; Topalian, S.L.; Schadendorf, D.; Larkin, J.; Sznol, M.; Long, G.; Li, H.; Waxman, I.M.; et al. Safety profile of nivolumab monotherapy: A pooled analysis of patients with advanced melanoma. *J. Clin. Oncol.* **2017**, *35*, 785–792. [CrossRef]
19. Conroy, M.; Naidoo, J. Immune-related adverse events and the balancing act of immunotherapy. *Nat. Commun.* **2022**, *13*, 392. [CrossRef]
20. Fan, Y.; Xie, W.; Huang, H.; Wang, Y.; Li, G.; Geng, Y.; Hao, Y.; Zhang, Z. Association of Immune Related Adverse Events With Efficacy of Immune Checkpoint Inhibitors and Overall Survival in Cancers: A Systemic Review and Meta-analysis. *Front. Oncol.* **2021**, *11*, 633032. [CrossRef]
21. Wang, D.Y.; Salem, J.-E.; Cohen, J.V.; Chandra, S.; Menzer, C.; Ye, F.; Zhao, S.; Das, S.; Beckermann, K.E.; Ha, L.; et al. Fatal Toxic Effects Associated With Immune Checkpoint Inhibitors. *JAMA Oncol.* **2018**, *4*, 1721–1728. [CrossRef] [PubMed]

22. Liu, T.; Jin, B.; Chen, J.; Wang, H.; Lin, S.; Dang, J.; Li, G. Comparative risk of serious and fatal treatment-related adverse events caused by 19 immune checkpoint inhibitors used in cancer treatment: A network meta-analysis. *Ther. Adv. Med Oncol.* **2020**, *12*, 1758835920940927. [CrossRef] [PubMed]
23. Kaur, A.; Doberstein, T.; Amberker, R.R.; Garje, R.; Field, E.H.; Singh, N. Immune-related adverse events in cancer patients treated with immune checkpoint inhibitors: A single-center expe-rience. *Medicine* **2019**, *98*, e17348. [CrossRef]
24. Möhn, N.; Beutel, G.; Gutzmer, R.; Ivanyi, P.; Satzger, I.; Skripuletz, T. Neurological Immune Related Adverse Events Associated with Nivolumab, Ipilimumab, and Pembrolizumab Therapy—Review of the Literature and Future Outlook. *J. Clin. Med.* **2019**, *8*, 1777. [CrossRef]
25. Xu, C.; Chen, Y.-P.; Du, X.-J.; Liu, J.-Q.; Huang, C.-L.; Chen, L.; Zhou, G.-Q.; Li, W.-F.; Mao, Y.-P.; Hsu, C.; et al. Comparative safety of immune checkpoint inhibitors in cancer: Systematic review and network meta-analysis. *BMJ* **2018**, *363*, k4226. [CrossRef]
26. Di Giacomo, A.M.; Grimaldi, A.M.; Ascierto, P.A.; Queirolo, P.; Del Vecchio, M.; Ridolfi, R.; De Rosa, F.; De Galitiis, F.; Testori, A.; Cognetti, F.; et al. Correlation between efficacy and toxicity in pts with pretreated advanced melanoma treated within the Italian cohort of the ipilimumab expanded access programme (EAP). *J. Clin. Oncol.* **2013**, *31*, 9065. [CrossRef]
27. Das, S.; Johnson, D.B. Immune-related adverse events and anti-tumor efficacy of immune checkpoint inhibitors. *J. Immunother. Cancer* **2019**, *7*, 306. [CrossRef]
28. Eggermont, A.M.M.; Kicinski, M.; Blank, C.U.; Mandala, M.; Long, G.; Atkinson, V.; Dalle, S.; Haydon, A.; Khattak, A.; Carlino, M.S.; et al. Association Between Immune-Related Adverse Events and Recurrence-Free Survival Among Patients With Stage III Melanoma Randomized to Receive Pembrolizumab or Placebo A Secondary Analysis of a Randomized Clinical Trial Supplemental content. *JAMA Oncol.* **2020**, *6*, 519–527. [CrossRef]
29. Maher, V.E.; Fernandes, L.L.; Weinstock, C.; Tang, S.; Agarwal, S.; Brave, M.; Ning, Y.-M.; Singh, H.; Suzman, D.; Xu, J.; et al. Analysis of the Association Between Adverse Events and Outcome in Patients Receiving a Programmed Death Protein 1 or Programmed Death Ligand 1 Antibody. *J. Clin. Oncol.* **2019**, *37*, 2730–2737. [CrossRef]
30. Xu, S.; Lai, R.; Zhao, Q.; Zhao, P.; Zhao, R.; Guo, Z. Correlation Between Immune-Related Adverse Events and Prognosis in Hepatocellular Carcinoma Patients Treated With Immune Checkpoint Inhibitors. *Front. Immunol.* **2021**, *12*, 5204. [CrossRef]
31. Haratani, K.; Hayashi, H.; Chiba, Y.; Kudo, K.; Yonesaka, K.; Kato, R.; Kaneda, H.; Hasegawa, Y.; Tanaka, K.; Takeda, M.; et al. Association of Immune-Related Adverse Events With Nivolumab Efficacy in Non–Small-Cell Lung Cancer. *JAMA Oncol.* **2018**, *4*, 374–378. [CrossRef] [PubMed]
32. Soneji, S.; Tanner, N.T.; Silvestri, G.A.; Lathan, C.S.; Black, W. Racial and Ethnic Disparities in Early-Stage Lung Cancer Survival. *Chest* **2017**, *152*, 587–597. [CrossRef] [PubMed]
33. Khan, S.; Gerber, D.E. Autoimmunity, checkpoint inhibitor therapy and immune-related adverse events: A review. *Semin. Cancer Biol.* **2020**, *64*, 93–101. [CrossRef] [PubMed]
34. Santos, E. Most common autoimmune diseases in relatives of portuguese patients with systemic lupus erythematosus. *RFML* **2007**, *12*, 181–186.
35. Kuusisalo, S.; Koivunen, J.P.; Iivanainen, S. Association of Rare Immune-Related Adverse Events to Survival in Advanced Cancer Patients Treated with Immune Checkpoint Inhibitors: A Real-World Single-Center Cohort Study. *Cancers* **2022**, *14*, 2276. [CrossRef] [PubMed]
36. Chin, I.S.; Khan, A.; Olsson-Brown, A.; Papa, S.; Middleton, G.; Palles, C. Germline genetic variation and predicting immune checkpoint inhibitor induced toxicity. *NPJ Genom. Med.* **2022**, *7*, 73. [CrossRef] [PubMed]

Review

Best Supportive Care of the Patient with Oesophageal Cancer

Rita Carrilho Pichel [1,*], Alexandra Araújo [1], Vital Da Silva Domingues [2,3], Jorge Nunes Santos [3,4], Elga Freire [2,3,5], Ana Sofia Mendes [1], Raquel Romão [1] and António Araújo [1,3,6]

1 Department of Medical Oncology, Centro Hospitalar Universitário do Porto (CHUPorto), 4099-001 Porto, Portugal
2 Department of Internal Medicine, Centro Hospitalar Universitário do Porto (CHUPorto), 4099-001 Porto, Portugal
3 Instituto de Ciências Biomédicas Abel Salazar (ICBAS), Universidade do Porto (UP), 4050-346 Porto, Portugal
4 Department of General Surgery, Centro Hospitalar Universitário do Porto (CHUPorto), 4099-001 Porto, Portugal
5 Equipa Intra-hospitalar de Suporte em Cuidados Paliativos, Centro Hospitalar Universitário do Porto (CHUPorto), 4099-001 Porto, Portugal
6 Unidade Multidisciplinar de Investigação Biomédica (UMIB), 4050-346 Porto, Portugal
* Correspondence: rita.pichel@chporto.min-saude.pt

Simple Summary: Oesophageal cancer is the sixth leading culprit of cancer-related mortality, and the majority of the patients with advanced disease are treated with best supportive care intent. Several management alternatives have been developed in recent years to address palliation of oesophageal cancer, such as self-expandable metallic stent placement and hypofractionated radiotherapy. Yet, optimal management is not standardized, and best supportive care decisions should be discussed on a case-by-case basis by multidisciplinary teams. This evidence-based review aimed for defining recommendations on the management of oesophageal cancer main symptoms and complications, such as dysphagia, malnutrition, pain, nausea and vomiting, fistula and bleeding. The late goal of our review is to improve (toward the "best") supportive care and decision making for oesophageal cancer patients.

Abstract: Background: Oesophageal cancer patients have poor survival, and most are unfit for curative or systemic palliative treatment. This article aims to review the best supportive care for oesophageal cancer, focusing on the management of its most frequent or distinctive symptoms and complications. Methods: Evidence-based review on palliative supportive care of oesophageal cancer, based on Pubmed search for relevant clinical practice guidelines, reviews and original articles, with additional records collected from related articles suggestions, references and societies recommendations. Results: We identified 1075 records, from which we screened 138 records that were related to oesophageal cancer supportive care, complemented with 48 additional records, finally including 60 records. This review summarizes the management of oesophageal cancer-related main problems, including dysphagia, malnutrition, pain, nausea and vomiting, fistula and bleeding. In recent years, several treatments have been developed, while optimal management is not yet standardized. Conclusion: This review contributes toward improving supportive care and decision making for oesophageal cancer patients, presenting updated summary recommendations for each of their main symptoms. A robust body of evidence is still lacking, and the best supportive care decisions should be individualized and shared.

Keywords: palliative supportive care; oesophageal cancer; dysphagia; malnutrition; oesophageal fistula

Citation: Pichel, R.C.; Araújo, A.; Domingues, V.D.S.; Santos, J.N.; Freire, E.; Mendes, A.S.; Romão, R.; Araújo, A. Best Supportive Care of the Patient with Oesophageal Cancer. *Cancers* **2022**, *14*, 6268. https://doi.org/10.3390/cancers14246268

Received: 17 October 2022
Accepted: 16 December 2022
Published: 19 December 2022

1. Introduction

Worldwide, oesophageal cancer is the eighth-most common cancer and the sixth leading cause of cancer-related mortality [1,2]. In Portugal and in Europe, it is far less incident, being the twentieth-most common cancer and the thirteenth leading cause of cancer-related mortality [1,3]. It remains, however, a highly fatal disease, with a 5-year overall survival rate of 20% for all stages, and 5% if considered metastatic disease alone [4].

From 1990 to 2017, there was an important decrease in age-standardized incidence (22.0%), mortality (29.0%), and disability-adjusted life years (33.4%) globally [2]. Still, the prognosis of oesophageal cancer remains poor. Oesophageal cancer is usually diagnosed late in the course of the disease and patients often have other relevant comorbidities, rendering them unfit for curative modalities of treatment [5].

Current guidelines on oesophageal cancer still suggest that, if there are distant metastases, or if the tumour is unresectable/invades adjacent structures/is too extensive and cannot be treated with surgery or curative-intent chemoradiotherapy, the patient should be treated with palliative intent, aimed at relieving cancer-related symptoms, improving quality of life (QOL) and prolonging survival. A poor performance status or relevant comorbidities preclude systemic treatment, so that only one fifth of incurable gastric or oesophageal cancer patients receive palliative chemotherapy (ChT) [6–8]. The advent of immunotherapy and targeted therapies applied to digestive tract tumours may allow us to treat more patients with advanced oesophageal cancer. Yet, for the majority of them, the most appropriate option remains the best supportive care.

Therefore, and trying to address the lack of therapeutic guidance [9], and our own clinical practice challenges, we aim to review the best palliative supportive care of oesophageal cancer, focusing on the management of its most frequent or distinctive symptoms and complications.

2. Methods

Authors performed an evidence-based review on palliative supportive care of oesophageal cancer, searching for relevant clinical practice guidelines, systematic reviews and the latest original articles. Our main search strategy included searching MEDLINE via PubMed, with the following search headings: ((supportive care) OR (palliative)) AND ((oesophag *) OR (esophag *)) AND ((cancer) OR (carcinoma)). Pubmed search was last undertaken on 25 August 2022. We also performed a secondary search, via PubMed, Cochrane and societies' sites, searching for "per symptom" epidemiology and treatment. We have also included articles collected from related articles suggestions and references sections. Selection flowchart is represented on Figure 1.

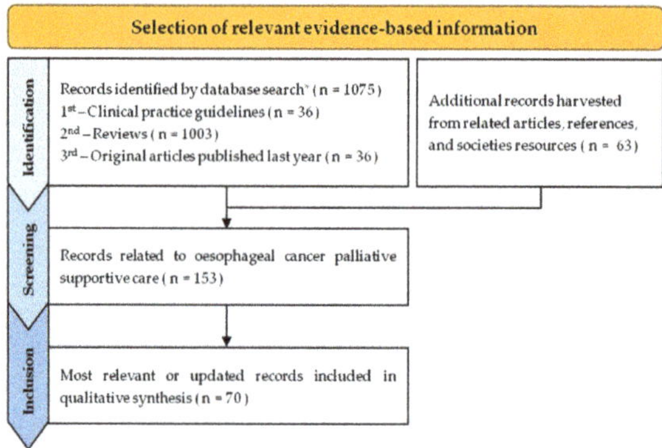

Figure 1. Flowchart of selection of relevant evidence-based information on palliative supportive care of oesophageal cancer. * PubMed search heading: ((supportive care) OR (palliative)) AND ((oesophag *) OR (esophag *)) AND ((cancer) OR (carcinoma)), last executed on 25 August 2022.

3. Results and Discussion

Main database search returned 1075 records, complemented with 63 additional records from "per symptom" searching, related articles, references and societies resources. From these, we selected 153 records that were related to oesophageal cancer supportive care, from which we included 70 in the qualitative synthesis.

3.1. Management of Oesophageal Cancer Symptoms or Complications

Mostly from empirical knowledge, we may say that dysphagia, weight loss and malnutrition, chest pain, nausea and vomiting, digestive bleeding and fistulisation to cardiopulmonary system are some of the main symptoms or complications related to oesophageal cancer, which we addressed in the following sections. Table 1 presents a summary of recommendations as explained on the text.

Table 1. Summary of recommendations.

	Dysphagia	Malnutrition	Pain	Nausea and Vomiting	Fistula	Bleeding and Anaemia
Do	Hypofractioned EBRT Brachytherapy Covered SEMS placement Pain control Dietary changes Adequate drug formulations (ex: liquid, powder, td, transmucosal, sc)	Malnutrition screening Early nutrition intervention Dysphagia treatment As-needed enteral tube feeding and parenteral nutrition (if expected survival of months) Assessment of end-of-life concerns on nutrition and hydration	(Re-)Assessment of the pain cause and its characteristics Stepwise approach Paracetamol [a] NSAID [a,b] Opioids (preferred oral morphine, or td fentanyl; or iv route for rapid pain control) Neuropathic-pain agents Corticosteroids Palliative antalgic RT Acupuncture and acupressure	Metoclopramide Sedative antiemetic drugs (ex: haloperidol, olanzapine) 5-HT3 agonists Anxiolytics Treat mechanical obstruction, if feasible Dexametasone, as part of the treatment of brain metastases or malignant bowel obstruction	Immediate treatment SEMS placement	Hemospray Palliative haemostatic RT Advanced care planning for the scenario of massive bleeding Assessment and treatment of causes of anaemia Red blood cell transfusion for symptomatic anaemia
Don't	Rigid plastic tube insertion Dilatation Laser ablation Photodynamic therapy	End-of-life invasive clinical assisted nutrition	Forget about side effects		Surgery in the palliative setting	RT for massive bleeding (due to great vessels fistula) Erythropoiesis stimulating agents
Don't know	RT and SEMS combination			Add dexametasone [c]		Argon plasma coagulation Arterial embolization

EBRT: external bean radiotherapy; NSAID: Non-steroidal anti-inflammatory drugs; RT: radiotherapy; SEMS: self-expandable metal stent; sc: subcutaneous; td: transdermal. [a] First step of the WHO analgesic ladder and widely used, despite no proven effectiveness for cancer pain. [b] Assess clinical benefit within two weeks. [c] Corticosteroids have no proven antiemetic effect in advance cancer, but must be used in case of brain metastases-related oedema or malignant bowel obstruction. Based on Lefroy J et al. "Guidelines: the do's, don'ts and don't knows of feedback for clinical education".

3.1.1. Dysphagia or Obstruction

Malignant dysphagia is the predominant symptom in more than 70% of patients with oesophageal cancer [10]. Usually, it arises from oesophageal lumen obstruction, but it can also be caused by tumour related dysmotility.

Regarding dysphagia palliation, there is no standard of care. Choosing the best treatment modality for each patient must rely on a multidisciplinary discussion, considering the symptoms burden and vital prognosis. In general, the first option for immediate relief of dysphagia is the insertion of a (fully or partially)-covered self-expandable metallic stent (SEMS). Palliative radiotherapy (RT) is another common option, more appropriate for patients with a longer life-expectancy, greater than three months [11–13].

SEMS placement is more effective and quicker in palliating dysphagia compared to other endoscopic procedures, which is of particular interest in cases of severe dysphagia and short life expectancy [14]. Complications from oesophageal stents are not negligible [15]. Different stents have been developed and were compared elsewhere [16].

Palliative RT, when compared to stent placement, has been shown to improve survival and QOL, and is being increasingly adopted [14,17]. Additionally, it correlates with significantly lower risk of toxicity, better pain control and equivalent relief of moderate to severe dysphagia [18]. Although intraluminal brachytherapy proved to be effective [14], it is more invasive and less widely implemented than external beam RT [9,19,20], with comparable efficacy [21]. Furthermore, patient-reported outcomes seem to favour external beam RT [22]. Short course palliative RT (20Gy in 5 fractions in consecutive workdays) was better tolerated with equal palliative effects than longer course [23].

Although it was theoretically promising, a recent trial failed to show additional benefit from adding palliative RT to SEMS insertion [24].

Laser ablation and photodynamic therapy require greater expertise than SEMS placement and more often need reintervention. Rigid plastic tube insertion and dilatation, either alone or in combination with other modalities, are no longer recommended due to high rate of delayed complications and recurrent dysphagia [14].

Complementarily, it is important to control pain, such as odynophagia, and also to educate patients for dietary changes, such as smaller and more frequent meals, softer food, liquid diets and oral nutritional supplements.

3.1.2. Malnutrition

Malnutrition (as synonym of undernutrition) and caquexia (as synonym of chronic disease-related malnutrition with inflammation) are common in advanced cancer disease and lead to poor QOL and performance status and to decreased survival [25–28]. Cancer caquexia is defined by either weight loss >5% alone, or weight loss >2% if body mass index is <20 kg/m^2 or fat free mass is reduced [29].

Patients with cancer of the upper digestive tract are prone to malnutrition, often present at diagnosis [26]. Oesophageal cancer raises specific nutritional concerns, as the disease predisposes to dysphagia, cancer-related caquexia and invasive therapies.

More recently, studies and best-practice guidelines increasingly advocate for early identification of malnutrition and implementation of nutrition interventions, aimed at maintaining or improving QOL [30–33]. These may also increase survival [34]. However, when life expectancy is shortened to few months or weeks, the intensity of such nutritional interventions should be decreased and focus should be directed towards immediate symptomatic relief and patient comfort (addressing thirst, eating-related distress, favourite foodstuffs and flavours) while dealing with family or carer concerns on end-of-life nutrition and hydration [33,35].

Since it is largely underdiagnosed and undertreated, routine nutritional status should be assessed with appropriate screening tools. The Patient-Generated Subjective Global Assessment (PGSGA), the Subjective Global Assessment (SGA), and Nutrition Risk Index (NRI) are examples of malnutrition screening tools validated for cancer patients with adequate specificity and sensitivity [31].

Oesophageal obstruction treatment as described above should be attempted, since even little oral feeding can achieve better QOL compared to enteral feeding. Nutritional counselling and support remains a cornerstone of successful management of patients with incurable oesophageal cancer, maintaining weight and performance status, even in the elderly. Clinical assisted nutrition, such as enteral feeding and parenteral nutrition should be based on a careful evaluation of the potential benefits, as QOL and functional expectant gain, and on the patient's and family's wishes [32,36,37]. Specifically about parenteral nutrition, it may benefit a limited percentage of patients when oral or enteric feeding is not possible, allowing them to survive longer [38,39]. When available, home parenteral nutrition allows good acceptance and stable QOL [39]. The decision to discontinue should attend the predetermined goals for each case and must be reconsidered in end-of-life situations.

3.1.3. Pain

Pain is one of the most common cancer-related symptoms and has a significant negative impact on QOL. Almost every patient with oesophageal cancer will experience pain at some point of the disease course, particularly in end stage disease [40].

The pain of oesophageal cancer can arise from different aetiologies, such as direct compression or invasion from tumour, and effects from treatments. Pain from the oesophagus usually manifests as odynophagia (with swallowing and eating), or chest and back pain at rest [40].

Following a stepwise approach, initial management with paracetamol (acetaminophen) is recommended for patients presenting with mild pain, although its effectiveness for cancer pain was not proven [41,42]. Non-steroidal anti-inflammatory drugs (NSAID) can also be considered, but their clinical benefit should be assessed within two weeks and balanced with the risks of gastroduodenal and cardiovascular toxicity [43,44]. Selective COX-2 inhibitors may be preferred to NSAID in this population. Metamizole (dipyrone) is an alternative non-opioid analgesic drug [45]. For moderate or severe cancer pain, opioids are the mainstay of pain management [46,47]. It is recommended to start with as-needed dosing of a short-acting opioid. Opioid dosing and titration should follow cancer-pain guidelines issued by World Health Organization and by associations dedicated to the management of this matter [42,44]. In the setting of dysphagia or odynophagia, liquid formulations or sublingual preparations of short-acting opioids should be preferred. Regarding long-acting (sustained or extended-release) opioids, there are few alternatives to tablets, such as fentanyl and buprenorphine transdermal systems, and some specific oxycodone and morphine sulfate capsules that can be opened so the content can be swallowed or administered through a feeding tube [40]. Yet, these options are not available in every country or institution. Alternatively, subcutaneous morphine may be administered by a portable pump, in perfusion or boluses (in patient-controlled analgesia systems). When prescribing opioids, patients have to be educated on the expected side effects, such as opioid-induced constipation, and prophylactic laxatives, such as senna and polyethylene glycol, should be prescribed [40,44,48].

Oesophageal cancer patients may also experience neuropathic pain, from tumour invasion of nerve plexus or as a side effect of platinum or taxane-based ChT previously used. For these patients, use of gabapentin, pregabalin, duloxetine or venlafaxine may be helpful [40,44]. When inflammatory pain is present, corticosteroids may also be used for pain control.

RT is also an effective palliative treatment for localized pain from the primary tumour or its metastasis. Hypofractionated RT (specifically 8Gy single dose) results in excellent pain relief from bone metastasis, and re-irradiation can be effective in recurrent pain [44,49]. The best response to pain relief occurs two to four weeks after completion of radiation, so, this treatment should be coupled with medical management and reserved to patients who have an expected survival of several weeks to months, in order to benefit from it [40].

Recent evidence favours the use of acupuncture and acupressure, as they significantly reduced pain intensity and opioid use [50].

3.1.4. Nausea and Vomiting

Nausea in patients with oesophageal cancer can have many causes, including mechanical obstruction, mucositis from ChT and radiation, pain medication, dehydration and anxiety [40].

For most advanced cancer patients, the best antiemetic choice are prokinetic drugs such as metoclopramide [51,52]. Alternative sedative antiemetic drugs include haloperidol, chlorpromazine or olanzapine. Anxiolytics are helpful for anxiety-driven nausea. Corticosteroids are synergistic with metoclopramide and 5-HT3 antagonist against chronic nausea, and may be most useful in nausea and vomiting induced by intracranial disease or bowel obstruction. Whenever feasible, mechanical obstruction nausea is more effectively managed with interventional procedures [40,52].

3.1.5. Fistula

The unique anatomy of the oesophagus and close proximity to the cardiopulmonary system favours the risk of major morbidity from cancer. Oesophageal fistula develops in 5 to 20% of oesophageal cancers, most commonly between the oesophagus and the respiratory tract (Figure 2a), and occasionally to pleural space, the aorta or other mediastinal structures (Figure 2b). Apart from tumour invasion, fistulas may also develop secondarily to radiation or endoscopic therapies. A history of worsening dysphagia and dyspnea, and coughing temporally related to drinking and eating is highly suggestive of an oesophagorespiratory fistula [53,54].

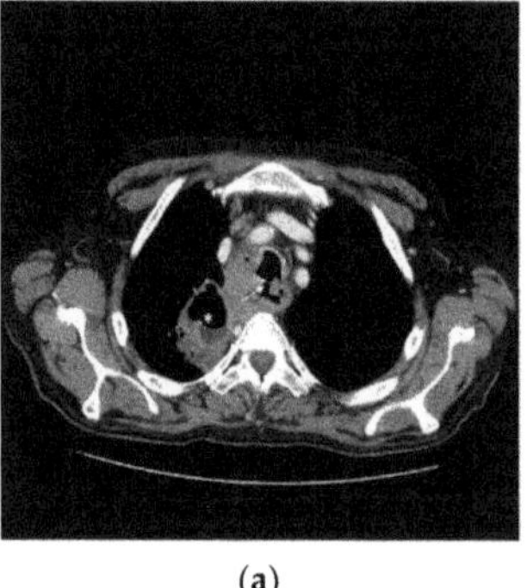

(a)

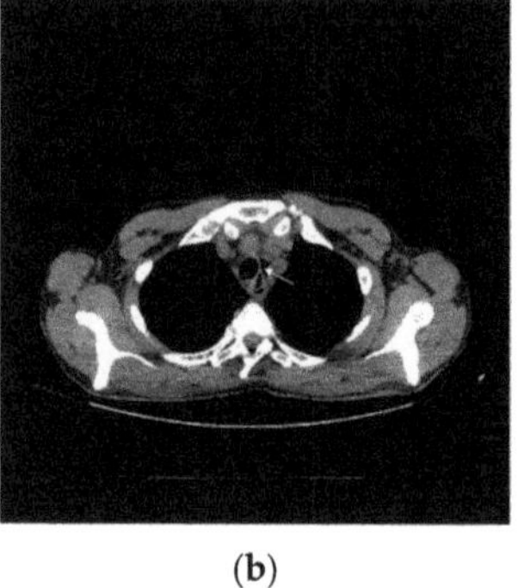

(b)

Figure 2. Oesophageal fistula. (**a**) Tracheo-oesophageal fistula (arrow) associated with pneumonia and lung abscess (*) (**b**) Squamous carcinoma of cervical oesophagus complicated with pneumomediastinum (arrow).

Oesophagorespiratory fistula are life-threatening, and immediate treatment should be attempted, as they can lead to aspiration pneumonia and poor nutritional intake, resulting in poorer survival (1 to 6 weeks with supportive care alone) [55].

SEMS placement is the treatment of choice for an oesophagorespiratory fistula [56]. In most published reports, complete sealing of the fistula was established in 90% of patients, while only half of these patients achieve long-term fistula closure [53,57,58]. In the palliative setting, surgical options, such as closure or resection of the fistula or bypass surgery, are associated with high morbidity and mortality and are not recommended [59].

3.1.6. Bleeding and Anaemia

Gastrointestinal (GI) bleeding in the cancer patient is a common and challenging clinical problem. Optimal clinical care is granted by access to a multidisciplinary team, including a gastroenterologist, an interventional radiologist, a radiation oncologist and a surgeon. It widens treatment options, to reach the most appropriate solution for each case [60].

The oesophagus is in close contact with the aorta and its arterial blood supply arises mainly from aortic small branches, but also from left gastric artery branches to the distal oesophagus [61].

In advanced oesophageal cancer, arterial bleeding can be massive and life-threatening, specially arising from aortic fistula. For instance, advanced care planning must include scope of treatment orders in such scenario: either to adopt an emergent, aggressive and invasive potentially life-saving treatment, or to allow natural death and improve comfort, including palliative sedation.

Endoscopic therapies are well established treatments of GI bleeding, but have been shown to be less effective to control malignancy-related bleeding [60,62]. Nevertheless, endoscopy is preponderant for diagnosing the source of bleeding and endoscopic interventions may avert the need for emergent surgery or temporarily control bleeding as a bridge for treatment with RT. Reflecting the lack of data on this matter, there are no widely accepted guidelines for optimal endoscopic therapeutic approach for tumour bleeding [60]. One of the caveats is that often tumour haemorrhage is diffuse and local endoscopic therapy fails to control the whole of it. Argon plasma coagulation is the method that has more evidence supporting its use, yet the majority of data indicate that control of bleeding is not reliably achieved [63]. Hemospray is a promising haemostatic agent for endoscopic application: for upper GI bleeding related to malignancy, the overall initial haemostasis rate was 94.9%, and the rebleeding rate was 30.3% [64].

RT has a pivotal role in the treatment of GI tumours. However, data describing RT for the treatment of bleeding are limited for oesophageal cancer [65]. In the setting of chronic blood loss, RT is suitable and may provide benefit as shown for gastric carcinoma. Common fractionation schedules use 1, 5 or 10 fractions, according to the expert radiation oncology physician judgment and discretion, based on many factors [66]. In case of active hemodynamically significant haemorrhage, RT is not useful for initial emergent bleeding control but it may latter complement the endoscopic or angiographic interventions.

Angiography and arterial embolization are feasible for the territory of the celiac artery and its branches. Here, the left gastric artery, which provides branches to distal oesophagus, is of particular interest. Considering oesophageal anatomic relations and blood supply, angiographic therapy can be only attempted in distal oesophagus bleeding as rescue, when endoscopic treatment fails. Yet, data supporting its use are limited.

Anaemia can adversely affect health related QOL of many cancer patients. Oesophageal cancer anaemia may particularly arise from tumour bleeding or malnutrition (addressed above). After assessing and treating iron deficiency or other reversible causes of anaemia, red blood cell transfusion can be used to treat symptomatic anaemia, as part of best supportive care [67]. Erythropoiesis stimulating agents are not recommended in patients who are not on ChT [68]. General recommendations on cancer anaemia management and palliative care experts' advisory can be applied to oesophageal cancer patients, taking into account their personal characteristics and options.

3.2. *Special Considerations*

Evidence relating specifically to elderly patients is limited, but available data suggest they derive similar benefit from treatment as younger patients [69].

Continuous updating of clinical management of patients with oesophageal cancer demands specialized multidisciplinary teams, including surgical, medical and radiation-oncology specialists, well as gastroenterologists, palliative care specialists, nutritionists, and nurses [7,70]. Referral to a palliative care specialist is recommended for oesophageal cancer patients with symptoms refractory to initial management and any cases that are particularly complex or difficult to manage.

4. Conclusions

Oesophageal cancer has a great personal and global impact. Often it presents as an advanced disease, with patients requiring best supportive care. However, apart from trials on the use of stents, there is a lack of high-level evidence regarding the several treatment options available. More well-designed trials and observational studies are needed to determine which approaches achieve the greatest benefit for these patients. Optimal management is not yet standardized, and best supportive care decisions should be discussed on a case-by-case basis by multidisciplinary teams.

Author Contributions: Conceptualization, V.D.S.D. and R.C.P.; Methodology, R.C.P. and A.A. (António Araújo); Literature Searching, R.C.P. and J.N.S.; Writing—Original Draft Preparation, R.C.P.; Writing—Review & Editing, R.C.P., A.A. (Alexandra Araújo), E.F., A.S.M., R.R. and A.A. (António Araújo); Supervision, A.A. (Alexandra Araújo); Publication Funding, A.A. (António Araújo). All authors named in this work meet the authorship criteria established by the International Committee of Medical Journal Editors (ICMJE) and they take responsibility for the integrity of the work and for the final version approved for publishing. All authors have read and agreed to the published version of the manuscript.

Funding: This research received no external funding.

Acknowledgments: The authors thank Olga Sousa (Radiation Oncology, Instituto Português de Oncologia do Porto) and Maria João Ramos (Medical Oncology, CHUPorto) for comments on the manuscript.

Conflicts of Interest: The authors declare no conflict of interests.

References

1. World Health Organization (WHO); International Agency for Research on Cancer (IARC). Global Cancer Observatory (GLOBOCAN): Cancer Today. 2020. Available online: http://gco.iarc.fr/today (accessed on 25 August 2022).
2. Kamangar, F.; Nasrollahzadeh, D.; Safiri, S.; Sepanlou, S.G.; Fitzmaurice, C.; Ikuta, K.S.; Bisignano, C.; Islami, F.; Roshandel, G.; Lim, S.S.; et al. The global, regional, and national burden of oesophageal cancer and its attributable risk factors in 195 countries and territories, 1990–2017: A systematic analysis for the Global Burden of Disease Study 2017. *Lancet Gastroenterol. Hepatol.* **2020**, *5*, 582–597. [CrossRef] [PubMed]
3. Ana da Costa Miranda, A.M.-d.-S.; Glória, L.; Brito, C. *Registo Oncológico Nacional de Todos os Tumores na População Residente em Portugal, em 2018*; Reprografia do IPO de Lisboa de Francisco Gentil; E.P.E.: Lisboa, Portugal, 2021.
4. Siegel, R.L.; Miller, K.D.; Fuchs, H.E.; Jemal, A. Cancer statistics, 2022. *CA A Cancer J. Clin.* **2022**, *72*, 7–33. [CrossRef] [PubMed]
5. Dandara, C.; Robertson, B.; Dzobo, K.; Moodley, L.; Parker, M.I. Patient and tumour characteristics as prognostic markers for oesophageal cancer: A retrospective analysis of a cohort of patients at Groote Schuur Hospital. *Eur. J. Cardio-Thorac. Surg.* **2015**, *49*, 629–634. [CrossRef] [PubMed]
6. Obermannová, R.; Alsina, M.; Cervantes, A.; Leong, T.; Lordick, F.; Nilsson, M.; van Grieken, N.; Vogel, A.; Smyth, E. Oesophageal cancer: ESMO Clinical Practice Guideline for diagnosis, treatment and follow-up. *Ann. Oncol.* **2022**, *33*, 992–1004. [CrossRef] [PubMed]
7. Ajani, J.A.; D'Amico, T.A.; Bentrem, D.J.; Chao, J.; Corvera, C.; Das, P.; Denlinger, C.S.; Enzinger, P.C.; Fanta, P.; Farjah, F.; et al. Esophageal and Esophagogastric Junction Cancers, Version 2.2019, NCCN Clinical Practice Guidelines in Oncology. *J. Natl. Compr. Cancer Netw.* **2019**, *17*, 855–883. [CrossRef]
8. Kitagawa, Y.; Uno, T.; Oyama, T.; Kato, K.; Kato, H.; Kawakubo, H.; Kawamura, O.; Kusano, M.; Kuwano, H.; Takeuchi, H.; et al. Esophageal cancer practice guidelines 2017 edited by the Japan esophageal society: Part 2. *Esophagus Off. J. Jpn. Esophageal Soc.* **2019**, *16*, 25–43. [CrossRef]
9. Opstelten, J.L.; De Wijkerslooth, L.R.H.; Leenders, M.; Bac, D.J.; Brink, M.A.; Loffeld, B.C.A.J.; Meijnen-Bult, M.J.F.; Minderhoud, I.M.; Verhagen, M.A.M.T.; Van Oijen, M.G.H.; et al. Variation in palliative care of esophageal cancer in clinical practice: Factors associated with treatment decisions. *Dis. Esophagus* **2016**, *30*, 1–7. [CrossRef]
10. Brierley, J.D.; Oza, A. Radiation and Chemotherapy in the Management of Malignant Esophageal Strictures. *Gastrointest. Endosc. Clin. N. Am.* **1998**, *8*, 451–463. [CrossRef]
11. Levy, A.; Wagner, A.D.; Chargari, C.; Moehler, M.; Verheij, M.; Durand-Labrunie, J.; Kissel, M.; Chirat, E.; Burtin, P.; Ducreux, M.; et al. Palliation of dysphagia in metastatic oesogastric cancers: An international multidisciplinary position. *Eur. J. Cancer* **2020**, *135*, 103–112. [CrossRef]
12. Spaander, M.C.W.; van der Bogt, R.D.; Baron, T.H.; Albers, D.; Blero, D.; de Ceglie, A.; Conio, M.; Czakó, L.; Everett, S.; Garcia-Pagán, J.-C.; et al. Esophageal stenting for benign and malignant disease: European Society of Gastrointestinal Endoscopy (ESGE) Guideline—Update 2021. *Endoscopy* **2021**, *53*, 751–762. [CrossRef]

13. Ahmed, O.; Lee, J.H.; Thompson, C.C.; Faulx, A. AGA Clinical Practice Update on the Optimal Management of the Malignant Alimentary Tract Obstruction: Expert Review. *Clin. Gastroenterol. Hepatol. Off. Clin. Pract. J. Am. Gastroenterol. Assoc.* **2021**, *19*, 1780–1788. [CrossRef] [PubMed]
14. Yang, C.; Li, C.; Fu, W.; Xu, W.; Yang, S. Interventions for dysphagia in oesophageal cancer. *Cochrane Database Syst. Rev.* **2014**, *10*, CD005048. [CrossRef]
15. Reijm, A.N.; Didden, P.; Schelling, S.J.C.; Siersema, P.D.; Bruno, M.J.; Spaander, M.C.W. Self-expandable metal stent placement for malignant esophageal strictures—Changes in clinical outcomes over time. *Endoscopy* **2018**, *51*, 18–29. [CrossRef]
16. Van Rossum, P.S.N.; Mohammad, N.H.; Vleggaar, F.P.; Van Hillegersberg, R. Treatment for unresectable or metastatic oesophageal cancer: Current evidence and trends. *Nat. Rev. Gastroenterol. Hepatol.* **2017**, *15*, 235–249. [CrossRef] [PubMed]
17. Homs, M.Y.; Steyerberg, E.W.; Eijkenboom, W.M.; Tilanus, H.W.; Stalpers, L.J.; Bartelsman, J.F.; van Lanschot, J.J.; Wijrdeman, H.K.; Mulder, C.J.; Reinders, J.G.; et al. Single-dose brachytherapy versus metal stent placement for the palliation of dysphagia from oesophageal cancer: Multicentre randomised trial. *Lancet* **2004**, *364*, 1497–1504. [CrossRef] [PubMed]
18. Martin, E.J.; Bruggeman, A.R.; Nalawade, V.V.; Sarkar, R.R.; Qiao, E.M.; Rose, B.S.; Murphy, J.D. Palliative Radiotherapy Versus Esophageal Stent Placement in the Management of Patients With Metastatic Esophageal Cancer. *J. Natl. Compr. Cancer Netw.* **2020**, *18*, 569–574. [CrossRef]
19. Sinha, S.; Varagunam, M.; Park, M.; Maynard, N.; Trudgill, N.; Crosby, T.; Cromwell, D. Brachytherapy in the Palliation of Oesophageal Cancer: Effective but Impractical? *Clin. Oncol.* **2019**, *31*, e87–e93. [CrossRef]
20. Merchant, S.J.; Kong, W.; Mahmud, A.; Booth, C.M.; Hanna, T.P. Palliative Radiotherapy for Esophageal and Gastric Cancer: Population-Based Patterns of Utilization and Outcomes in Ontario, Canada. *J. Palliat. Care* **2022**. [CrossRef]
21. Jeene, P.M.; Vermeulen, B.D.; Rozema, T.; Braam, P.M.; Lips, I.; Muller, K.; van Kampen, D.; Homs, M.Y.; Oppedijk, V.; Berbée, M.; et al. Short-Course External Beam Radiotherapy Versus Brachytherapy for Palliation of Dysphagia in Esophageal Cancer: A Matched Comparison of Two Prospective Trials. *J. Thorac. Oncol.* **2020**, *15*, 1361–1368. [CrossRef]
22. van Rossum, P.S.; Jeene, P.M.; Rozema, T.; Braam, P.M.; Lips, I.M.; Muller, K.; van Kampen, D.; Vermeulen, B.D.; Homs, M.Y.; Oppedijk, V.; et al. Patient-reported outcomes after external beam radiotherapy versus brachytherapy for palliation of dysphagia in esophageal cancer: A matched comparison of two prospective trials. *Radiother. Oncol.* **2020**, *155*, 73–79. [CrossRef]
23. Ólafsdóttir, H.S.; Klevebro, F.; Ndegwa, N.; von Döbeln, G.A. Short-course compared to long-course palliative radiotherapy for oesophageal cancer: A single centre observational cohort study. *Radiat. Oncol.* **2021**, *16*, 153. [CrossRef] [PubMed]
24. Adamson, D.; Byrne, A.; Porter, C.; Blazeby, J.; Griffiths, G.; Nelson, A.; Sewell, B.; Jones, M.; Svobodova, M.; Fitzsimmons, D.; et al. Palliative radiotherapy after oesophageal cancer stenting (ROCS): A multicentre, open-label, phase 3 randomised controlled trial. *Lancet Gastroenterol. Hepatol.* **2021**, *6*, 292–303. [CrossRef] [PubMed]
25. Cederholm, T.; Barazzoni, R.; Austin, P.; Ballmer, P.; Biolo, G.; Bischoff, S.C.; Compher, C.; Correia, I.; Higashiguchi, T.; Holst, M.; et al. ESPEN guidelines on definitions and terminology of clinical nutrition. *Clin. Nutr.* **2017**, *36*, 49–64. [CrossRef]
26. Hébuterne, X.; Lemarié, E.; Michallet, M.; De Montreuil, C.B.; Schneider, S.; Goldwasser, F. Prevalence of Malnutrition and Current Use of Nutrition Support in Patients With Cancer. *J. Parenter. Enter. Nutr.* **2014**, *38*, 196–204. [CrossRef] [PubMed]
27. Ottery, F.D. Supportive nutrition to prevent cachexia and improve quality of life. *Semin. Oncol.* **1995**, *22*, 98–111.
28. Nitenberg, G.; Raynard, B. Nutritional support of the cancer patient: Issues and dilemmas. *Crit. Rev. Oncol. Hematol.* **2000**, *34*, 137–168. [CrossRef]
29. Fearon, K.; Strasser, F.; Anker, S.D.; Bosaeus, I.; Bruera, E.; Fainsinger, R.L.; Jatoi, A.; Loprinzi, C.; MacDonald, N.; Mantovani, G.; et al. Definition and classification of cancer cachexia: An international consensus. *Lancet Oncol.* **2011**, *12*, 489–495. [CrossRef]
30. Silvers, M.A.; Savva, J.; Huggins, C.E.; Truby, H.; Haines, T. Potential benefits of early nutritional intervention in adults with upper gastrointestinal cancer: A pilot randomised trial. *Support. Care Cancer* **2014**, *22*, 3035–3044. [CrossRef]
31. Huhmann, M.B.; August, D. Review of American Society for Parenteral and Enteral Nutrition (A.S.P.E.N.) Clinical Guidelines for Nutrition Support in Cancer Patients: Nutrition Screening and Assessment. *Nutr. Clin. Pract.* **2008**, *23*, 182–188. [CrossRef]
32. Alderman, B.; Allan, L.; Amano, K.; Bouleuc, C.; Davis, M.; Lister-Flynn, S.; Mukhopadhyay, S.; Davies, A. Multinational Association of Supportive Care in Cancer (MASCC) expert opinion/guidance on the use of clinically assisted nutrition in patients with advanced cancer. *Support. Care Cancer* **2021**, *30*, 2983–2992. [CrossRef]
33. Arends, J.; Strasser, F.; Gonella, S.; Solheim, T.; Madeddu, C.; Ravasco, P.; Buonaccorso, L.; de van der Schueren, M.; Baldwin, C.; Chasen, M.; et al. Cancer cachexia in adult patients: ESMO Clinical Practice Guidelines ☆. *ESMO Open* **2021**, *6*, 100092. [CrossRef] [PubMed]
34. Furness, K.; Silvers, M.A.; Savva, J.; Huggins, C.E.; Truby, H.; Haines, T. Long-term follow-up of the potential benefits of early nutritional intervention in adults with upper gastrointestinal cancer: A pilot randomised trial. *Support. Care Cancer* **2017**, *25*, 3587–3593. [CrossRef] [PubMed]
35. Crawford, G.; Dzierżanowski, T.; Hauser, K.; Larkin, P.; Luque-Blanco, A.; Murphy, I.; Puchalski, C.; Ripamonti, C. Care of the adult cancer patient at the end of life: ESMO Clinical Practice Guidelines. *ESMO Open* **2021**, *6*, 100225. [CrossRef]
36. Cereda, E.; Pedrolli, C.A.S.P.E.N. Recommendations for Enteral Nutrition: Practice Is the Result of Potential Benefits, Harms, Clinical Judgment, and Ethical Issues. *J. Parenter. Enter. Nutr.* **2009**, *34*, 103. [CrossRef] [PubMed]
37. Nunes, G.; Fonseca, J.; Barata, A.T.; Dinis-Ribeiro, M.; Pimentel-Nunes, P. Nutritional Support of Cancer Patients without Oral Feeding: How to Select the Most Effective Technique? *GE-Port. J. Gastroenterol.* **2019**, *27*, 172–184. [CrossRef] [PubMed]

38. Ukleja, A.; Freeman, K.L.; Gilbert, K.; Kochevar, M.; Kraft, M.D.; Russell, M.K.; Shuster, M.H. Standards for Nutrition Support. *Nutr. Clin. Pract.* **2010**, *25*, 403–414. [CrossRef]
39. Bozzetti, F.; Cozzaglio, L.; Biganzoli, E.; Chiavenna, G.; DE Cicco, M.; Donati, D.; Gilli, G.; Percolla, S.; Pironi, L. Quality of life and length of survival in advanced cancer patients on home parenteral nutrition. *Clin. Nutr.* **2002**, *21*, 281–288. [CrossRef]
40. Guyer, D.L.; Almhanna, K.; McKee, K.Y. Palliative care for patients with esophageal cancer: A narrative review. *Ann. Transl. Med.* **2020**, *8*, 1103. [CrossRef]
41. Wiffen, P.J.; Derry, S.; Moore, R.A.; McNicol, E.D.; Bell, R.F.; Carr, D.B.; McIntyre, M.; Wee, B. Oral paracetamol (acetaminophen) for cancer pain. *Cochrane Database Syst. Rev.* **2017**, *2020*, CD012637. [CrossRef]
42. World Health Organization. WHO Guidelines Approved by the Guidelines Review Committee. In *WHO Guidelines for the Pharmacological and Radiotherapeutic Management of Cancer Pain in Adults and Adolescents*; World Health Organization: Geneva, Switzerland, 2018.
43. Derry, S.; Wiffen, P.J.; Moore, R.A.; McNicol, E.D.; Bell, R.F.; Carr, D.B.; McIntyre, M.; Wee, B. Oral nonsteroidal anti-inflammatory drugs (NSAIDs) for cancer pain in adults. *Cochrane Database Syst. Rev.* **2017**, *2020*, CD012638. [CrossRef]
44. Fallon, M.; Giusti, R.; Aielli, F.; Hoskin, P.; Rolke, R.; Sharma, M.; Ripamonti, C.; ESMO Guidelines Committee. Management of cancer pain in adult patients: ESMO Clinical Practice Guidelines. *Ann. Oncol.* **2018**, *29*, iv166–iv191. [CrossRef] [PubMed]
45. Gaertner, J.; Stamer, U.M.; Remi, C.; Voltz, R.; Bausewein, C.; Sabatowski, R.; Wirz, S.; Müller-Mundt, G.; Simon, S.T.; Pralong, A.; et al. Metamizole/dipyrone for the relief of cancer pain: A systematic review and evidence-based recommendations for clinical practice. *Palliat. Med.* **2016**, *31*, 26–34. [CrossRef] [PubMed]
46. Wiffen, P.J.; Wee, B.; Moore, R.A. Oral morphine for cancer pain. *Cochrane Database Syst. Rev.* **2016**, *2021*, CD003868. [CrossRef]
47. Wiffen, P.J.; Wee, B.; Derry, S.; Bell, R.F.; Moore, R.A. Opioids for cancer pain—An overview of Cochrane reviews. *Cochrane Database Syst. Rev.* **2017**, *2020*, CD012592. [CrossRef]
48. Larkin, P.; Cherny, N.; La Carpia, D.; Guglielmo, M.; Ostgathe, C.; Scotté, F.; Ripamonti, C. Diagnosis, assessment and management of constipation in advanced cancer: ESMO Clinical Practice Guidelines. *Ann. Oncol.* **2018**, *29*, iv111–iv125. [CrossRef]
49. Lutz, S.; Balboni, T.; Jones, J.; Lo, S.; Petit, J.; Rich, S.E.; Wong, R.; Hahn, C. Palliative radiation therapy for bone metastases: Update of an ASTRO Evidence-Based Guideline. *Pract. Radiat. Oncol.* **2016**, *7*, 4–12. [CrossRef]
50. He, Y.; Guo, X.; May, B.H.; Zhang, A.L.; Liu, Y.; Lu, C.; Mao, J.J.; Xue, C.C.; Zhang, H. Clinical Evidence for Association of Acupuncture and Acupressure With Improved Cancer Pain: A Systematic Review and Meta-Analysis. *JAMA Oncol.* **2020**, *6*, 271. [CrossRef]
51. Bruera, E.; Belzile, M.; Neumann, C.; Harsanyi, Z.; Babul, N.; Darke, A. A Double-Blind, Crossover Study of Controlled-Release Metoclopramide and Placebo for the Chronic Nausea and Dyspepsia of Advanced Cancer. *J. Pain Symptom Manag.* **2000**, *19*, 427–435. [CrossRef]
52. Davis, M.; Hui, D.; Davies, A.; Ripamonti, C.; Capela, A.; DeFeo, G.; Del Fabbro, E.; Bruera, E. MASCC antiemetics in advanced cancer updated guideline. *Support. Care Cancer* **2021**, *29*, 8097–8107. [CrossRef]
53. Siersema, P.D.; Vleggaar, F.P. Esophageal strictures, tumors, and fistulae: Alternative techniques for palliating primary esophageal cancer. *Tech. Gastrointest. Endosc.* **2010**, *12*, 203–209. [CrossRef]
54. Di Mauro, P.; Ferrari, V. Bronchoesophageal Fistula. *New Engl. J. Med.* **2020**, *383*, e137. [CrossRef] [PubMed]
55. Reed, M.F.; Mathisen, D.J. Tracheoesophageal fistula. *Chest Surg. Clin. N. Am.* **2003**, *13*, 271–289. [CrossRef] [PubMed]
56. Pennathur, A.; Gibson, M.K.; Jobe, B.A.; Luketich, J.D. Oesophageal carcinoma. *Lancet* **2013**, *381*, 400–412. [CrossRef] [PubMed]
57. May, A.; Ell, C. Palliative Treatment of Malignant Esophagorespiratory Fistulas With Gianturco-Z Stents: A Prospective Clinical Trial and Review of the Literature on Covered Metal Stents. *Am. J. Gastroenterol.* **1998**, *93*, 532–535. [CrossRef]
58. Shin, J.H.; Song, H.-Y.; Ko, G.-Y.; Lim, J.-O.; Yoon, H.-K.; Sung, K.-B. Esophagorespiratory Fistula: Long-term Results of Palliative Treatment with Covered Expandable Metallic Stents in 61 Patients. *Radiology* **2004**, *232*, 252–259. [CrossRef] [PubMed]
59. Vleggaar, F.; Siersema, P. Expandable Stents for Malignant Esophageal Disease. *Gastrointest. Endosc. Clin. N. Am.* **2011**, *21*, 377–388. [CrossRef]
60. Heller, S.J.; Tokar, J.L.; Nguyen, M.T.; Haluszka, O.; Weinberg, D.S. Management of bleeding GI tumors. *Gastrointest. Endosc.* **2010**, *72*, 817–824. [CrossRef]
61. Schwartz, S.I.; Brunicardi, F.C. *Schwartz's Principles of Surgery*; McGraw-Hill: New York, NY, USA, 2010.
62. Imbesi, J.J.; Kurtz, R.C. A multidisciplinary approach to gastrointestinal bleeding in cancer patients. *J. Support. Oncol.* **2005**, *3*, 101–110.
63. Akhtar, K.; Byrne, J.; Bancewicz, J.; Attwood, S.E.A. Argon beam plasma coagulation in the management of cancers of the esophagus and stomach. *Surg. Endosc.* **2000**, *14*, 1127–1130. [CrossRef]
64. de Rezende, D.T.; Brunaldi, V.O.; Bernardo, W.M.; Ribeiro, I.B.; Mota, R.C.L.; Baracat, F.I.; de Moura, D.T.H.; Baracat, R.; Matuguma, S.E.; de Moura, E.G.H. Use of hemostatic powder in treatment of upper gastrointestinal bleeding: A systematic review and meta-analysis. *Endosc. Int. Open* **2019**, *07*, E1704–E1713. [CrossRef]
65. Hughes, C.; Radhakrishna, G. Haemostatic radiotherapy for bleeding cancers of the upper gastrointestinal tract. *Br. J. Hosp. Med.* **2019**, *80*, 579–583. [CrossRef] [PubMed]
66. Kim, M.M.; Rana, V.; Janjan, N.A.; Das, P.; Phan, A.T.; Delclos, M.E.; Mansfield, P.F.; Ajani, J.A.; Crane, C.H.; Krishnan, S. Clinical benefit of palliative radiation therapy in advanced gastric cancer. *Acta Oncol.* **2008**, *47*, 421–427. [CrossRef]

67. Chin-Yee, N.; Taylor, J.; Downar, J.; Tanuseputro, P.; Saidenberg, E. Red Blood Cell Transfusion in Palliative Care: A Survey of Palliative Care Physicians. *J. Palliat. Med.* **2019**, *22*, 1139–1142. [CrossRef]
68. Aapro, M.; Beguin, Y.; Bokemeyer, C.; Dicato, M.; Gascón, P.; Glaspy, J.; Hofmann, A.; Link, H.; Littlewood, T.; Ludwig, H.; et al. Management of anaemia and iron deficiency in patients with cancer: ESMO Clinical Practice Guidelines. *Ann. Oncol.* **2018**, *29*, iv96–iv110. [CrossRef] [PubMed]
69. Bracken-Clarke, D.; Farooq, A.R.; Horgan, A.M. Management of Locally Advanced and Metastatic Esophageal Cancer in the Older Population. *Curr. Oncol. Rep.* **2018**, *20*, 99. [CrossRef] [PubMed]
70. Allum, W.H.; Blazeby, J.M.; Griffin, S.M.; Cunningham, D.; Jankowski, J.A.; Wong, R. Guidelines for the management of oesophageal and gastric cancer. *Gut* **2011**, *60*, 1449–1472. [CrossRef] [PubMed]

Article

Relationships among Inflammatory Biomarkers and Self-Reported Treatment-Related Symptoms in Patients Treated with Chemotherapy for Gynecologic Cancer: A Controlled Comparison

Aasha I. Hoogland [1], Brent J. Small [2], Laura B. Oswald [1], Crystal Bryant [1], Yvelise Rodriguez [1], Brian D. Gonzalez [1], Xiaoyin Li [1], Michelle C. Janelsins [3], Hailey W. Bulls [4], Brian W. James [5], Bianca Arboleda [5], Claudia Colon-Echevarria [6], Mary K. Townsend [6], Shelley S. Tworoger [6], Paulo C. Rodriguez [7], Julienne E. Bower [8], Sachin M. Apte [9], Robert M. Wenham [10] and Heather S. L. Jim [1,*]

1 Department of Health Outcomes and Behavior, Moffitt Cancer Center, Tampa, FL 33612, USA; aasha.hoogland@moffitt.org (A.I.H.)
2 School of Aging Studies, University of South Florida, Tampa, FL 33612, USA
3 Department of Surgery and Neuroscience, University of Rochester Medical Center, Rochester, NY 14642, USA
4 Department of Medicine, University of Pittsburgh, Pittsburgh, PA 15213, USA
5 Morsani College of Medicine, University of South Florida, Tampa, FL 33602, USA
6 Department of Cancer Epidemiology, Moffitt Cancer Center, Tampa, FL 33612, USA
7 Department of Immunology, Moffitt Cancer Center, Tampa, FL 33612, USA
8 Department of Psychology, University of California Los Angeles, Los Angeles, CA 90095, USA
9 Department of Obstetrics and Gynecology, Huntsman Cancer Institute, Salt Lake City, UT 84132, USA
10 Department of Gynecologic Oncology, Moffitt Cancer Center, Tampa, FL 33612, USA
* Correspondence: heather.jim@moffitt.org

Simple Summary: Treatment-related symptoms such as fatigue, depression, and disruptions in sleep and physical activity are common and distressing in gynecologic cancer patients. The aim of our study was to examine whether higher levels of inflammation are associated with worse symptomatology, and if these associations are stronger for patients with gynecologic cancer ($n = 121$) than age-matched women without a cancer history (i.e., controls; $n = 105$). Elevated levels of C-reactive protein were associated with depression and disrupted physical activity, but there were no other significant associations between inflammation and treatment-related symptoms. Findings suggest that inflammation may not play a significant role in the development of fatigue or sleep disturbance among gynecologic cancer patients but may contribute to depression and physical inactivity.

Abstract: Previous research suggests that inflammation triggers cancer-treatment-related symptoms (i.e., fatigue, depression, and disruptions in sleep and physical activity), but evidence is mixed. This study examined relationships between inflammatory biomarkers and symptoms in patients with gynecologic cancer compared to age-matched women with no cancer history (i.e., controls). Patients ($n = 121$) completed assessments before chemotherapy cycles 1, 3, and 6, and 6 and 12 months later. Controls ($n = 105$) completed assessments at similar timepoints. Changes in inflammation and symptomatology were evaluated using random-effects mixed models, and cross-sectional differences between patients and controls in inflammatory biomarkers and symptoms were evaluated using least squares means. Associations among inflammatory biomarkers and symptoms were evaluated using random-effects fluctuation mixed models. The results indicated that compared to controls, patients typically have higher inflammatory biomarkers (i.e., TNF-alpha, TNFR1, TNFR2, CRP, IL-1ra) and worse fatigue, depression, and sleep ($ps < 0.05$). Patients reported lower levels of baseline physical activity ($p = 0.02$) that became more similar to controls over time. Significant associations were observed between CRP, depression, and physical activity ($ps < 0.05$), but not between inflammation and other symptoms. The results suggest that inflammation may not play a significant role in fatigue or sleep disturbance among gynecologic cancer patients but may contribute to depression and physical inactivity.

Citation: Hoogland, A.I.; Small, B.J.; Oswald, L.B.; Bryant, C.; Rodriguez, Y.; Gonzalez, B.D.; Li, X.; Janelsins, M.C.; Bulls, H.W.; James, B.W.; et al. Relationships among Inflammatory Biomarkers and Self-Reported Treatment-Related Symptoms in Patients Treated with Chemotherapy for Gynecologic Cancer: A Controlled Comparison. *Cancers* **2023**, *15*, 3407. https://doi.org/10.3390/cancers15133407

Academic Editor: António Araújo

Received: 15 May 2023
Revised: 24 June 2023
Accepted: 27 June 2023
Published: 29 June 2023

Keywords: chemotherapy; cytokines; gynecologic cancer; quality of life in cancer patients

1. Introduction

Fatigue, depression, and disruptions in sleep and physical activity, known collectively as treatment-related symptoms, tend to be common and distressing in gynecologic cancer patients [1–5], due in part to the arduous platinum- and taxane-based chemotherapy regimens prescribed as first- and second-line treatments [6]. Previous research has shown that symptoms occur in a cascade pattern during chemotherapy to treat gynecologic cancer, with disrupted sleep contributing to next-day increases in fatigue, and fatigue contributing to next-day increases in depression [7,8]. Interventions that target treatment-related symptoms early in the cascade may mitigate later symptoms, but the underlying biology that drives this cascade is yet to be elucidated.

There is strong evidence for a causal role of inflammation in symptomatology outside the context of cancer. Therapeutic and experimental administration of pro-inflammatory cytokines induces fatigue, depression, and altered sleep and activity patterns in both humans and animals [9–21]. In addition, inflammatory biomarkers are known to activate inflammatory processes in the brain, which in turn regulate neurotransmitters involved in mood and behavior, such as serotonin, norepinephrine, and dopamine [22–26]. However, the existing literature examining the relationship between inflammatory biomarkers and symptomatology in cancer patients is mixed and interpretation is limited by methodological challenges [27].

Previous studies of the relationships between inflammatory biomarkers and symptomatology in cancer have typically been characterized by cross-sectional study designs; a lack of non-cancer comparison groups to facilitate the interpretation of results; and patients that are heterogeneous in terms of cancer type, treatment received, and time since treatment completion. The majority of work has examined cancer survivors several months or years after the completion of treatment [27], with few studies examining inflammatory biomarkers and symptomatology both during and shortly after treatment [28–30]. Cross-sectional studies have found that post-treatment breast cancer survivors who report high levels of fatigue and depression also demonstrate higher levels of interleukin 1 receptor antagonist (sIL-1ra), interleukin 6 (IL-6), soluble IL-6 receptor (sIL-6r), soluble tumor necrosis factor receptor type II (sTNFR-II), and C-reactive protein (CRP) compared to those with low levels of symptomatology [31–33], although other studies of multiple cancer types have found no association [34–36]. In some studies of breast cancer survivors, cross-sectional associations have been found between inflammation and fatigue, but not depression or sleep problems [37]. Research examining these relationships over time has found that treatment-related increases in IL-6, sTNF-R1, and CRP are associated with worsening fatigue in patients with multiple cancer types [28,30,38]. Further, studies of breast cancer patients in the first 18 months after completing treatment suggest that psychological risk factors such as perceived stress may moderate the relationship between inflammation (i.e., CRP in one study, and the combination of CRP, IL-6, and sTNF-RII in another study) and depression [39,40]. Collectively, these findings raise the question of whether changes in inflammation levels affect symptomatology.

According to the 3P Factors model [41], inflammation may be a predisposing factor that increases the likelihood of worse symptomatology, a precipitating factor that hastens the onset of symptomatology, and/or a perpetuating factor that worsens or prolongs symptoms. For example, research on inflammatory biomarkers in cancer patients suggests that inflammation is altered prior to starting adjuvant chemotherapy, by the cancer itself and/or surgery. Advanced ovarian cancer is associated with greater circulating pro-inflammatory cytokines before surgery than early-stage disease [42]. Surgery is also known to trigger inflammatory processes [43]. In addition, high levels of symptom severity have been observed prior to the start of chemotherapy [44,45]. Research on breast cancer patients before treatment with radiotherapy or chemotherapy has also shown that fatigue is associated with higher circulating levels of pro-inflammatory cytokines [46].

Some studies suggest that inflammatory biomarkers are triggered by chemotherapy. In preclinical studies, platinum- and taxane-based chemotherapy elicit expression of interleukin 1 beta (IL-1β), IL-6, interleukin-8 (IL-8), interleukin-12 (IL-12), and tumor necrosis factor alpha (TNF-α) in primary human monocytes, macrophages, plasma, and various breast and ovarian cancer cell lines [47–52]. Interestingly however, there are few data regarding the effects of chemotherapy on circulating pro-inflammatory cytokines in humans [53–56]. There are even fewer data examining the relationship between pro-inflammatory cytokines and symptoms during chemotherapy; thus, it is unclear if inflammation precipitates symptoms. In patients with breast cancer, IL-8 was associated with flulike symptoms measured over three occasions during the first chemotherapy cycle [51], and soluble intercellular adhesion molecule 1 (sICAM-1) was associated with sleepiness before the first and fourth chemotherapy infusions [57]. In patients receiving chemoradiation for non-small cell lung cancer (NSCLC), advanced colorectal cancer, or advanced esophageal cancer, increases in sTNF-R1 and serum IL-6 have been associated with increases in fatigue and sleep disturbance [30,58]. However, other research in lung cancer patients undergoing chemotherapy found no association between IL-6, IL-8, or CRP with symptoms, including fatigue, sleep disturbance, and depression [59].

There is additional research in cancer patients suggesting that inflammatory biomarkers are triggered after chemotherapy is complete (i.e., late effects of treatment); data indicate that symptoms are worse in post-treatment cancer patients than individuals without cancer [60–62]. While it was previously assumed that these differences reflected persistent, acute effects of treatment, research by our group and others indicates that chemotherapy may be associated with late-onset symptoms not present at the end of treatment [63–66]. Thus, it may be that inflammation perpetuates symptoms during or after treatment. These observations are intriguing and merit additional investigation.

The goal of this study was to examine longitudinal relationships among changes in circulating pro- and anti-inflammatory biomarkers (i.e., IL-10, IL-1b, TNF-alpha, IL-6, IL-1ra, TNFR1, TNFR2, and CRP) and patient-reported symptoms in women with gynecologic cancer throughout the chemotherapy treatment trajectory. Specifically, our aim was to examine whether higher levels of inflammation were associated with worse symptomatology, and if these associations were stronger for patients than age-matched women without a cancer history (i.e., controls). Because inflammatory biomarkers can fluctuate very rapidly and individual measurements of inflammatory biomarkers may not reflect systemic or stable levels of inflammation, our analytic focus was on examining inflammation relative to participants' personal average levels of inflammation, and relative to other participants' average levels of inflammation. It was hypothesized that (1) patients would demonstrate higher levels of inflammation and worse symptomatology than controls throughout the treatment trajectory, (2) participants with higher levels of inflammation would report worse symptomatology than those with lower levels of inflammation (i.e., between-person effects), and (3) at times when participants had higher levels of inflammation than their own average, they would also report worse symptomatology than usual for them (i.e., within-person effects). We also explored whether there were differences in relationships among circulating inflammatory biomarkers and patient-reported symptomatology in patients vs. controls. We previously reported on symptoms among patients during active treatment [67]. For this paper, we expand on those findings to report on symptoms over one year post-treatment.

2. Materials and Methods

2.1. Participants

Study participants were recruited as part of a larger study examining the side effects of chemotherapy in patients with gynecologic cancer. As described previously [67], 6935 patients were screened for participation, 264 were approached, and 150/264 (57%) provided consent. Control participants were recruited from women identified through a national marketing company. A total of 355 controls were screened for participation, 346 were approached, and 150/346 (43%) provided consent. A sample size of 150 per group was chosen in order to have 80% power to detect between-group differences in slopes of at least 20% [68], assuming an alpha of 0.05 (two-tailed), intra-class correlation (ICC) of 0.50, and eight measurement occasions (in the larger study, data were collected before and after the first, third, and sixth infusion, and 6 and 12 months after completing chemotherapy). Eligible participants for both patient and control groups were (a) 18–89 years of age, (b) without psychiatric or neurological disorders that could interfere with study participation (e.g., dementia, psychosis), (c) without reported or documented diagnosis of an immune-related disease (e.g., HIV, systemic lupus erythematosus, rheumatoid arthritis), (d) not pregnant, (e) able to speak and read English, and (f) able to provide informed consent. Additional eligibility criteria for patients were (g) diagnosed with gynecologic cancer (i.e., ovarian, endometrial, peritoneal, fallopian, cervical, uterine, or vaginal), and (h) scheduled to receive intravenous or intraperitoneal chemotherapy at Moffitt Cancer Center (Tampa, FL, USA). Additional eligibility criteria for non-cancer controls were (i) not diagnosed with any form of cancer (except non-melanoma skin cancer), (j) age within five years of the patient participant to whom they were being matched, and (k) having a mailing address, working telephone number, and internet access. Participants were recruited between August 2013 and October 2018. This study was approved by the University of South Florida Institutional Review Board (Pro00005797).

Potential patient participants were identified by the research team in collaboration with their treating physician and were contacted via phone or in person to determine initial eligibility and interest in the study. Eligible and interested patients provided written informed consent to participate. Patient participants completed assessments before their first, third (i.e., approximately middle), and sixth (i.e., approximately last) chemotherapy infusions, and 6 and 12 months after their sixth chemotherapy infusion. Control participants provided written informed consent and completed assessments at equivalent time points.

2.2. Measures

2.2.1. Demographic and Clinical Data

Patients completed a baseline questionnaire that assessed sociodemographic characteristics prior to starting chemotherapy and controls completed at time of enrollment, including date of birth, race, ethnicity, marital status, education level, household income, and menopausal status. Medical comorbidities were ascertained via a self-report version [69] of the Charlson Comorbidity Index [70]. Clinical characteristics for gynecologic cancer patients were obtained by a medical record review at baseline and included cancer type, stage, and previous chemotherapy.

2.2.2. Circulating Inflammatory Biomarkers

Participants provided blood samples at baseline (i.e., before starting chemotherapy for patients), before chemo cycles 3 and 6, and again at 6 and 12 months after completing chemotherapy. Each serum or plasma blood sample was evaluated for the presence of inflammatory biomarkers, including IL-10, IL-1beta, TNF-alpha, TNFR1, TNFR2, CRP, IL-6, and IL-1ra. These inflammatory biomarkers were selected because they have shown significant associations with symptomatology in previous research and are readily detectable using existing laboratory methodology [31–33,71–73]. Blood samples were typically drawn at the same time of day (i.e., 8:00 a.m.–12:00 p.m.). Participants were asked to refrain from exercise, alcohol consumption, caffeine use, and non-prescription medications for the 24 h prior to blood draws [74,75]. Blood samples were sent to the Cancer Control and Psychoneuroimmunology Lab at the University of Rochester for analysis. All samples were assayed in one run using a multiplexed cytokine bead assay (i.e., IL-10, IL-1beta, TNF-alpha [HSTCMAG-28SK-04], TNFR1, TNFR2 [HSCRMAG-32K-02], IL-6, and IL-1ra [HCYTOMAG-60K-01]) or enzyme-linked immunosorbent assays (i.e., CRP; R&D Systems Human Quantikine ELISA; Minneapolis, MN, DCRP00) per the manufacturer's instructions. All kits were from the same lot. For Luminex, the median concentration was taken from 50 beads per well. For ELISA, the average was taken from duplicates. All data and internal controls were inspected for a CV < 20%, with all kits run with a standard curve with an $r^2 > 0.98$. All sample collections from the same participant were run on the same plate. The lower limits of detection of the assays, with sample dilution taken into account, were IL-10 = 0.30; IL-1β = 0.14; IL-6 = 0.04; TNFα = 0.08; IL-1Rα = 7.41; TNFR-1 = 10.60; TNFR-2 = 10.18; CRP = 5.0 pg/mL.

2.2.3. Treatment-Related Symptoms

Fatigue was measured using the Fatigue Symptom Inventory [76]. The average of 4 items assessing the highest, lowest, average, and current levels of fatigue over the previous week was used in analyses. Scores ranged from 0 to 10 with higher scores indicating greater fatigue. A score of 3 and above indicated clinically meaningful fatigue [77].

Depression was assessed using the 7-item depression subscale of the Hospital Anxiety and Depression Scale designed to detect depressive symptoms in medically ill patients, including people with cancer [78,79]. Participants rated each item on a 4-point scale from 0 (absence) to 3 (extreme presence). All items were summed with possible scores ranging from 0 to 21. Scores of 8 or higher indicated clinically meaningful depressive symptoms [80].

Sleep was evaluated using the 19-item Pittsburgh Sleep Quality Index (PSQI) that assesses types and frequency of sleep disturbance experienced over the last month. All items were summed to derive an overall sleep quality score ranging from 0 to 21. Scores of 5 or higher indicated clinically meaningful sleep difficulties [81].

Physical activity was measured using the International Physical Activity Questionnaire-Short Form (IPAQ). The IPAQ assesses the frequency (days per week) and duration (minutes per day) of physical activity in the last 7 days [82]. Values were weighted by energy requirements for activities of varying intensities, which were defined by metabolic equivalents (METs) as follows: walking = 3.3 METs/min, moderate physical activity = 4.0 METs/min, and vigorous physical activity = 8.0 METs/min [83]. Total physical activity was calculated as the sum of METs per week. Patients were considered to be meeting the American Cancer Society guidelines [84] if they reported 600 or more MET minutes per week [85].

2.3. *Data Analyses*

Inflammatory biomarkers with estimates below the limit of detection were assigned a value of the limit of detection divided by the square root of 2. Indeterminate inflammatory biomarker concentrations were set to missing. Inflammatory biomarker values that were three standard deviations or more from the sample mean for each group were treated as outliers and set to missing. Raw inflammatory biomarkers with non-normal distributions (i.e., IL-6, IL-1ra) were natural-log-transformed to normalize their distributions, and the natural log values were used in analyses. To facilitate interpretation, both non-transformed data and transformed data are presented in selected tables. All cytokines that were not natural-log-transformed (i.e., IL-10, IL-1b, TNF-alpha, TNFR1, TNFR2, CRP) were mean-centered to facilitate interpretation.

Participants who provided a blood sample and questionnaire data at one or more timepoints were included in analyses. Means, standard deviations, frequencies, and percentages were used to describe sociodemographic and clinical characteristics of the sample. Differences in sociodemographic and clinical characteristics by group (i.e., patients vs. controls) were examined using independent sample t-tests, chi-square tests, and Fisher's tests. Variables significant at $p < 0.10$ were included as covariates in longitudinal analyses. Inflammatory biomarkers and symptoms were described at each time point using means and standard deviations. Changes in inflammatory biomarkers and symptoms over time (for patients only and separately for controls only), and group differences (patients vs. controls) in these changes were evaluated using linear and quadratic random-effects mixed models and time coded as the number of months since baseline enrollment. When quadratic models were not significant, linear models were used instead. Cross-sectional group differences in inflammatory biomarkers and symptoms at each time point were examined using between-subject comparisons of least squares means from the quadratic random-effects mixed models, as these account for quadratic changes over time. Proportions of patients and controls with clinically meaningful fatigue, depression, problems with sleep, and physical inactivity were assessed at each timepoint.

Associations among inflammatory biomarkers and symptoms aggregated over the five assessments were evaluated using random-effects fluctuation mixed models [86]. Because of the time-varying nature of the inflammatory biomarkers, each model included both between-person predictors and within-person predictors. Between-person predictors tested whether participants' average levels of each inflammatory biomarker differed from other participants' average levels, and whether these differences were associated with symptoms. Within-person predictors tested whether participants' average levels of each inflammatory biomarker differed from their own average level, and whether these differences were associated with symptoms. Interactions between fluctuations in inflammatory biomarkers and the group with symptoms were included in the fluctuation models to evaluate whether these associations between inflammation and symptoms were stronger for patients than controls. Significant interactions were further evaluated within each group using separate random-effects fluctuation models with the group effect and covariates removed. All statistical analyses were conducted using SAS version 9.4 (Cary, NC, USA).

3. Results

The sociodemographic and clinical characteristics of patients ($n = 121$) and controls ($n = 105$) are presented in Table 1. On average, patients were 60 years of age, married (70%), White (93%), and without a college education (63%). Most patients were diagnosed with stage 3 or 4 cancer (70%) of the ovaries or endometrium (75%). Controls were, on average, 58 years of age, White (89%), and college-educated (72%). Patients were less likely than controls to be college graduates ($p < 0.01$). Patients also reported more comorbidities ($p = 0.04$), on average, and were more likely to be post-menopausal than controls ($p = 0.01$). Age, education, comorbidities, and menopausal status were included as covariates in all multivariable analyses. Body mass index was not included, as it was not collected for controls. As expected, in this population where controls were age-matched to patients, there was not a statistically significant group difference in age. Nonetheless, age was included as a covariate because it is often associated with inflammation [87,88].

Table 1. Participant characteristics of patients with a diagnosis of gynecologic cancer and non-cancer controls.

Variable	Patients (n = 121)	Controls (n = 105)	p-Values
Age: M (SD)	60.4 (10.7)	58.0 (12.8)	0.12
Race: n (%) White	112 (93)	93 (89)	0.11
Education: n (%) college graduate	45 (37)	76 (72)	<0.0001
Income: n (%) USD 40 k or more	66 (69)	74 (80)	0.08
Comorbidities: mean (SD) [range]	2.4 (0.8) [2–7]	2.2 (0.5) [2–5]	0.04
Menopausal status: n (%)			0.01
Pre-menopausal	15 (13)	27 (26)	
Post-menopausal	104 (87)	78 (74)	
Cancer type: n (%)		-	-
Cervical	4 (3)		
Endometrial	27 (22)		
Fallopian	5 (4)		
Ovarian	64 (53)		
Peritoneal	7 (6)		
Uterine	10 (8)		
Vulvar	1 (1)		
Other	2 (2)		
Stage: n (%)		-	-

Table 1. *Cont.*

Variable	Patients (n = 121)	Controls (n = 105)	p-Values
1	22 (19)		
2	13 (11)		
3	61 (54)		
4	18 (16)		
Prior lines of chemotherapy: n (%) 3 or more	14 (12)	-	-

3.1. Group Differences in Inflammatory Biomarkers

Approximately 5% of inflammatory biomarker levels were below the lower limit of detection and divided by the square root of 2. Within each group, inflammatory biomarker levels that were more than three standard deviations from the group mean were set to missing (approximately 1% of all values, from 20 patients and 18 controls). Raw and log-transformed (IL-6 and IL-1ra only, per visual inspection of variable distributions) means for inflammatory biomarkers are displayed in Table 2. Adjusted means of circulating inflammatory biomarkers from the quadratic random-effects mixed models stratified by group are displayed in Figure 1A–H. Adjusted parameter estimates from the random-effects mixed models examining changes in inflammation over time are presented in Supplementary Table S1. Between-subject comparisons of adjusted means (i.e., least squares means) from the quadratic random-effects mixed models at each timepoint demonstrated that IL-1b was significantly lower in patients than controls at baseline only (p = 0.02). TNF-alpha and TNFR2 were significantly higher in patients than controls during active treatment (baseline, pre-chemo 3, and pre-chemo 6) and at the 6 month follow-up (ps < 0.04). TNFR1 and IL-1ra were significantly higher in patients than controls during active treatment (baseline, pre-chemo 3, and pre-chemo 6) (ps < 0.02). CRP was significantly higher in patients than controls at all timepoints (ps < 0.02). IL-6 was significantly higher in patients than controls at pre-chemo 3 only (p = 0.05).

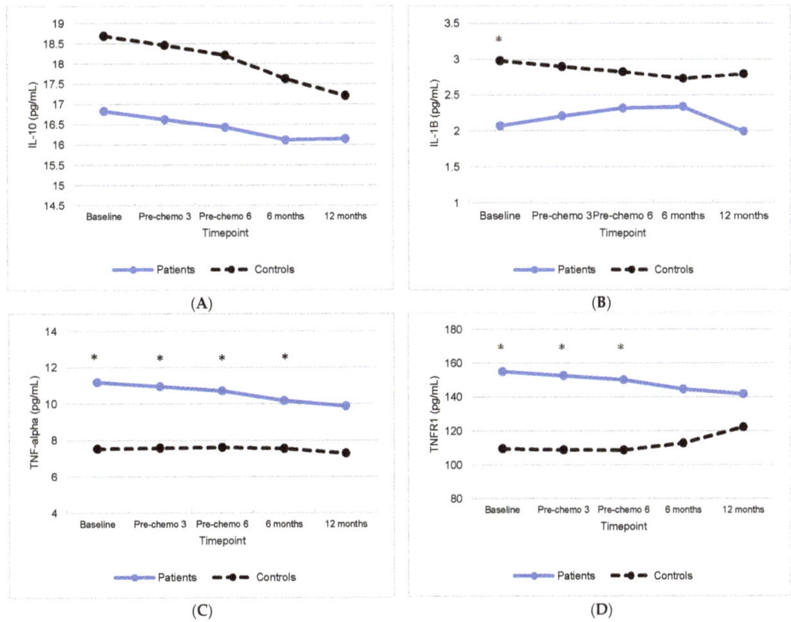

Figure 1. *Cont.*

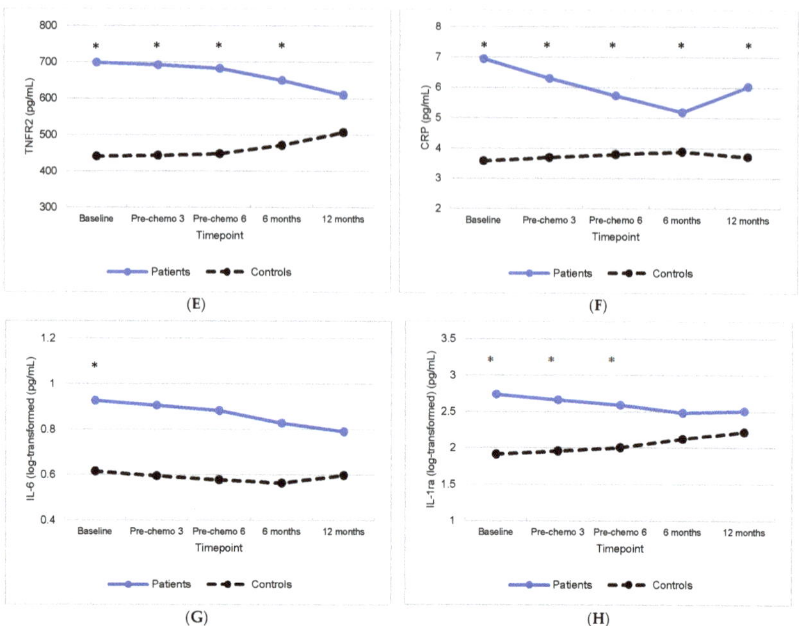

Figure 1. Least square mean levels over time for patients with gynecologic cancer and non-cancer controls for (**A**) IL-10, (**B**) IL-1B, (**C**) TNF-alpha, (**D**) TNFR1, (**E**) TNFR2, (**F**) CRP, (**G**) log-transformed IL-6, and (**H**) log-transformed IL-1ra. Asterisks indicate significant group differences in least squares means ($p < 0.05$).

Adjusted parameter estimates from the random-effects mixed models are presented in Supplementary Table S2. The results of the quadratic mixed models revealed a significant interaction between group and the quadratic effect of time for CRP ($p < 0.001$) such that, on average, levels of circulating CRP decreased from baseline to 6 months after completing chemotherapy and increased thereafter ($p < 0.01$) in patients, whereas controls' levels of CRP remained unchanged over time ($p = 0.29$). The results of the linear mixed models revealed no significant changes in inflammation over time (*p*s > 0.05). There was a significant interaction between group and time for TNFR1 ($p = 0.03$), but the effect of time was not significant for patients or controls when each group was examined separately (*p*s > 0.05). There were also significant interactions between group and time for TNFR2 ($p < 0.01$) and IL-1ra ($p = 0.03$) such that levels of circulating TNFR2 and IL-1ra increased over time for controls (*p*s = 0.03) but not patients (*p*s > 0.05).

3.2. Group Differences in Treatment-Related Symptoms and Associations with Inflammation

The unadjusted mean levels of treatment-related symptoms (i.e., fatigue, depression, sleep, physical activity) over time are displayed in Table 2. The adjusted means of symptoms from the quadratic random-effects mixed models stratified by group are displayed in Figure 2A–D. The adjusted parameter estimates from the random-effects mixed models examining change in symptoms over time are presented in Supplementary Table S2. The results of the random-effects fluctuation models examining associations between inflammation and symptoms are presented in Supplementary Table S3.

Table 2. Mean scores for biomarkers of inflammation and treatment-related symptoms over time for patients with gynecologic cancer and non-cancer controls.

	Baseline		Pre-Chemotherapy Cycle 3		Pre-Chemotherapy Cycle 6		6-Month Follow-Up		12-Month Follow-Up	
	Patients Mean (SD)	Controls Mean (SD)	Patients Mean (SD)	Controls Mean (SD)	Patients Mean (SD)	Controls Mean (SD)	Patients Mean (SD)	Controls Mean (SD)	Patients Mean (SD)	Controls Mean (SD)
Biomarkers of Inflammation (pg/mL):										
IL-10	24.4 (21.1)	23.0 (16.7)	22.3 (18.4)	23.3 (14.8)	22.7 (18.0)	24.2 (14.8)	23.7 (18.5)	22.7 (15.9)	25.7 (20.1)	23.2 (15.6)
IL-1b	3.4 (2.5)	3.9 (2.9)	3.4 (2.6)	4.0 (2.7)	3.5 (2.6)	4.1 (2.8)	3.4 (2.1)	4.0 (2.8)	3.4 (2.6)	4.4 (3.2)
TNF-alpha	16.3 (7.9)	12.0 (6.6)	15.4 (7.9)	12.3 (5.7)	15.8 (7.4)	12.4 (7.3)	16.0 (7.1)	12.6 (6.8)	16.6 (7.5)	12.7 (6.3)
TNFR1	181.6 (107.7)	126.0 (78.4)	178.7 (106.5)	126.1 (82.8)	185.3 (105.9)	133.7 (90.7)	187.9 (105.1)	132.6 (74.9)	185.4 (100.1)	133.5 (80.5)
TNFR2	1015.4 (580.2)	733.2 (313.8)	1078.3 (626.3)	748.3 (363.6)	1071.7 (560.6)	757.5 (326.5)	1077.0 (581.1)	793.9 (348.0)	1021.3 (474.4)	758.5 (334.4)
CRP	6.2 (4.6)	2.9 (2.9)	6.2 (4.7)	3.0 (3.1)	5.6 (4.9)	3.0 (2.8)	5.0 (4.6)	2.9 (2.8)	5.4 (5.1)	2.5 (2.2)
IL-6	6.4 (4.8)	4.4 (3.0)	5.6 (3.5)	4.5 (2.7)	5.5 (3.4)	4.5 (2.8)	6.3 (5.2)	4.6 (2.8)	5.7 (4.3)	4.4 (2.8)
IL-6 (log-transformed)	1.5 (1.1)	1.0 (1.4)	1.4 (1.2)	1.2 (1.2)	1.4 (0.9)	1.1 (1.4)	1.4 (1.4)	1.1 (1.4)	1.5 (0.9)	1.0 (1.4)
IL-1ra	62.9 (94.2)	55.1 (101.5)	59.6 (84.6)	59.7 (103.7)	49.9 (80.1)	50.3 (97.2)	74.5 (132.1)	59.3 (118.9)	93.8 (139.1)	54.4 (100.8)
IL-1ra (log-transformed)	3.4 (1.4)	2.5 (1.8)	3.2 (1.4)	2.6 (1.8)	3.0 (1.4)	2.3 (1.8)	3.3 (1.5)	2.5 (1.8)	3.4 (1.9)	2.4 (1.8)
Sickness Behaviors:										
Fatigue [1]	3.4 (1.9)	2.5 (1.6)	3.7 (1.7)	2.6 (1.8)	4.3 (2.2)	2.4 (1.7)	3.4 (2.0)	2.4 (1.9)	3.2 (2.0)	2.6 (1.9)
Depression [2]	4.7 (3.7)	2.1 (2.2)	5.0 (3.9)	2.2 (2.7)	5.6 (3.9)	2.1 (2.6)	3.9 (3.4)	2.1 (2.5)	3.9 (3.4)	2.2 (2.4)
Overall Sleep Quality [3]	7.2 (3.5)	4.9 (2.8)	7.4 (4.0)	4.9 (2.7)	6.8 (3.4)	5.2 (3.0)	6.9 (3.8)	4.1 (2.9)	6.5 (4.2)	4.7 (3.1)
Physical Activity (Total METs) [4]	1771 (2254)	2744 (2440)	1984 (2205)	2700 (2825)	2003 (3281)	2408 (2783)	2590 (3051)	2455 (2808)	2618 (3155)	2336 (2639)

Note: IL-10 = interleukin-10; IL-1b = interleukin 1 beta; TNF-alpha = tumor necrosis factor–alpha; TNFR1 = tumor necrosis factor receptor 1; TNFR2 = tumor necrosis factor receptor 2; CRP = c-reactive protein; IL-6 = interleukin-6; IL-1ra = interleukin 1 receptor antagonist. [1] Measured using the Fatigue Symptom Inventory; scores of 3 or above indicate clinically meaningful fatigue. [2] Measured using the Depression subscale of the Hospital Anxiety and Depression Scale; scores of 8 or higher indicate clinically meaningful depressive symptoms. [3] Measured using the total score from the Pittsburgh Sleep Quality Index; scores of 5 and above indicate clinically meaningful sleep problems. [4] Measured using the International Physical Activity Questionnaire-Short Form; fewer than 600 MET-minutes per week indicated non-adherence to American Cancer Society physical activity guidelines.

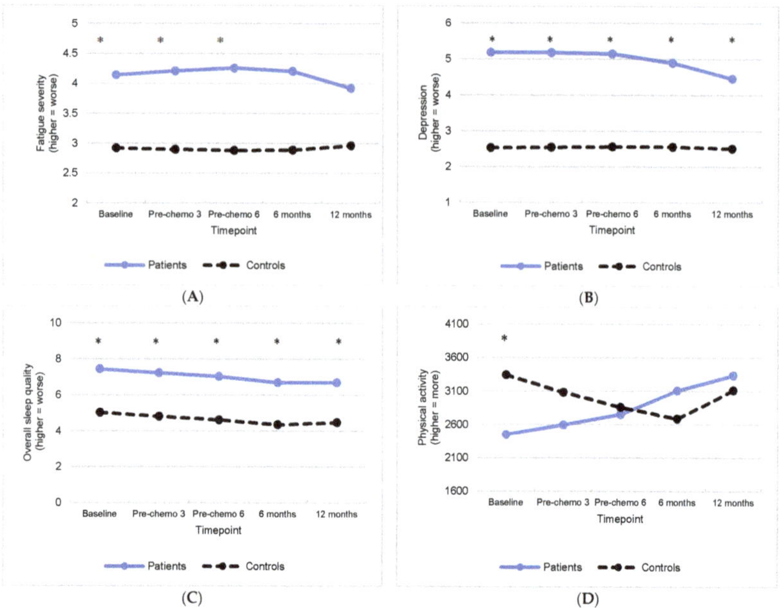

Figure 2. Least square mean levels over time for patients with gynecologic cancer and non-cancer controls for (**A**) fatigue severity (from the Fatigue Symptom Inventory), (**B**) depression (from the Hospital Anxiety and Depression Scale), (**C**) overall sleep quality (from the Pittsburgh Sleep Quality Index total score), and (**D**) total METS (from the International Physical Activity-Short Form). Asterisks indicate significant group differences in least squares means ($p < 0.05$).

3.2.1. Fatigue

Between-subject comparisons of adjusted means at each timepoint revealed that patients reported significantly more fatigue than controls during active treatment (baseline, pre-chemo 3, and pre-chemo 6) ($ps < 0.03$). At baseline, 60% of patients reported clinically meaningful fatigue (i.e., Fatigue Severity Inventory scores of 3+), which increased to 72% by pre-chemo 6, compared to 39% and 36% of controls at the same timepoints. By 12 months after completing chemotherapy, 55% of patients and 45% of controls reported clinically meaningful fatigue. There was no significant change in fatigue over time, nor were there differences between groups in change in fatigue over time ($ps > 0.05$). There were no significant effects of circulating levels of inflammation or interactions between inflammation and group on fatigue ($ps > 0.05$).

3.2.2. Depression

Patients reported significantly more depression than controls at all timepoints ($ps < 0.01$). At baseline, 23% of patients reported clinically meaningful depression (i.e., HADS depression scores of 8+), which increased slightly to 26% by pre-chemo 6 and decreased to 18% by 12 months after completing chemotherapy. In contrast, between 2% and 6% of controls reported clinically meaningful depression at any timepoint. Over time, there was no change in depression ($p > 0.05$), but there was a significant interaction between group and time ($p = 0.05$) such that depressive symptoms significantly improved over time in patients ($p = 0.03$) but remained unchanged for controls ($p = 0.90$). There were significant interactions of between-person variance in circulating TNF-alpha ($p = 0.05$) and IL-1ra ($p = 0.04$) with group on depression, but these effects were not significant for patients or controls when each group was examined separately ($ps > 0.05$). However, across groups, participants with higher circulating CRP levels had significantly greater depressive symptomatology ($p = 0.01$) than participants with lower CRP levels (i.e., main effect of between-person variance in circulating CRP). There were no other significant effects of circulating levels of inflammation or interactions between inflammation and group on depression.

3.2.3. Sleep

Patients reported significantly worse overall sleep quality than controls at all timepoints (*p*s < 0.01). In total, 80% of patients reported clinically meaningful problems with sleep at baseline (i.e., PSQI total scores of 5+), which remained stable by pre-chemo 6 (78%) and decreased to 61% by 12 months after chemotherapy. In contrast, 50% of controls reported clinically meaningful problems with sleep at baseline, which remained stable by pre-chemo 6 (47%) and declined slightly to 41% at 12 months after chemotherapy. There was no significant change in sleep over time, nor were there differences between groups in change in sleep over time (*p*s > 0.05). There were no significant effects of circulating levels of inflammation or interactions between inflammation and group on sleep (*p*s > 0.05).

3.2.4. Physical Activity

Patients reported significantly less physical activity than controls prior to chemotherapy only ($p = 0.02$). At baseline, 33% of patients did not meet the American Cancer Society guidelines for weekly METs expenditure (i.e., 600+ METs), compared to 11% of controls. By pre-chemo 6, 39% of patients reported under 600 METs per week, compared to 22% of controls, and 12 months after completing chemotherapy 21% of patients and controls reported under 600 METs per week. Over time, there was a main effect of group ($p < 0.01$) that was qualified by a significant interaction between group and time ($p = 0.04$) such that patients reported a linear increase in physical activity over time ($p < 0.01$), but controls' physical activity decreased from baseline through the 6-month follow-up but increased thereafter ($p < 0.01$). There was a significant interaction of within-person fluctuations in circulating levels of TNFR1 with group on physical activity ($p = 0.03$), but this effect was not significant for patients or controls when each group was examined separately (*p*s > 0.05). Similarly, there was a significant interaction of between-person variance in circulating levels of IL-1ra by group on physical activity ($p = 0.04$), but this effect was not significant for patients or controls when each group was examined separately (*p*s > 0.05). However, at times when participant levels of circulating CRP were higher than their normal, they reported significantly less physical activity ($p = 0.03$) (i.e., main effect of within-person variance in circulating CRP).

4. Discussion

This study is among the first to examine relationships between circulating inflammatory biomarkers and treatment-related symptoms before, during, and after chemotherapy for gynecologic cancer. As hypothesized, inflammatory biomarkers were generally higher and fatigue, depression, and sleep disturbance were typically worse in patients with gynecologic cancer than controls. Physical activity was lower in patients before treatment and became more similar to controls over time. However, only CRP was associated with greater depression and less physical activity among both patients and controls. There were no associations with other inflammatory markers and fatigue or sleep disturbance.

Consistent with the previous literature [42,44,45], our results demonstrated that several inflammatory biomarkers (i.e., IL-1B, TNF-alpha, TNFR1, TNFR2, CRP, and IL-1ra) were elevated in patients prior to chemotherapy as compared to controls. The majority of these remained elevated in patients compared to controls before the third and sixth infusion of chemotherapy, and at 6 months and 12 months after completing chemotherapy, consistent with other studies of cancer patients [47–52]. For most of these cytokines (i.e., IL-10, IL-1b, TNFR2, or IL-6), there were no significant associations between symptoms and inflammation. In particular, between-person differences in inflammatory biomarkers were not associated with fatigue. While this finding is concordant with research on IL-6 and CRP in lung cancer patients undergoing chemotherapy [59] and in prostate cancer patients treated with androgen deprivation therapy [28], it is in contrast to research examining cytokines and fatigue in other cancer patient populations [30,58].

For patients only, there were quadratic changes in CRP levels over time such that CRP levels declined for patients only during chemotherapy, and subsequently increased by 12 months after treatment. Levels of CRP remained unchanged for controls assessed at similar timepoints. The results also indicated that both patients and controls with higher levels of circulating CRP than other participants, on average, reported worse depression. This is consistent with previous research identifying a significant association between CRP levels and depression severity in patients with other types of cancer [89–92]. In addition, women in our sample with higher levels of circulating CRP than their personal average reported less physical activity. This finding is consistent with prior research identifying higher CRP levels among post-treatment women with breast cancer who were less physically active than their personal average [93] and who exhibited poor physical fitness post treatment [94]. However, the associations between CRP and depression and physical activity were similar for patients and controls. Considering patients consistently reported worse symptomatology than controls, and patients demonstrated fluctuations in CRP over time that were not found in controls, one possible explanation for the lack of a group-based difference in the association between CRP and depression is that there are multiple triggering factors of symptoms, such as oxidative stress, genetic risk factors, metabolic dysregulation, or host and microbiome genomic risk factors [41,95]. For patients, symptoms may also be perpetuated by other symptoms, irrespective of systemic inflammation [41].

This study had several strengths, including a longitudinal study design, the inclusion of on-treatment and post-treatment pro-inflammatory cytokine and symptom data, and a non-cancer comparison group. However, study limitations should also be noted. The sample was relatively homogeneous in terms of race and ethnicity, which could limit the generalizability of these findings. We did not collect height and weight data from controls; thus, we could not include body mass index as a covariate in longitudinal analyses. We also had limited statistical power to identify moderate associations between fluctuations in inflammatory biomarkers and symptoms throughout the treatment period. Of note, some patients may have been taking anti-inflammatory drugs (e.g., NSAIDS); thus, we may have underestimated inflammation levels. Additional research is needed to replicate these findings in larger and more diverse samples.

5. Conclusions

In conclusion, our results suggest that several inflammatory biomarkers and symptoms are elevated or worse in gynecologic cancer patients prior to, during, and after chemotherapy compared to non-cancer controls. For both patients and controls, levels of CRP were associated with worse symptomatology (i.e., depression and physical activity). These associations were no stronger for patients than for controls, suggesting that there may be other causal mechanisms of symptoms in gynecologic cancer patients beyond inflammation. Additional research is needed to identify additional biological mechanisms of symptoms and examine whether reductions in inflammation (e.g., circulating levels of CRP) can reduce the severity of symptoms in patients treated with chemotherapy. Such interventions may considerably improve the quality of life of patients with gynecologic cancer treated with chemotherapy. Further, reductions in inflammation may improve quality of life in individuals without cancer.

Supplementary Materials: The following supporting information can be downloaded at: https://www.mdpi.com/article/10.3390/cancers15133407/s1, Table S1. Adjusted parameter estimates from mixed models examining group differences in changes in biomarkers of inflammation among patients with gynecologic cancer treated with chemotherapy and non-cancer controls; Table S2. Adjusted parameter estimates from mixed models examining group differences in changes in treatment-related symptoms among patients with gynecologic cancer treated with chemotherapy and non-cancer controls; Table S3. Adjusted parameter estimates from mixed models examining associations between fluctuations in biomarkers of inflammation with symptoms among patients with gynecologic cancer treated with chemotherapy and non-cancer controls.

Author Contributions: Conceptualization, A.I.H. and H.S.L.J.; data curation, A.I.H., C.B., B.D.G., H.W.B., B.W.J. and B.A.; formal analysis, A.I.H. and B.J.S.; funding acquisition, H.S.L.J.; investigation, B.J.S.; methodology, A.I.H.; project administration, Y.R.; supervision, M.C.J.; visualization, A.I.H.; writing—original draft, A.I.H. and H.S.L.J.; writing—review and editing, A.I.H., L.B.O., C.B., Y.R., B.D.G., X.L., M.C.J., H.W.B., B.W.J., B.A., C.C.-E., M.K.T., S.S.T., P.C.R., J.E.B., S.M.A., R.M.W. and H.S.L.J. All authors have read and agreed to the published version of the manuscript.

Funding: This work was supported by the National Cancer Institute (R01CA164109; U54CA163071, U54CA163068) and the Participant Research, Interventions, and Measurements Core Facility at the H. Lee Moffitt Cancer Center & Research Institute, an NCI-designated Comprehensive Cancer Center (P30CA076292).

Institutional Review Board Statement: The study was conducted according to the guidelines of the Declaration of Helsinki and approved by the Institutional Review Board of the University of South Florida (protocol #: Pro00005797).

Informed Consent Statement: Informed consent was obtained from all subjects involved in the study.

Data Availability Statement: The data presented in this study can be shared upon reasonable request.

Conflicts of Interest: B.D.G. reports fees unrelated to this work from SureMed Compliance and Elly Health, Inc. H.S.L.J. is a paid consultant for RedHill BioPharma, Janssen Scientific Affairs, and Merck. All other authors have no relevant financial or non-financial interests to disclose.

References

1. Beesley, V.L.; Price, M.A.; Butow, P.N.; Green, A.C.; Olsen, C.M.; Australian Ovarian Cancer Study Group; Australian Ovarian Cancer Study—Quality of Life Study; Webb, P.M. Physical activity in women with ovarian cancer and its association with decreased distress and improved quality of life. *Psychooncology* **2011**, *20*, 1161–1169. [PubMed]
2. Stevinson, C.; Faught, W.; Steed, H.; Tonkin, K.; Ladha, A.B.; Vallance, J.K.; Capstick, V.; Schepansky, A.; Courneya, K.S. Associations between physical activity and quality of life in ovarian cancer survivors. *Gynecol. Oncol.* **2007**, *106*, 244–250. [CrossRef] [PubMed]
3. Stevinson, C.; Steed, H.; Faught, W.; Tonkin, K.; Vallance, J.K.; Ladha, A.B.; Schepansky, A.; Capstick, V.; Courneya, K.S. Physical activity in ovarian cancer survivors: Associations with fatigue, sleep, and psychosocial functioning. *Int. J. Gynecol. Cancer* **2009**, *19*, 73–78. [CrossRef]
4. Pozzar, R.A.; Hammer, M.J.; Paul, S.M.; Cooper, B.A.; Kober, K.M.; Conley, Y.P.; Levine, J.D.; Miaskowski, C. Distinct sleep disturbance profiles among patients with gynecologic cancer receiving chemotherapy. *Gynecol. Oncol.* **2021**, *163*, 419–426. [CrossRef]
5. Poort, H.; de Rooij, B.H.; Uno, H.; Weng, S.; Ezendam, N.P.M.; van de Poll-Franse, L.; Wright, A.A. Patterns and predictors of cancer-related fatigue in ovarian and endometrial cancers: 1-year longitudinal study. *Cancer* **2020**, *126*, 3526–3533. [CrossRef]
6. Armstrong, D.K.; Alvarez, R.D.; Bakkum-Gamez, J.N.; Barroilhet, L.; Behbakht, K.; Berchuck, A.; Chen, L.M.; Cristea, M.; DeRosa, M.; Eisenhauer, E.L.; et al. Ovarian Cancer, Version 2.2020, NCCN Clinical Practice Guidelines in Oncology. *J. Natl. Compr. Cancer Netw.* **2021**, *19*, 191–226. [CrossRef]
7. Jim, H.S.; Jacobsen, P.B.; Phillips, K.M.; Wenham, R.M.; Roberts, W.; Small, B.J. Lagged relationships among sleep disturbance, fatigue, and depressed mood during chemotherapy. *Health Psychol.* **2013**, *32*, 768–774. [CrossRef]
8. Jim, H.S.; Small, B.; Faul, L.A.; Franzen, J.; Apte, S.; Jacobsen, P.B. Fatigue, depression, sleep, and activity during chemotherapy: Daily and intraday variation and relationships among symptom changes. *Ann. Behav. Med.* **2011**, *42*, 321–333. [CrossRef]
9. Bluthe, R.M.; Pawlowski, M.; Suarez, S.; Parnet, P.; Pittman, Q.; Kelley, K.W.; Dantzer, R. Synergy between tumor necrosis factor alpha and interleukin-1 in the induction of sickness behavior in mice. *Psychoneuroendocrinology* **1994**, *19*, 197–207. [CrossRef] [PubMed]
10. Raison, C.L.; Borisov, A.S.; Majer, M.; Drake, D.F.; Pagnoni, G.; Woolwine, B.J.; Vogt, G.J.; Massung, B.; Miller, A.H. Activation of central nervous system inflammatory pathways by interferon-alpha: Relationship to monoamines and depression. *Biol. Psychiatry* **2009**, *65*, 296–303. [CrossRef] [PubMed]
11. Capuron, L.; Ravaud, A.; Dantzer, R. Early depressive symptoms in cancer patients receiving interleukin 2 and/or interferon alfa-2b therapy. *J. Clin. Oncol.* **2000**, *18*, 2143–2151. [CrossRef]
12. Capuron, L.; Ravaud, A.; Dantzer, R. Timing and specificity of the cognitive changes induced by interleukin-2 and interferon-alpha treatments in cancer patients. *Psychosom. Med.* **2001**, *63*, 376–386. [CrossRef] [PubMed]
13. Kelley, K.W.; Bluthe, R.M.; Dantzer, R.; Zhou, J.H.; Shen, W.H.; Johnson, R.W.; Broussard, S.R. Cytokine-induced sickness behavior. *Brain Behav. Immun.* **2003**, *17* (Suppl. 1), S112–S118. [CrossRef]
14. Malik, U.R.; Makower, D.F.; Wadler, S. Interferon-mediated fatigue. *Cancer* **2001**, *92* (Suppl. S6), 1664–1668. [CrossRef]
15. Swiergiel, A.H.; Burunda, T.; Patterson, B.; Dunn, A.J. Endotoxin- and interleukin-1-induced hypophagia are not affected by adrenergic, dopaminergic, histaminergic, or muscarinic antagonists. *Pharmacol. Biochem. Behav.* **1999**, *63*, 629–637. [CrossRef]
16. Yirmiya, R. Endotoxin produces a depressive-like episode in rats. *Brain Res.* **1996**, *711*, 163–174. [CrossRef]

17. Bower, J.E.; Ganz, P.A.; Irwin, M.R.; Castellon, S.; Arevalo, J.; Cole, S.W. Cytokine genetic variations and fatigue among patients with breast cancer. *J. Clin. Oncol.* **2013**, *31*, 1656–1661. [CrossRef] [PubMed]
18. Jeon, S.W.; Kim, Y.K. Inflammation-induced depression: Its pathophysiology and therapeutic implications. *J. Neuroimmunol.* **2017**, *313*, 92–98. [CrossRef] [PubMed]
19. Karshikoff, B.; Sundelin, T.; Lasselin, J. Role of Inflammation in Human Fatigue: Relevance of Multidimensional Assessments and Potential Neuronal Mechanisms. *Front. Immunol.* **2017**, *8*, 21. [CrossRef]
20. Irwin, M.R. Sleep and inflammation: Partners in sickness and in health. *Nat. Rev. Immunol.* **2019**, *19*, 702–715. [CrossRef]
21. Kiecolt-Glaser, J.K.; Derry, H.M.; Fagundes, C. Inflammation: Depression fans the flames and feasts on the heat. *Am. J. Psychiatry* **2015**, *172*, 1075–1091. [CrossRef] [PubMed]
22. Miller, A.H.; Ancoli-Israel, S.; Bower, J.E.; Capuron, L.; Irwin, M.R. Neuroendocrine-immune mechanisms of behavioral comorbidities in patients with cancer. *J. Clin. Oncol.* **2008**, *26*, 971–982. [CrossRef] [PubMed]
23. Anisman, H. Cascading effects of stressors and inflammatory immune system activation: Implications for major depressive disorder. *J. Psychiatry Neurosci.* **2009**, *34*, 4–20. [PubMed]
24. Muller, N.; Schwarz, M.J. The immune-mediated alteration of serotonin and glutamate: Towards an integrated view of depression. *Mol. Psychiatry* **2007**, *12*, 988–1000. [CrossRef]
25. Pucak, M.L.; Kaplin, A.I. Unkind cytokines: Current evidence for the potential role of cytokines in immune-mediated depression. *Int. Rev. Psychiatry* **2005**, *17*, 477–483. [CrossRef]
26. Dunn, A.J.; Wang, J.; Ando, T. Effects of cytokines on cerebral neurotransmission. Comparison with the effects of stress. *Adv. Exp. Med. Biol.* **1999**, *461*, 117–127.
27. Schubert, C.; Hong, S.; Natarajan, L.; Mills, P.J.; Dimsdale, J.E. The association between fatigue and inflammatory marker levels in cancer patients: A quantitative review. *Brain Behav. Immun.* **2007**, *21*, 413–427. [CrossRef]
28. Hoogland, A.I.; Jim, H.S.L.; Gonzalez, B.D.; Small, B.J.; Gilvary, D.; Breen, E.C.; Bower, J.E.; Fishman, M.; Zachariah, B.; Jacobsen, P.B. Systemic inflammation and symptomatology in patients with prostate cancer treated with androgen deprivation therapy: Preliminary findings. *Cancer* **2021**, *127*, 1476–1482. [CrossRef]
29. Bower, J.E.; Ganz, P.A.; Tao, M.L.; Hu, W.; Belin, T.R.; Sepah, S.; Cole, S.; Aziz, N. Inflammatory biomarkers and fatigue during radiation therapy for breast and prostate cancer. *Clin. Cancer Res.* **2009**, *15*, 5534–5540. [CrossRef]
30. Wang, X.S.; Williams, L.A.; Krishnan, S.; Liao, Z.; Liu, P.; Mao, L.; Shi, Q.; Mobley, G.M.; Woodruff, J.F.; Cleeland, C.S. Serum sTNF-R1, IL-6, and the development of fatigue in patients with gastrointestinal cancer undergoing chemoradiation therapy. *Brain Behav. Immun.* **2012**, *26*, 699–705. [CrossRef]
31. Bower, J.E.; Ganz, P.A.; Aziz, N.; Fahey, J.L. Fatigue and proinflammatory cytokine activity in breast cancer survivors. *Psychosom. Med.* **2002**, *64*, 604–611. [CrossRef]
32. Bower, J.E.; Ganz, P.A.; Aziz, N.; Fahey, J.L.; Cole, S.W. T-cell homeostasis in breast cancer survivors with persistent fatigue. *J. Natl. Cancer Inst.* **2003**, *95*, 1165–1168. [CrossRef]
33. Collado-Hidalgo, A.; Bower, J.E.; Ganz, P.A.; Cole, S.W.; Irwin, M.R. Inflammatory biomarkers for persistent fatigue in breast cancer survivors. *Clin. Cancer Res.* **2006**, *12*, 2759–2766. [CrossRef] [PubMed]
34. Shafqat, A.; Einhorn, L.H.; Hanna, N.; Sledge, G.W.; Hanna, A.; Juliar, B.E.; Monahan, P.; Bhatia, S. Screening studies for fatigue and laboratory correlates in cancer patients undergoing treatment. *Ann. Oncol.* **2005**, *16*, 1545–1550. [CrossRef] [PubMed]
35. Gelinas, C.; Fillion, L. Factors related to persistent fatigue following completion of breast cancer treatment. *Oncol. Nurs. Forum* **2004**, *31*, 269–278. [CrossRef]
36. Dimeo, F.; Schmittel, A.; Fietz, T.; Schwartz, S.; Kohler, P.; Boning, D.; Thiel, E. Physical performance, depression, immune status and fatigue in patients with hematological malignancies after treatment. *Ann. Oncol.* **2004**, *15*, 1237–1242. [CrossRef]
37. Bower, J.E.; Ganz, P.A.; Irwin, M.R.; Kwan, L.; Breen, E.C.; Cole, S.W. Inflammation and behavioral symptoms after breast cancer treatment: Do fatigue, depression, and sleep disturbance share a common underlying mechanism? *J. Clin. Oncol.* **2011**, *29*, 3517–3522. [CrossRef] [PubMed]
38. Xiao, C.; Beitler, J.J.; Higgins, K.A.; Conneely, K.; Dwivedi, B.; Felger, J.; Wommack, E.C.; Shin, D.M.; Saba, N.F.; Ong, L.Y.; et al. Fatigue is associated with inflammation in patients with head and neck cancer before and after intensity-modulated radiation therapy. *Brain Behav. Immun.* **2016**, *52*, 145–152. [CrossRef] [PubMed]
39. Manigault, A.W.; Ganz, P.A.; Irwin, M.R.; Cole, S.W.; Kuhlman, K.R.; Bower, J.E. Moderators of inflammation-related depression: A prospective study of breast cancer survivors. *Transl. Psychiatry* **2021**, *11*, 615. [CrossRef]
40. Manigault, A.W.; Kuhlman, K.R.; Irwin, M.R.; Cole, S.W.; Ganz, P.A.; Crespi, C.M.; Bower, J.E. Vulnerability to inflammation-related depressive symptoms: Moderation by stress in women with breast cancer. *Brain Behav. Immun.* **2021**, *94*, 71–78. [CrossRef] [PubMed]
41. Sleight, A.G.; Crowder, S.L.; Skarbinski, J.; Coen, P.; Parker, N.H.; Hoogland, A.I.; Gonzalez, B.D.; Playdon, M.C.; Cole, S.; Ose, J.; et al. A New Approach to Understanding Cancer-Related Fatigue: Leveraging the 3P Model to Facilitate Risk Prediction and Clinical Care. *Cancers* **2022**, *14*, 1982. [CrossRef] [PubMed]
42. Lutgendorf, S.K.; Weinrib, A.Z.; Penedo, F.; Russell, D.; DeGeest, K.; Costanzo, E.S.; Henderson, P.J.; Sephton, S.E.; Rohleder, N.; Lucci, J.A., 3rd; et al. Interleukin-6, cortisol, and depressive symptoms in ovarian cancer patients. *J. Clin. Oncol.* **2008**, *26*, 4820–4827. [CrossRef] [PubMed]
43. Witek-Janusek, L.; Gabram, S.; Mathews, H.L. Psychologic stress, reduced NK cell activity, and cytokine dysregulation in women experiencing diagnostic breast biopsy. *Psychoneuroendocrinology* **2007**, *32*, 22–35. [CrossRef]

44. Palesh, O.G.; Collie, K.; Batiuchok, D.; Tilston, J.; Koopman, C.; Perlis, M.L.; Butler, L.D.; Carlson, R.; Spiegel, D. A longitudinal study of depression, pain, and stress as predictors of sleep disturbance among women with metastatic breast cancer. *Biol. Psychol.* **2007**, *75*, 37–44. [CrossRef] [PubMed]
45. Berger, A.M.; Farr, L.A.; Kuhn, B.R.; Fischer, P.; Agrawal, S. Values of sleep/wake, activity/rest, circadian rhythms, and fatigue prior to adjuvant breast cancer chemotherapy. *J. Pain Symptom Manag.* **2007**, *33*, 398–409. [CrossRef] [PubMed]
46. Bower, J.E. Cancer-related fatigue—mechanisms, risk factors, and treatments. *Nat. Rev. Clin. Oncol.* **2014**, *11*, 597–609. [CrossRef] [PubMed]
47. Collins, T.S.; Lee, L.F.; Ting, J. Paclitaxel up-regulates interleukin-8 synthesis in human lung carcinoma through an NF-kappaB- and AP-1-dependent mechanism. *Cancer Immunol. Immunother.* **2000**, *49*, 78–84. [CrossRef]
48. White, C.M.; Martin, B.K.; Lee, L.F.; Haskill, J.S.; Ting, J. Effects of paclitaxel on cytokine synthesis by unprimed human monocytes, T lymphocytes, and breast cancer cells. *Cancer Immunol. Immunother.* **1998**, *46*, 104–112. [CrossRef]
49. Bogdan, C.; Ding, A. Taxol, a microtubule-stabilizing antineoplastic agent, induces expression of tumor necrosis factor alpha and interleukin-1 in macrophages. *J. Leukoc. Biol.* **1992**, *52*, 119–121. [CrossRef]
50. Mullins, D.W.; Burger, C.J.; Elgert, K.D. Paclitaxel enhances macrophage IL-12 production in tumor-bearing hosts through nitric oxide. *J. Immunol.* **1999**, *162*, 6811–6818. [CrossRef]
51. Pusztai, L.; TMendoza, R.; Reuben, J.M.; Martinez, M.M.; Willey, J.S.; Lara, J.; Syed, A.; Fritsche, H.A.; Bruera, E.; Booser, D.; et al. Changes in plasma levels of inflammatory cytokines in response to paclitaxel chemotherapy. *Cytokine* **2004**, *25*, 94–102. [CrossRef]
52. Wang, T.H.; Chan, Y.-H.; Chen, C.W.; Kung, W.H.; Lee, Y.S.; Wang, S.T.; Chang, T.C.; Wang, H.S. Paclitaxel (Taxol) upregulates expression of functional interleukin-6 in human ovarian cancer cells through multiple signaling pathways. *Oncogene* **2006**, *25*, 4857–4866. [CrossRef]
53. Penson, R.T.; Kronish, K.; Duan, Z.; Feller, A.J.; Stark, P.; Cook, S.E.; Duska, L.R.; Fuller, A.F.; Goodman, A.K.; Nikrui, N.; et al. Cytokines IL-1beta, IL-2, IL-6, IL-8, MCP-1, GM-CSF and TNFalpha in patients with epithelial ovarian cancer and their relationship to treatment with paclitaxel. *Int. J. Gynecol. Cancer* **2000**, *10*, 33–41. [CrossRef]
54. Tsavaris, N.; Kosmas, C.; Vadiaka, M.; Kanelopoulos, P.; Boulamatsis, D. Immune changes in patients with advanced breast cancer undergoing chemotherapy with taxanes. *Br. J. Cancer* **2002**, *87*, 21–27. [CrossRef] [PubMed]
55. Janelsins, M.C.; Mustian, K.M.; Palesh, O.G.; Mohile, S.G.; Peppone, L.J.; Sprod, L.K.; Heckler, C.E.; Roscoe, J.A.; Katz, A.W.; Williams, J.P.; et al. Differential expression of cytokines in breast cancer patients receiving different chemotherapies: Implications for cognitive impairment research. *Support Care Cancer* **2012**, *20*, 831–839. [CrossRef] [PubMed]
56. Bower, J.E.; Ganz, P.A.; Irwin, M.R.; Cole, S.W.; Carroll, J.; Kuhlman, K.R.; Petersen, L.; Garet, D.; Asher, A.; Hurvitz, S.A.; et al. Acute and Chronic Effects of Adjuvant Therapy on Inflammatory Markers in Breast Cancer Patients. *JNCI Cancer Spectr.* **2022**, *6*, pkac052. [CrossRef] [PubMed]
57. Mills, P.J.; Parker, B.; Dimsdale, J.E.; Sadler, G.R.; Ancoli-Israel, S. The relationship between fatigue and quality of life and inflammation during anthracycline-based chemotherapy in breast cancer. *Biol. Psychol.* **2005**, *69*, 85–96. [CrossRef]
58. Wang, X.S.; Shi, Q.; Williams, L.A.; Mao, L.; Cleeland, C.S.; Komaki, R.R.; Mobley, G.M.; Liao, Z. Inflammatory cytokines are associated with the development of symptom burden in patients with NSCLC undergoing concurrent chemoradiation therapy. *Brain Behav. Immun.* **2010**, *24*, 968–974. [CrossRef]
59. Chou, H.L.; Chao, T.Y.; Chen, T.C.; Chu, C.M.; Hsieh, C.H.; Yao, C.T.; Janckila, A.J. The Relationship Between Inflammatory Biomarkers and Symptom Distress in Lung Cancer Patients Undergoing Chemotherapy. *Cancer Nurs.* **2017**, *40*, E1–E8. [CrossRef]
60. Jacobsen, P.B.; Donovan, K.A.; Small, B.J.; Jim, H.S.; Munster, P.N.; Andrykowski, M.A. Fatigue after treatment for early stage breast cancer: A controlled comparison. *Cancer* **2007**, *110*, 1851–1859. [CrossRef]
61. Lee, E.S.; Lee, M.K.; Kim, S.H.; Ro, J.S.; Kang, H.S.; Kim, S.W.; Lee, K.S.; Yun, Y.H. Health-related quality of life in survivors with breast cancer 1 year after diagnosis compared with the general population: A prospective cohort study. *Ann. Surg.* **2011**, *253*, 101–108. [CrossRef] [PubMed]
62. Arndt, V.; Merx, H.; Stegmaier, C.; Ziegler, H.; Brenner, H. Quality of life in patients with colorectal cancer 1 year after diagnosis compared with the general population: A population-based study. *J. Clin. Oncol.* **2004**, *22*, 4829–4836. [CrossRef]
63. Andrykowski, M.A.; Donovan, K.A.; Laronga, C.; Jacobsen, P.B. Prevalence, predictors, and characteristics of off-treatment fatigue in breast cancer survivors. *Cancer* **2010**, *116*, 5740–5748. [CrossRef] [PubMed]
64. Wefel, J.S.; Saleeba, A.K.; Buzdar, A.U.; Meyers, C.A. Acute and late onset cognitive dysfunction associated with chemotherapy in women with breast cancer. *Cancer* **2010**, *116*, 3348–3356. [CrossRef]
65. Goedendorp, M.M.; Andrykowski, M.A.; Donovan, K.A.; Jim, H.; Phillips, K.; Small, B.J.; Laronga, C.; Jacobsen, J. Prolonged impact of chemotherapy on fatigue in breast cancer survivors: A longitudinal comparison with radiotherapy treated breast cancer survivors and non-cancer controls. *Cancer* **2011**, *118*, 3833–3841. [CrossRef]
66. Phillips, K.; Jim, H.; Small, B.; Laronga, C.; Andrykowski, M.A.; Jacobsen, P. Cognitive functioning after cancer treatment: A three-year longitudinal comparison of breast cancer survivors treated with chemotherapy or radiation and non-cancer controls. *Cancer* **2011**, *118*, 1925–1932. [CrossRef] [PubMed]
67. Oswald, L.B.; Eisel, S.L.; Tometich, D.B.; Bryant, C.; Hoogland, A.I.; Small, B.J.; Apte, S.M.; Chon, H.S.; Shahzad, M.M.; Gonzalez, B.D.; et al. Cumulative burden of symptomatology in patients with gynecologic malignancies undergoing chemotherapy. *Health Psychol.* **2022**, *41*, 864–873. [CrossRef]
68. Diggle, P.J.; Heagerty, P.; Liang, K.-Y.; Zeger, S.L. *Analysis of Longitudinal Data*; Oxford University Press: New York, NY, USA, 2022.
69. Chaudhry, S.; Jin, L.; Meltzer, D. Use of a self-report-generated Charlson Comorbidity Index for predicting mortality. *Med. Care* **2005**, *43*, 607–615. [CrossRef]

70. Charlson, M.E.; Pompei, P.; Ales, K.L.; MacKenzie, C.R. A new method of classifying prognostic comorbidity in longitudinal studies: Development and validation. *J. Chronic Dis.* **1987**, *40*, 373–383. [CrossRef]
71. Greenberg, D.B.; Gray, J.L.; Mannix, C.M.; Eisenthal, S.; Carey, M. Treatment-related fatigue and serum interleukin-1 levels in patients during external beam irradiation for prostate cancer. *J. Pain Symptom Manag.* **1993**, *8*, 196–200. [CrossRef]
72. Wratten, C.; Kilmurray, J.; Nash, S.; Seldon, M.; Hamilton, C.S.; O'Brien, P.C.; Denham, J.W. Fatigue during breast radiotherapy and its relationship to biological factors. *Int. J. Radiat. Oncol. Biol. Phys.* **2004**, *59*, 160–167. [CrossRef]
73. Bower, J.E. Cancer-related fatigue: Links with inflammation in cancer patients and survivors. *Brain Behav. Immun.* **2007**, *21*, 863–871. [CrossRef] [PubMed]
74. O'Connor, M.F.; Bower, J.E.; Cho, H.J.; Creswell, J.D.; Dimitrov, S.; Hamby, M.E.; Hoyt, M.A.; Martin, J.L.; Robles, T.F.; Sloan, E.K.; et al. To assess, to control, to exclude: Effects of biobehavioral factors on circulating inflammatory markers. *Brain Behav. Immun.* **2009**, *23*, 887–897. [CrossRef] [PubMed]
75. Lutgendorf, S.K.; Logan, H.; Costanzo, E.; Lubaroff, D. Effects of acute stress, relaxation, and a neurogenic inflammatory stimulus on interleukin-6 in humans. *Brain Behav. Immun.* **2004**, *18*, 55–64. [CrossRef] [PubMed]
76. Hann, D.M.; Jacobsen, P.B.; Azzarello, L.M.; Martin, S.C.; Curran, S.L.; Fields, K.K.; Greenberg, H.; Lyman, G. Measurement of fatigue in cancer patients: Development and validation of the Fatigue Symptom Inventory. *Qual. Life Res.* **1998**, *7*, 301–310. [CrossRef] [PubMed]
77. Donovan, K.A.; Jacobsen, P.B.; Small, B.J.; Munster, P.N.; Andrykowski, M.A. Identifying clinically meaningful fatigue with the Fatigue Symptom Inventory. *J. Pain Symptom Manag.* **2008**, *36*, 480–487. [CrossRef]
78. Moorey, S.; Greer, S.; Watson, M.; Gorman, C.; Rowden, L.; Tunmore, R.; Robertson, B.; Bliss, J. The factor structure and factor stability of the hospital anxiety and depression scale in patients with cancer. *Br. J. Psychiatry* **1991**, *158*, 255–259. [CrossRef]
79. Zigmond, A.S.; Snaith, R. The hospital anxiety and depression scale. *Acta Psychiatr. Scand.* **1983**, *67*, 361–370. [CrossRef]
80. Hipkins, J.; Whitworth, M.; Tarrier, N.; Jayson, G. Social support, anxiety and depression after chemotherapy for ovarian cancer: A prospective study. *Br. J. Health Psychol.* **2004**, *9*, 569–581. [CrossRef] [PubMed]
81. Buysse, D.J.; Reynolds, C.F., 3rd; Monk, T.H.; Berman, S.R.; Kupfer, D.J. The Pittsburgh Sleep Quality Index: A new instrument for psychiatric practice and research. *Psychiatry Res.* **1989**, *28*, 193–213. [CrossRef]
82. Lee, P.H.; Macfarlane, D.J.; Lam, T.H.; Stewart, S.M. Validity of the International Physical Activity Questionnaire Short Form (IPAQ-SF): A systematic review. *Int. J. Behav. Nutr. Phys. Act.* **2011**, *8*, 115. [CrossRef]
83. Forde, C. *Scoring the International Physical Activity Questionnaire (IPAQ)*; University of Dublin: Dublin, Ireland, 2018; Volume 3.
84. Rock, C.L.; Thomson, C.; Gansler, T.; Gapstur, S.M.; McCullough, M.L.; Patel, A.V.; Andrews, K.S.; Bandera, E.V.; Spees, C.K.; Robien, K.; et al. American Cancer Society guideline for diet and physical activity for cancer prevention. *CA Cancer J. Clin.* **2020**, *70*, 245–271. [CrossRef] [PubMed]
85. Park, S.H.; Knobf, M.T.; Kerstetter, J.; Jeon, S. Adherence to American Cancer Society Guidelines on Nutrition and Physical Activity in Female Cancer Survivors: Results From a Randomized Controlled Trial (Yale Fitness Intervention Trial). *Cancer Nurs.* **2019**, *42*, 242–250. [CrossRef] [PubMed]
86. Hoffman, L. *Longitudinal Analysis: Modeling Within-Person Fluctuation and Change*; Routledge: East Sussex, UK, 2015.
87. Franceschi, C.; Campisi, J. Chronic inflammation (inflammaging) and its potential contribution to age-associated diseases. *J. Gerontol. A Biol. Sci. Med. Sci.* **2014**, *69* (Suppl. 1), S4–S9. [CrossRef] [PubMed]
88. Bottazzi, B.; Riboli, E.; Mantovani, A. Aging, inflammation and cancer. *Semin. Immunol.* **2018**, *40*, 74–82. [CrossRef]
89. Howren, M.B.; Lamkin, D.M.; Suls, J. Associations of depression with C-reactive protein, IL-1, and IL-6: A meta-analysis. *Psychosom. Med.* **2009**, *71*, 171–186. [CrossRef] [PubMed]
90. Renna, M.E.; Shrout, M.R.; Madison, A.A.; Alfano, C.M.; Povoski, S.P.; Lipari, A.M.; Carson, W.E., 3rd; Malarkey, W.B.; Kiecolt-Glaser, J.K. Depression and anxiety in colorectal cancer patients: Ties to pain, fatigue, and inflammation. *Psychooncology* **2022**, *31*, 1536–1544. [CrossRef]
91. Kuhlman, K.R.; Irwin, M.R.; Ganz, P.A.; Cole, S.W.; Manigault, A.W.; Crespi, C.M.; Bower, J.E. Younger women are more susceptible to inflammation: A longitudinal examination of the role of aging in inflammation and depressive symptoms. *J. Affect. Disord.* **2022**, *310*, 328–336. [CrossRef]
92. McFarland, D.C.; Doherty, M.; Atkinson, T.M.; O'Hanlon, R.; Breitbart, W.; Nelson, C.J.; Miller, A.H. Cancer-related inflammation and depressive symptoms: Systematic review and meta-analysis. *Cancer* **2022**, *128*, 2504–2519. [CrossRef]
93. Sabiston, C.M.; Wrosch, C.; Castonguay, A.L.; Sylvester, B.D. Changes in physical activity behavior and C-reactive protein in breast cancer patients. *Ann. Behav. Med.* **2018**, *52*, 545–551. [CrossRef]
94. Romero-Elías, M.; Álvarez-Bustos, A.; Cantos, B.; Maximiano, C.; Méndez, M.; Méndez, M.; de Pedro, C.G.; Rosado-García, S.; Sanchez-Lopez, A.J.; García-González, D.; et al. C-Reactive Protein Is Associated with Physical Fitness in Breast Cancer Survivors. *J. Clin. Med.* **2022**, *12*, 65. [CrossRef] [PubMed]
95. Crowder, S.L.; Hoogland, A.I.; Welniak, T.L.; LaFranchise, E.A.; Carpenter, K.M.; Li, D.; Rotroff, D.M.; Mariam, A.; Pierce, C.M.; Fischer, S.M.; et al. Metagenomics and chemotherapy-induced nausea: A roadmap for future research. *Cancer* **2022**, *128*, 461–470. [CrossRef] [PubMed]

Article

Relationships among Inflammatory Biomarkers and Objectively Assessed Physical Activity and Sleep during and after Chemotherapy for Gynecologic Malignancies

Danielle B. Tometich [1,*], Aasha I. Hoogland [1], Brent J. Small [2], Michelle C. Janelsins [3], Crystal Bryant [1], Yvelise Rodriguez [1], Brian D. Gonzalez [1], Xiaoyin Li [1], Hailey W. Bulls [4], Brian W. James [5], Bianca Arboleda [5], Claudia Colon-Echevarria [6], Mary K. Townsend [6], Shelley S. Tworoger [6], Paulo Rodriguez [7], Laura B. Oswald [1], Julienne E. Bower [8], Sachin M. Apte [9], Robert M. Wenham [10], Hye Sook Chon [10], Mian M. Shahzad [10] and Heather S. L. Jim [1]

1 Department of Health Outcomes and Behavior, Moffitt Cancer Center, Tampa, FL 33612, USA
2 School of Aging Studies, University of South Florida, Tampa, FL 33620, USA
3 Department of Surgery and Neuroscience, University of Rochester Medical Center, Rochester, NY 14642, USA
4 Section of Palliative Care and Medical Ethics, Division of General Internal Medicine, Department of Medicine, University of Pittsburgh, Pittsburgh, PA 15260, USA
5 Morsani College of Medicine, University of South Florida, Tampa, FL 33620, USA
6 Department of Cancer Epidemiology, Moffitt Cancer Center, Tampa, FL 33612, USA
7 Department of Immunology, Moffitt Cancer Center, Tampa, FL 33612, USA
8 Department of Psychology, University of California Los Angeles, Los Angeles, CA 90095, USA
9 Huntsman Cancer Institute, Department of Obstetrics and Gynecology, University of Utah, Salt Lake City, UT 84112, USA
10 Department of Gynecologic Oncology, Moffitt Cancer Center, Tampa, FL 33612, USA
* Correspondence: danielle.tometich@moffitt.org

Citation: Tometich, D.B.; Hoogland, A.I.; Small, B.J.; Janelsins, M.C.; Bryant, C.; Rodriguez, Y.; Gonzalez, B.D.; Li, X.; Bulls, H.W.; James, B.W.; et al. Relationships among Inflammatory Biomarkers and Objectively Assessed Physical Activity and Sleep during and after Chemotherapy for Gynecologic Malignancies. *Cancers* **2023**, *15*, 3882. https://doi.org/10.3390/cancers15153882

Academic Editors: Christian Singer and António Araújo

Received: 31 May 2023
Revised: 28 June 2023
Accepted: 29 July 2023
Published: 30 July 2023

Simple Summary: Physical inactivity and sleep problems are commonly reported by women going through chemotherapy for gynecologic cancer. Inflammation from cancer and its treatment might contribute to these issues, but existing findings are limited. We examined relationships between biomarkers of inflammation and data from wearable devices to objectively measure physical activity and sleep. We collected data from women with gynecologic cancer during chemotherapy and followed up with them for a year after completing chemotherapy. We also compared their results to women without cancer who were assessed at similar time intervals. We found that women with cancer were less active and had more sleep problems than controls even a year after completing chemotherapy. Greater inflammation was also related to less physical activity and more sleep problems. Future research should test whether interventions aimed at reducing inflammation can help women with cancer to be more active and have fewer sleep problems.

Abstract: Little is known regarding associations between inflammatory biomarkers and objectively measured physical activity and sleep during and after chemotherapy for gynecologic cancer; thus, we conducted a longitudinal study to address this gap. Women with gynecologic cancer (patients) and non-cancer controls (controls) completed assessments before chemotherapy cycles 1, 3, and 6 (controls assessed contemporaneously), as well as at 6- and 12-month follow-ups. Physical activity and sleep were measured using wrist-worn actigraphs and sleep diaries, and blood was drawn to quantify circulating levels of inflammatory markers. Linear and quadratic random-effects mixed models and random-effects fluctuation mixed models were used to examine physical activity and sleep over time, as well as the associations with inflammatory biomarkers. On average, patients ($n = 97$) and controls ($n = 104$) were 62 and 58 years old, respectively. Compared to controls, patients were less active, more sedentary, had more time awake after sleep onset, and had lower sleep efficiency (p-values < 0.05). Across groups, higher levels of TNF-α were associated with more sedentary time and less efficient sleep (p-values ≤ 0.05). Higher levels of IL-1β, TNF-α, and IL-6 were associated with lower levels of light physical activity (p-values < 0.05). Associations between inflammatory biomarkers, physical activity, and sleep did not differ between patients and controls. Given these results, we speculate that inflammation may contribute to less physical activity and more sleep problems that persist even 12 months after completing chemotherapy.

Keywords: gynecologic cancer; physical activity; sleep; inflammation

1. Introduction

Over 1.4 million people in the United States are living with gynecologic cancers, including primary cancers of the ovary, uterus, and cervix [1]. Survival rates are improving for ovarian cancer but have remained relatively stagnant for uterine and cervical cancer for the past 50 years [1,2]. Currently, 5-year relative survival rates for gynecologic cancer range from 49% for ovarian to 81% for uterine cancer. Standard of care treatment includes surgery and chemotherapy for locally advanced ovarian cancer, surgery and/or combination chemotherapy/radiation for cervical cancers, and surgery followed by observation or chemotherapy and/or radiation for uterine cancers [1,3,4]. Although systemic chemotherapy treatment is often necessary for effective treatment, chemotherapy can result in distressing side effects that interfere with quality of life [5–9].

Inflammation is a potential mechanism to explain the relationship of cancer and its treatment with well-established 'sickness behaviors' associated with cancer, such as reduced physical activity and changes in sleep patterns [10,11]. Sleep disturbance reportedly affects up to 80% of people with gynecologic cancer during treatment [12], and prior research has found evidence for inflammation-associated sleep disturbance lasting at least one year after treatment for patients with ovarian cancer [13]. Physical activity can also be affected in those with gynecologic cancer, as fewer than 20% of ovarian cancer survivors report meeting recommendations for 150 min of moderate to strenuous physical activity per week [5]. Existing research on physical activity, sleep, and/or inflammation among individuals with gynecologic cancers is limited by the use of self-reported measures of physical activity and sleep [7,12–15], few assessments during chemotherapy [13,15] or short-term follow-ups of a few weeks up to four months after treatment [12,16], and/or a lack of control comparison [7,12–16]. There is currently a paucity of research designed to examine associations between inflammation, sleep, and physical activity during and following chemotherapy for gynecologic cancer. Collecting this information will inform biobehavioral symptom management interventions.

The goal of this study was to examine longitudinal relationships between circulating biomarkers of inflammation (i.e., IL-10, IL-1β, TNF-α, IL-6, IL-1Ra, TNFR1, TNFR2, and CRP) and objectively measured physical activity and sleep in individuals with gynecologic cancer during and after chemotherapy (i.e., patients) as compared to frequency age-matched individuals without cancer (i.e., controls), assessed contemporaneously. We hypothesized that (1) patients would have worse sleep and less physical activity than controls before, during, and after chemotherapy; (2) participants with higher levels of inflammation would report worse sleep and less physical activity than those with lower levels of inflammation (between-person effects); (3) at times when participants had higher levels of inflammation than their own average, they would also report worse sleep and less physical activity (within-person effects). We also explored differences in relationships among circulating biomarkers of inflammation and sleep and physical activity by group.

2. Materials and Methods

2.1. Participants

The study methodology has been described in detail elsewhere [17–19]. Briefly, participants were recruited between August 2013 and July 2018 prior to starting a new chemotherapy regimen for gynecologic malignancies at the Moffitt Cancer Center. Inclusion criteria were (1) 18–89 years of age; (2) able to speak and read English; (3) diagnosed with a gynecologic malignancy (e.g., ovarian, endometrial, uterine, cervical, vulvar, fallopian tube, or peritoneal); (4) scheduled to start intravenous or intraperitoneal chemotherapy; (5) no current or prior history of immune-related diseases (e.g., HIV, rheumatoid arthritis, systemic lupus erythematosus); (6) no documented psychiatric, sleep, or neurological disorders that could interfere with study participation (e.g., psychosis, sleep apnea, dementia); (7) no receipt of chemotherapy or radiation in the month prior to enrollment; (8) not pregnant; and (9) able to provide informed consent. Eligibility criteria for controls were the same apart from the presence of a gynecologic malignancy and chemotherapy. Additionally, controls were required not to have a history of any form of cancer except non-melanoma skin cancer, and they had to be within five years of age of the patient participant to whom they were being matched. This study was approved by the University of South Florida Institutional Review Board.

Recruitment procedures for patients included physician referral, screening of clinic schedules, and in-person or telephone screening by a trained research coordinator to determine eligibility and obtain informed consent. Controls were recruited from a contact list from a national marketing company. Patients who consented to participate completed assessments at eight timepoints: pre-chemotherapy cycle 1 (i.e., 1 week before beginning treatment) and post-chemotherapy cycle 1 (i.e., 1 week after first treatment); pre- and post-chemotherapy cycle 3 (i.e., 1 week before and 1 week after cycle 3); pre- and post-chemotherapy cycle 6 (i.e., 1 week before and 1 week after cycle 6); 6-month follow-up (i.e., 1 week 6 months after cycle 6); and 12-month follow-up (i.e., 1 week 12 months after cycle 6). Controls were assessed contemporaneously (i.e., two-week assessments where first week coincides with pre-chemotherapy cycle and second week with post-cycle, 6 weeks between first two and second two assessments, 9 weeks between second two and third two assessments, and one-week assessments at 6 and 12 months after third two assessments). Participants were compensated USD 25 per assessment after returning the survey and actigraph. The current project used data from five timepoints: pre-chemotherapy cycle 1, pre-chemotherapy cycle 3, pre-chemotherapy cycle 6, 6-month follow-up, and 12-month follow-up.

2.2. Measures

2.2.1. Sociodemographic and Clinical Variables

Demographics were self-reported by participants at baseline (i.e., age, race, education, household income, comorbidities, and menopausal status). Medical record review for cases only was conducted for clinical data (i.e., cancer type, cancer stage, prior lines of chemotherapy).

2.2.2. Physical Activity and Sleep

Wrist-worn actigraph devices (ActiGraph GT3X, MiniMitter, Bend, OR, USA) were provided to participants, along with instructions to continuously wear the device on their non-dominant wrist for each assessment period (1 week per assessment). Prior research has found actigraphy to provide reliable and valid data for the estimation of sleep and physical activity in cancer populations [20–22]. Participants also completed daily sleep diaries of bedtimes and rising times. Sleep diary data were integrated with sustained periods of inactivity measured by actigraphy, and Cole/Kripke scoring algorithms [23] were used to estimate sleep efficiency (i.e., percentage of time in bed spent asleep) and the amount of time (in hours) awake after sleep onset. Data were downloaded via the ActiLife software (version 6.13.4, ActiGraph, Pensacola, FL) and raw data were exported to GGIR (version 2.8-2) in R (version 3.6.3) for processing. Criteria for data inclusion were at least 50% wear time and 3 valid wear days [24]. Outliers were reviewed and data were excluded if sleep time or activity appeared inconsistent with actigrams (i.e., graphs that showed physical activity over the entire wear period of the assessment for each participant). This procedure excluded data from 8 assessments (i.e., one patient and one control at baseline, one patient at pre-cycle 3, one control at the assessment corresponding to pre-cycle 6, one control at 6-month follow-up, and one patient and two controls at 12-month follow-up). For physical activity, movement over one-minute epochs was used to measure activity. After data processing, the daily activity level based on Euclidean Norms Minus One (ENMO) and the duration of each activity level were obtained in minutes and converted into hours. Thresholds for activity levels were <40 ENMO for sedentary behavior, 40–99 ENMO for light activity, and ≥100 ENMO for moderate-to-vigorous physical activity (MVPA) [20,21]. For sleep, data were excluded if the average sleep time was <3 or >12 h per night as these values were likely errors [25]. This excluded data from 3 assessments (i.e., one control at baseline, and one patient and one control at 12-month follow-up).

2.2.3. Biomarkers of Inflammation

Participants provided blood samples at five timepoints: baseline (i.e., before starting chemotherapy for patients), before chemotherapy cycles 3 and 6 (and similar time intervals for controls), and again at 6 and 12 months after patients had completed chemotherapy. Each blood sample was evaluated for the presence of circulating markers of inflammation, which included interleukin 10 (IL-10), interleukin 1β (IL-1β), tumor necrosis factor α (TNF-α), interleukin 6 (IL-6), interleukin 1 receptor antagonist (IL-1Ra), tumor necrosis factor receptor 1 (TNFR1), tumor necrosis factor receptor 2 (TNFR2), and c-reactive protein (CRP). These inflammatory biomarkers were selected because previous studies have shown that they are associated with physical activity and sleep, they are relatively stable over time, and they are readily detectable using existing laboratory methodologies [10,11,13]. All blood samples were sent to the Cancer Control and Psychoneuroimmunology Lab at the University of Rochester for analysis. All samples were assayed in one run on a multiplexed cytokine bead assay (i.e., IL-10, IL-1β, TNF-α [HSTCMAG-28SK-04], TNFR1, TNFR2 [HSCRMAG-32K-02], IL-6, and IL-1Ra [HCYTOMAG-60K-01]) or using enzyme-linked immunosorbent assays (i.e., CRP; R&D Systems Human Quantikine ELISA; Minneapolis, MN, DCRP00; R&D Systems Human Quantikine ELISA; Minneapolis, MN, DCRP00) per the manufacturer's instructions. The same lot was used for all kits. The median concentrated was taken from 50 beads per well for Luminex, and the average was taken from duplicates for ELISA. All data and internal controls were inspected for a CV < 20%, with all kits run with a standard curve with an $r^2 > 0.98$. The same plate was used for all sample collections from the same participant. The lower limits of detection of the assays, with sample dilution taken into account, were IL-10 = 0.30; IL-1β = 0.14; IL-6 = 0.04; TNFα = 0.08; IL-1Rα = 7.41; TNFR1 = 10.60; TNFR2 = 10.18; CRP = 5.0 pg/mL.

2.3. Data Analyses

Circulating markers of inflammation with estimates below the lower limit of detection were divided by 1.4, and indeterminate inflammatory biomarker concentrations were set to missing. Inflammatory biomarker values that were three standard deviations from the sample mean for each group (cases and controls) were set to missing. Raw inflammatory biomarkers with non-normal distributions (i.e., IL-6, IL-1Ra) were natural log-transformed to normalize their distributions. All cytokines that were not natural log-transformed (i.e., IL-10, IL-1β, TNF-α, TNFR1, TNFR2, CRP) were mean-centered across all participants to facilitate interpretation.

Participants with non-missing data for at least one biomarker and actigraphy data at one or more timepoints were included in analyses. Sociodemographic and clinical characteristics of the sample were described using means, standard deviations, frequencies, and percentages. Independent-sample t-tests, chi-square tests, and Fisher's tests were used to evaluate differences in sociodemographic and clinical characteristics between groups (i.e., patients vs. controls). Independent-sample t-tests were also used to evaluate differences in physical activity and sleep between groups at each timepoint. Linear and quadratic random-effects mixed models were used to examine changes in physical activity and sleep over time between each group and within each group. Time was coded as the number of months since baseline. Random-effects fluctuation mixed models were used to examine associations of inflammatory biomarkers with physical activity and sleep aggregated over the five assessments [26]. Biomarkers of inflammation were included as between-person predictors (having an average level of inflammation that differed from other participants) and within-person predictors (having an average level of inflammation that differed from a participant's own average). To determine whether associations differed between patients and controls, interactions between the group and fluctuations in inflammatory biomarkers with physical activity and sleep were also included. Significant interactions between the group and fluctuations in inflammatory biomarkers were further probed within each group using separate random-effects fluctuation models with the group effect removed. Sensitivity analyses were completed by excluding patients with metastatic disease and examining fluctuations in inflammatory biomarkers with physical activity and sleep. All statistical analyses were conducted using SAS version 9.4 (Cary, NC, USA).

3. Results

The sociodemographic and clinical characteristics of patients (n = 97) and controls (n = 104) are displayed in Table 1. Patients were older, more likely to be White, and less likely to have a college education or a household income of USD 40,000 or more than controls (p-values < 0.05). Patients also reported more comorbidities on average (p = 0.05) and were more likely to be post-menopausal than controls (p < 0.01). Because of these group differences, we included age, race, education, comorbidities, and menopausal status as covariates. Income was not included as a covariate because it was highly correlated with education (Spearman's rho = 0.34, p < 0.0001). Among patients, cancer types included ovarian (51%), endometrial (36%), and other gynecologic malignancies (13%).

Table 1. Participant characteristics.

Variable	Patients (n = 97)	Controls (n = 104)	p-Values
Age: M (SD)	61.62 (10.07)	58.38 (12.44)	0.05
Race: n (%) White	92 (96)	92 (89)	0.05
Education: n (%) college graduate	33 (34)	74 (71)	<0.0001
Income: n (%) USD 40,000 or more	51 (68)	73 (81)	0.05
Comorbidities: M (SD, range)	2.40 (0.85, 2–7)	2.19 (0.53, 2–5)	0.05
Menopausal status: n (%)			<0.01
Pre-menopausal	10 (11)	26 (25)	
Post-menopausal	84 (89)	78 (75)	
Cancer Type: n (%)		-	-
Cervical	1 (1)		
Fallopian	3 (3)		
Ovarian	49 (51)		
Vulvar	1 (1)		
Endometrial	35 (36)		
Peritoneal	5 (5)		
Other	2 (2)		
Stage: n (%)		-	-
1	18 (20)		
2	9 (10)		
3	48 (53)		
4	16 (18)		
Prior lines of chemotherapy: n (%) 3 or more	9 (9)	-	-

3.1. Group Differences in Biomarkers of Inflammation

Raw and log-transformed (Il-6 and IL-1Ra only) means and longitudinal changes in biomarkers of inflammation in this study are reported elsewhere [27]. Patients generally had higher circulating levels of inflammatory biomarkers than controls over time. The results of the random-effects mixed models also demonstrated group differences over time such that the levels of some inflammatory biomarkers (i.e., TNFR1, TNFR2, IL-1Ra, CRP) generally decreased in patients but remained unchanged in controls.

3.2. Group Differences in Physical Activity and Sleep

Raw means for physical activity and sleep are presented in Table 2. Cross-sectional comparisons of physical activity (i.e., time in hours for MVPA, light activity, and sedentary time) and sleep (time in hours awake after sleep onset, sleep efficiency) by group indicated that patients spent significantly less time each day engaging in both light activity and MVPA than controls at all timepoints (patients engaged in 17.4 to 37.2 fewer minutes of MVPA and 18.6 to 44.4 min of light activity per day than controls). Patients were also significantly more sedentary throughout chemotherapy and at the 12-month follow-up compared to controls (patients were sedentary for 16.2 to 30 more minutes per day than controls). Patients had significantly more time awake after sleep onset than controls at all timepoints (patients spent 16.2 to 30 more minutes awake after sleep onset per night than controls) and had lower sleep efficiency than controls throughout chemotherapy and at 12-month follow-up (patients had 3–6% less efficient sleep than controls).

Table 2. Physical activity and sleep raw means.

	Baseline		Pre-Chemo 3		Pre-Chemo 6		6 Month Follow-Up		12 Month Follow-Up	
	Patients	Controls	Patients	Controls	Patients	Controls	Patients	Controls	Patients	Controls
Physical Activity										
MVPA [a]	**0.55 (0.42)**	**1.17 (0.52)**	**0.74 (0.52)**	**1.29 (0.6)**	**0.64 (0.48)**	**1.15 (0.53)**	**0.85 (0.5)**	**1.14 (0.53)**	**0.71 (0.49)**	**1.12 (0.52)**
Light Activity [b]	**2.1 (0.66)**	**2.84 (0.82)**	**2.34 (0.8)**	**3.04 (0.84)**	**2.21 (0.84)**	**2.82 (0.86)**	**2.6 (0.62)**	**2.91 (0.82)**	**2.38 (0.81)**	**2.9 (0.88)**
Sedentary Time [c]	11.88 (1.27)	11.45 (1.38)	**11.68 (1.45)**	**11.2 (1.5)**	11.73 (1.42)	11.32 (1.34)	11.26 (1.51)	11.14 (1.24)	**11.79 (1.57)**	**11.24 (1.22)**
Sleep Measures										
WASO [d]	**1.6 (0.76)**	**1.1 (0.56)**	**1.56 (0.78)**	**1.21 (0.73)**	**1.65 (0.95)**	**1.29 (0.73)**	**1.39 (0.75)**	**1.12 (0.64)**	**1.46 (0.77)**	**1.07 (0.62)**
Sleep Efficiency [e]	**0.79 (0.09)**	**0.85 (0.06)**	**0.8 (0.08)**	**0.83 (0.08)**	**0.8 (0.09)**	**0.83 (0.08)**	0.82 (0.08)	0.84 (0.07)	**0.81 (0.08)**	**0.84 (0.08)**

[a] MVPA = moderate-to-vigorous physical activity in hours per day. [b] Light activity in hours per day. [c] Sedentary time in hours per day. [d] WASO = wake after sleep onset in hours per night. [e] Sleep efficiency, or proportion of time in bed spent asleep. Significant group differences are in bold.

Adjusted means of physical activity and sleep from the quadratic random-effects mixed models controlling for group (adjusted for age, race, education, comorbidities, and menopausal status) are displayed in Figure 1a–e. There was a significant quadratic change in MVPA for patients only, such that the number of hours spent in MVPA gradually decreased during chemotherapy and increased thereafter (from 0.63 h per day at baseline to 0.54 h before cycle 3 and 0.78 h at 12-month follow-up, $p < 0.01$). The results of the mixed models revealed significant interactions between group and time for light activity ($p = 0.01$) and time awake after sleep onset ($p = 0.03$). Specifically, the estimated hours per day of light activity increased over time for patients (from 2.22 h per day at baseline to 2.49 h at 12-month follow-up, $p = 0.02$) but remained unchanged for controls (2.86 to 2.89 h per day, $p = 0.70$). Although the interaction between group and time was significant for time awake after sleep onset in the entire sample, the effect of time was not significant for patients or controls when each group was examined separately (p-values > 0.05). There were no other significant changes in physical activity or sleep over time.

3.3. Longitudinal Relationships among Biomarkers of Inflammation and Physical Activity and Sleep

3.3.1. Moderate-to-Vigorous Activity

The results of the random-effects fluctuation models are presented in Table 3. There was a significant interaction of within-person variance in IL-10 with group ($B = 0.007$, $p = 0.04$); there was no association between IL-10 and moderate-to-vigorous activity for controls ($B = 0.002$, $p = 0.27$). At times when patients had greater circulating IL-10, they tended to engage in less moderate-to-vigorous physical activity ($B = -0.005$, $p = 0.05$).

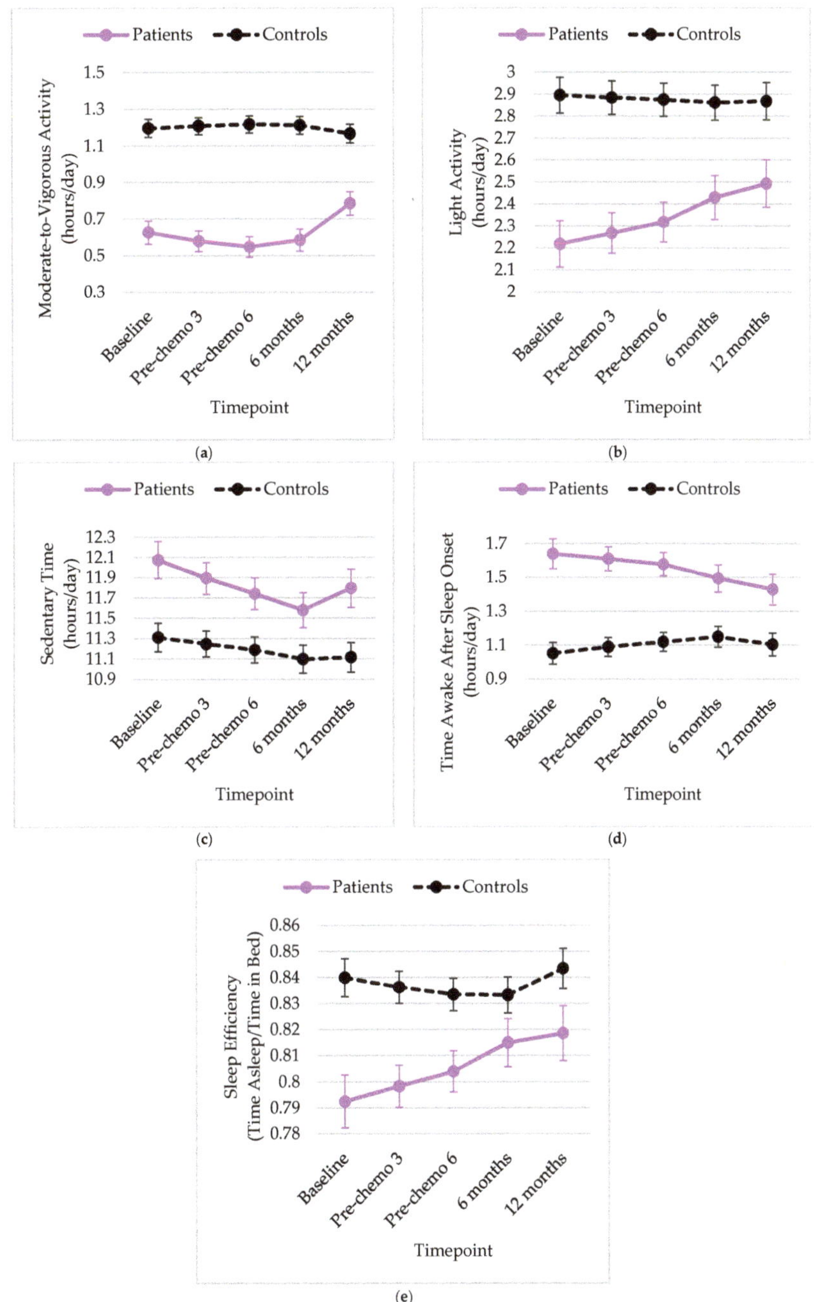

Figure 1. Least squares means over time by group for (**a**) moderate-to-vigorous activity hours per day, (**b**) light activity hours per day, (**c**) sedentary time hours per day, (**d**) average hours awake after sleep onset per night, (**e**) sleep efficiency, or proportion of time in bed spent sleeping.

Table 3. Associations of fluctuations in biomarkers of inflammation with sleep and physical activity among patients with gynecologic cancer treated with chemotherapy and noncancer controls.

Variable	Moderate to Vigorous Activity							
	IL-10	IL-1β	TNF-α	TNFR1	TNFR2	CRP	IL-6^	IL-1Ra^
Intercept	1.76 ***	1.84 ***	1.85 ***	1.72 ***	1.86 ***	1.90 ***	1.81 ***	1.75 ***
Group	−0.46 ***	−0.70 ***	−0.73 ***	−0.52 **	−0.68 ***	−0.61 ***	−0.57 ***	−0.59 ***
Between-person variance in cytokine	2.91×10^{-3}	-6.91×10^{-3}	-8.71×10^{-3}	5.26×10^{-4}	-8.00×10^{-5}	−0.03	0.04	7.88×10^{-3}
Within-person variance in cytokine	2.46×10^{-3}	-1.66×10^{-3}	-1.77×10^{-3}	-2.00×10^{-5}	1.90×10^{-5}	1.45×10^{-3}	−0.01	0.02
Group × between-person variance in cytokine	-5.37×10^{-3}	0.04	0.01	-4.20×10^{-4}	1.05×10^{-4}	0.02	−0.02	0.02
Group × within-person variance in cytokine	7.24×10^{-3} *	7.34×10^{-3}	-8.89×10^{-3}	8.10×10^{-5}	2.00×10^{-5}	-7.70×10^{-3}	−0.01	−0.04
Variable	**Light Activity**							
	IL-10	IL-1β	TNF-α	TNFR1	TNFR2	CRP	IL-6^	IL-1Ra^
Intercept	3.14 ***	3.19 ***	3.15 ***	2.79 ***	2.93 ***	3.20 ***	3.04 ***	3.05 ***
Group	−0.75 ***	−0.79 ***	−0.48	−0.05	−0.20	−0.51 **	−0.50 ***	−0.42 *
Between-person variance in cytokine	-6.05×10^{-3}	−0.03	−0.02	1.12×10^{-3}	4.00×10^{-5}	−0.05	−0.07	−0.04
Within-person variance in cytokine	-6.55×10^{-3}	−0.04 *	−0.02 *	-3.80×10^{-4}	-1.10×10^{-4}	-8.12×10^{-3}	−0.08 *	−0.03
Group × between-person variance in cytokine	0.01	0.09	5.36×10^{-3}	-2.44×10^{-3}	-2.50×10^{-4}	0.03	0.06	-3.37×10^{-3}
Group × within-person variance in cytokine	5.51×10^{-3}	0.06 *	3.78×10^{-3}	-1.16×10^{-3}	-9.00×10^{-5}	−0.04	0.06	-2.12×10^{-3}
Variable	**Sedentary Time**							
	IL-10	IL-1β	TNF-α	TNFR1	TNFR2	CRP	IL-6^	IL-1Ra^
Intercept	10.86 ***	10.81 ***	10.55 ***	11.26 ***	11.02 ***	10.72 ***	10.99 ***	10.97 ***
Group	0.63	0.91 *	1.28 *	7.05×10^{-3}	0.37	0.29	0.42	0.28
Between-person variance in cytokine	-1.34×10^{-3}	0.02	0.04 *	-1.91×10^{-3}	-7.00×10^{-5}	0.06	0.06	6.91×10^{-4}
Within-person variance in cytokine	0.01 *	0.05	6.83×10^{-3}	1.13×10^{-3}	1.17×10^{-4}	−0.01	0.04	7.44×10^{-3}
Group × between-person variance in cytokine	-4.13×10^{-3}	−0.12	−0.06	3.20×10^{-3}	1.48×10^{-4}	0.02	3.76×10^{-3}	0.05
Group × within-person variance in cytokine	8.95×10^{-4}	−0.02	2.26×10^{-4}	-1.96×10^{-3}	-6.60×10^{-4}	0.09 *	0.05	0.15
Variable	**Wake after Sleep Onset**							
	IL-10	IL-1β	TNF-α	TNFR1	TNFR2	CRP	IL-6^	IL-1Ra^
Intercept	1.50 ***	1.46 ***	1.38 ***	1.42 ***	1.45 ***	1.35 ***	1.49 ***	1.50 ***
Group	0.49 **	0.53 **	0.59 **	0.51 **	0.42	0.51 ***	0.44 ***	0.68 ***
Between-person variance in cytokine	-2.00×10^{-4}	1.99×10^{-4}	9.05×10^{-3}	3.39×10^{-4}	7.30×10^{-5}	0.03	-8.63×10^{-3}	−0.01
Within-person variance in cytokine	-2.96×10^{-3}	-1.22×10^{-3}	-6.25×10^{-3}	-4.10×10^{-4}	-6.00×10^{-5}	1.72×10^{-3}	−0.01	−0.05
Group × between-person variance in cytokine	-2.88×10^{-3}	−0.03	−0.01	-5.30×10^{-4}	-9.88×10^{-6}	−0.03	−0.14	−0.12
Group × within-person variance in cytokine	-1.88×10^{-3}	0.03	8.23×10^{-3}	4.00×10^{-5}	1.35×10^{-4}	-1.04×10^{-3}	0.04	−0.01
Variable	**Sleep Efficiency**							
	IL-10	IL-1β	TNF-α	TNFR1	TNFR2	CRP	IL-6^	IL-1Ra^
Intercept	0.80 ***	0.80 ***	0.82 ***	0.81 ***	0.81 ***	0.84 ***	0.81 ***	0.81 ***
Group	−0.04 *	−0.03	−0.05 *	−0.03	−0.03	−0.04 *	−0.04 **	−0.08 ***
Between-person variance in cytokine	4.70×10^{-4}	1.03×10^{-3}	-1.34×10^{-3}	-2.00×10^{-5}	-2.00×10^{-5}	-4.96×10^{-3}	3.93×10^{-3}	-1.02×10^{-3}
Within-person variance in cytokine	-5.20×10^{-4}	-3.78×10^{-3}	-1.92×10^{-3} *	9.00×10^{-5}	1.90×10^{-5}	-1.97×10^{-3}	-7.81×10^{-3}	-2.32×10^{-3}
Group × between-person variance in cytokine	2.04×10^{-4}	9.06×10^{-4}	1.53×10^{-3}	7.50×10^{-6}	1.54×10^{-6}	3.90×10^{-3}	9.58×10^{-3}	0.02 *
Group × within-person variance in cytokine	8.70×10^{-4}	3.99×10^{-3}	5.50×10^{-4}	-1.40×10^{-4}	-3.00×10^{-5}	3.41×10^{-3}	5.32×10^{-3}	0.01

Analyses controlled for age, education, comorbidities, and menopausal status. IL-6^ and IL-1Ra^ were natural log-transformed. *** = $p < 0.001$; ** = $p < 0.01$; * = $p < 0.05$.

3.3.2. Light Activity

Participants with higher levels of circulating IL-1β, TNF-α, and IL-6 (log-transformed) than their personal average spent significantly fewer hours in light activity each day (i.e., main effects of within-person variance in inflammation) (IL-1β: $B = -0.04$, $p = 0.04$; TNF-α: $B = -0.02$, $p = 0.04$; IL-6: $B = -0.08$, $p = 0.02$). There was also a significant interaction of within-person variance in IL-1β by group ($B = 0.06$, $p = 0.04$), such that there was no association between IL-1β and light activity for patients ($B = 0.02$, $p = 0.49$), but when controls had greater circulating IL-1β, they were engaged in less light activity ($B = -0.04$, $p = 0.02$).

3.3.3. Sedentary Time

Participants with higher levels of circulating TNF-α than other participants tended to be more sedentary each day (i.e., main effect of between-person variance in inflammation; $B = 0.04$, $p = 0.05$). Participants with higher levels of circulating IL-10 than their personal average tended to be more sedentary each day (i.e., main effect of within-person variance in inflammation; $B = 0.01$, $p = 0.05$). There was also a trend for an interaction of within-person variance in CRP by group ($B = 0.09$, $p = 0.05$); however, this association of sedentary time with CRP was not significant when examined in patients ($B = 0.03$, $p = 0.19$) and controls separately ($B = -0.005$, $p = 0.88$).

3.3.4. Wake after Sleep Onset

There were no significant associations between circulating markers of inflammation and time awake after sleep onset.

3.3.5. Sleep Efficiency

Participants with higher levels of circulating TNF-α than their personal average also tended to have less efficient sleep (i.e., main effect of within-person variance in inflammation; $B = -0.002$, $p = 0.05$). There was also a significant interaction of between-person levels of IL-1Ra (log-transformed) and group on sleep efficiency ($B = 0.02$, $p = 0.02$); however, the association of IL-1Ra and sleep efficiency was not significant when examined in patients ($B = 0.008$, $p = 0.17$) and controls separately ($B = -0.001$, $p = 0.64$).

3.3.6. Sensitivity Analyses

After excluding 16 patient participants with metastatic disease, the effects were similar, with a few exceptions (see Supplemental Table S1). For light activity, the main effect of group was no longer significant in the model with IL-1Ra (log-transformed; $B = -0.25$, $p = 0.30$), and the interaction effect of group and within-person levels of IL-1β was no longer significant ($B = 0.03$, $p = 0.35$). For sleep efficiency, there was an additional trend for a main effect of group in the model with IL-1β ($B = -0.04$, $p = 0.05$).

4. Discussion

Changes to physical activity and sleep are commonly reported by people with gynecologic cancer during chemotherapy [12,13], yet there is limited research on the objective measurement of these sickness behaviors and associations with inflammation during and following chemotherapy. To our knowledge, this is the first longitudinal study with a control group to examine inflammatory markers and objectively measured physical activity and sleep during and following chemotherapy for gynecologic cancer. Our findings provide evidence for affected physical activity and sleep during chemotherapy. Additionally, when participants had higher circulating levels of inflammatory cytokines relative to their own average, they also demonstrated greater sickness behaviors in the form of reduced physical activity and sleep efficiency.

Patients were overall less active and had more sleep problems than controls. Compared to controls, patients engaged in less light and moderate-to-vigorous physical activity at all timepoints, and patients were more sedentary during chemotherapy and at 12-month follow-up. Patients increased their activity over time, demonstrating some recovery of activity following chemotherapy; however, patient activity levels remained affected even at 12-month follow-up. Compared to controls, patients also had more time awake after sleep onset at all timepoints and lower sleep efficiency during chemotherapy. These findings are consistent with prior research showing that reduced physical activity and sleep problems are common and can be long-term side effects of chemotherapy [5,7,12–15]. Our findings extend prior research by showing these associations using objectively measured physical activity and sleep, and by demonstrating changes in physical activity and sleep over the course of six chemotherapy cycles (worsening) as well as 6- and 12-month follow-ups (improving) among patients with gynecologic cancer.

We found that several inflammatory markers were significantly associated with physical activity and sleep. When participants had higher circulating levels of IL-1β, TNF-α, and IL-6 than their usual average, they had lower light physical activity. The associations for TNF-α and IL-6 were similar for both cases and controls, suggesting common underlying relationships between inflammation and both activity and sleep. Further, participants who had higher circulating levels of TNF-α also were more sedentary. When participants had higher TNF-α than their own average, they also had less efficient sleep. IL-6, TNF-α, and IL-1β are proinflammatory cytokines [28]. Prior research has found that TNF-α is essential for the sleep–wake cycle, and both elevations and decrements can disrupt the homeostasis necessary for efficient sleep [11,29]. These results suggest that greater inflammation is associated with less light physical activity, more sedentary behavior, and less efficient sleep.

Interestingly, higher IL-10, which has anti-inflammatory properties, was previously found to be associated with less sedentary time and greater physical activity in a healthy sample without cancer [30]. However, in our study, when participants had higher IL-10, they were more sedentary and engaged in less moderate-to-vigorous physical activity. The previous study enrolled both men and women under the age of 55 years, with a mean age of 32, while our study had a much older population on average (60 years) and comprised only females. It is possible that there may be differences in associations in older groups, particularly postmenopausal women, who tend to have higher levels of inflammation [31]. There is also emerging evidence that IL-10 can be 'non-classical' and pro-inflammatory in the context of cancer [32]. Future research should evaluate potential differences in these associations across sex and age and determine the impact of cancer on functional changes in IL-10 action.

Strengths of our study include the unique contribution of objectively measured physical activity and sleep during chemotherapy, with a full one year of follow-up, and a comparison to controls. The inclusion of inflammatory biomarkers and associations with objective activity and sleep measurements are also strengths. However, our study also had limitations. The majority of participants were White, with middle to upper-level socioeconomic status, and the study was conducted at one cancer center, which may limit the generalizability of the findings to other populations. Prior research has found disparities in gynecological morbidity and mortality outcomes between Black and White women in the United States, and the likely main source of this disparity is inadequate access to screening, HPV vaccination, and cancer treatment [33]. To enhance the generalizability and investigate potential disparities in inflammation, sleep, and physical activity, multiple recruitment sites and community networking are recommended for future research [34]. There were also significant group differences in a few sample characteristics (i.e., education, menopausal status), and this limitation was addressed by including covariates in the analyses. Income is another potential confounding variable; however, income and education were highly correlated, and including both as covariates would have resulted in multicollinearity. A future project with a larger sample may be better able to control for additional confounders such as income, as well as variables that were unmeasured in the current study (e.g., body mass index, waist-to-hip ratio).

The findings point to several opportunities for future research. Interventions aimed at reducing inflammation, such as an anti-inflammatory diet or medications (such as bupropion), potentially could address sickness behaviors [35–37]. In addition, randomized trials of health behavior change interventions such as exercise and cognitive–behavioral therapy for insomnia have shown promise in reducing markers of inflammation among healthy middle-aged and older adults [38,39] and cancer survivors [40,41]. Thus, there may be bidirectional effects in the relationship between inflammation and sickness behaviors, and an intervention focused on one may have a cascading positive impact. Since many associations were similar between cases and non-cases, this suggests that interventions addressing sickness behaviors or reducing inflammation may work in both populations. For example, TNF-α inhibitors have been used to treat multiple auto-immune diseases [42], which also exhibit similar sickness behaviors to those seen in cancer patients. Future research should identify the health behaviors and specific inflammatory factors with the strongest effects, the factors most amenable to long-term change, and which interventions work best for which patients at which timepoints in the trajectory from treatment to survivorship. Future research may also improve the generalizability by recruiting diverse samples and utilizing multiple sites and community partnerships.

5. Conclusions

In conclusion, this study is among the first to examine the associations between inflammatory markers and objectively measured physical activity and sleep during and following chemotherapy for gynecologic cancer. The findings suggest that greater inflammation is associated with less light physical activity, more sedentary behavior, and more sleep problems in those with and without cancer, and that patients are less active and have more sleep problems than controls, possibly due to their higher inflammation levels. Clinical implications include a need to examine anti-inflammatory health behavior change interventions or drugs among people with gynecologic cancer to address patient-reported outcomes.

Supplementary Materials: The following supporting information can be downloaded at: https://www.mdpi.com/article/10.3390/cancers15153882/s1, Table S1: Sensitivity analyses: Associations of fluctuations in biomarkers of inflammation with sleep and physical activity among patients with gynecologic cancer treated with chemotherapy and noncancer controls excluding those with metastatic disease.

Author Contributions: Conceptualization, D.B.T., A.I.H., B.J.S. and H.S.L.J.; methodology, D.B.T., A.I.H., B.J.S., M.C.J., M.K.T., J.E.B. and H.S.L.J.; software, A.I.H., B.J.S. and X.L.; validation, D.B.T., A.I.H., B.J.S., M.C.J., M.K.T., J.E.B. and H.S.L.J.; formal analysis, A.I.H., B.J.S. and X.L.; investigation, C.B., Y.R., H.W.B., B.W.J., B.A. and C.C.-E.; resources, M.C.J., M.K.T., S.S.T., P.R., J.E.B., S.M.A., R.M.W., H.S.C., M.M.S. and H.S.L.J.; data curation, D.B.T., A.I.H., X.L., H.W.B., B.W.J., B.A. and C.C.-E.; writing—original draft preparation, D.B.T., A.I.H. and H.S.L.J.; writing—review and editing, all authors; visualization, D.B.T. and A.I.H.; supervision, B.J.S. and H.S.L.J.; project administration, C.B. and Y.R.; funding acquisition, H.S.L.J. All authors have read and agreed to the published version of the manuscript.

Funding: This work was supported by the National Cancer Institute (NCI), grant number R01CA164109, and the Participant Research, Interventions, and Measurements Core Facility at the H. Lee Moffitt Cancer Center & Research Institute, an NCI designated Comprehensive Cancer Center, grant number P30CA076292. D.B.T. was supported by the NCI, grant number K99CA270294. H.W.B. was supported by the NCI, grant number K08CA263317. C.C.-E. was supported by the NCI, grant numbers U54CA163071 and U54CA163068. The views expressed in this article are those of the authors and do not necessarily represent the official position of the National Cancer Institute.

Institutional Review Board Statement: The study was conducted in accordance with the Declaration of Helsinki and approved by the Institutional Review Board of the University of South Florida (protocol code Pro0005797).

Informed Consent Statement: Informed consent was obtained from all subjects involved in the study.

Data Availability Statement: The data presented in this study are available on request from the corresponding author. The data are not publicly available due to patient participant privacy.

Acknowledgments: This work would not have been possible without the contributions of the women who participated in the study.

Conflicts of Interest: The funders had no role in the design of the study; in the collection, analyses, or interpretation of data; in the writing of the manuscript; or in the decision to publish the results.

References

1. American Cancer Society. *Cancer Treatment & Survivorship Facts & Figures 2022–2024*; American Cancer Society: Atlanta, GA, USA, 2022.
2. National Cancer Institute. Cancer Stat Facts: New Cases, Deaths, and 5-Year Relative Survival 1975–2019. Available online: https://seer.cancer.gov/statfacts/html/all.html (accessed on 29 July 2023).
3. Kumar, L.; Harish, P.; Malik, P.S.; Khurana, S. Chemotherapy and targeted therapy in the management of cervical cancer. *Curr. Probl. Cancer* **2018**, *42*, 120–128. [CrossRef]
4. Mikuła-Pietrasik, J.; Witucka, A.; Pakuła, M.; Uruski, P.; Begier-Krasińska, B.; Niklas, A.; Tykarski, A.; Książek, K. Comprehensive review on how platinum- and taxane-based chemotherapy of ovarian cancer affects biology of normal cells. *Cell Mol. Life Sci.* **2019**, *76*, 681–697. [CrossRef] [PubMed]
5. Joly, F.; Ahmed-Lecheheb, D.; Kalbacher, E.; Heutte, N.; Clarisse, B.; Grellard, J.M.; Gernier, F.; Berton-Rigaud, D.; Tredan, O.; Fabbro, M.; et al. Long-term fatigue and quality of life among epithelial ovarian cancer survivors: A GINECO case/control VIVROVAIRE I study. *Ann. Oncol.* **2019**, *30*, 845–852. [CrossRef]
6. Lee, Y.C.; King, M.T.; Connell, R.L.; Lanceley, A.; Joly, F.; Hilpert, F.; Davis, A.; Roncolato, F.T.; Okamoto, A.; Bryce, J.; et al. Symptom burden and quality of life with chemotherapy for recurrent ovarian cancer: The Gynecologic Cancer InterGroup-Symptom Benefit Study. *Int. J. Gynecol. Cancer* **2022**, *32*, 761. [CrossRef]
7. Ross, T.L.; DeFazio, A.; Friedlander, M.; Grant, P.; Nagle, C.M.; Williams, M.; Webb, P.M.; Beesley, V.L. Insomnia and its association with quality of life in women with ovarian cancer. *Gynecol. Oncol.* **2020**, *158*, 760–768. [CrossRef]
8. Westin, S.N.; Sun, C.C.; Tung, C.S.; Lacour, R.A.; Meyer, L.A.; Urbauer, D.L.; Frumovitz, M.M.; Lu, K.H.; Bodurka, D.C. Survivors of gynecologic malignancies: Impact of treatment on health and well-being. *J. Cancer Surviv.* **2016**, *10*, 261–270. [CrossRef]
9. Wu, N.; Su, X.; Song, H.; Li, Y.; Gu, F.; Sun, X.; Li, X.; Cheng, G. A multi-institutional retrospective analysis of oncologic outcomes for patients with locally advanced cervical cancer undergoing platinum-based adjuvant chemotherapy after concurrent chemoradiotherapy. *Cancer Control* **2021**, *28*, 1–12. [CrossRef] [PubMed]
10. Bower, J.E. Cancer-related fatigue—Mechanisms, risk factors, and treatments. *Nat. Rev. Clin. Oncol.* **2014**, *11*, 597–609. [CrossRef] [PubMed]
11. Zhao, C.; Grubbs, A.; Barber, E.L. Sleep and gynecological cancer outcomes: Opportunities to improve quality of life and survival. *Int. J. Gynecol. Cancer* **2022**, *32*, 669. [CrossRef]
12. Pozzar, R.A.; Hammer, M.J.; Paul, S.M.; Cooper, B.A.; Kober, K.M.; Conley, Y.P.; Levine, J.D.; Miaskowski, C. Distinct sleep disturbance profiles among patients with gynecologic cancer receiving chemotherapy. *Gynecol. Oncol.* **2021**, *163*, 419–426. [CrossRef]
13. Clevenger, L.; Schrepf, A.; Christensen, D.; DeGeest, K.; Bender, D.; Ahmed, A.; Goodheart, M.J.; Penedo, F.; Lubaroff, D.M.; Sood, A.K.; et al. Sleep disturbance, cytokines, and fatigue in women with ovarian cancer. *Brain Behav. Immun.* **2012**, *26*, 1037–1044. [CrossRef]
14. Jones, T.; Sandler, C.; Vagenas, D.; Janda, M.; Obermair, A.; Hayes, S. Physical activity levels among ovarian cancer survivors: A prospective longitudinal cohort study. *Int. J. Gynecol. Cancer* **2021**, *31*, 553–561. [CrossRef] [PubMed]
15. Beesley, V.L.; Ross, T.L.; King, M.T.; Campbell, R.; Nagle, C.M.; Obermair, A.; Grant, P.; DeFazio, A.; Webb, P.M.; Friedlander, M.L. Evaluating patient-reported symptoms and late adverse effects following completion of first-line chemotherapy for ovarian cancer using the MOST (Measure of Ovarian Symptoms and Treatment concerns). *Gynecol. Oncol.* **2022**, *164*, 437–445. [CrossRef] [PubMed]
16. Gorzelitz, J.; Costanzo, E.S.; Spencer, R.J.; Rumble, M.; Rose, S.L.; Cadmus-Bertram, L. Longitudinal assessment of post-surgical physical activity in endometrial and ovarian cancer patients. *PLoS ONE* **2019**, *14*, e0223791. [CrossRef] [PubMed]
17. Oswald, L.B.; Eisel, S.L.; Tometich, D.B.; Bryant, C.; Hoogland, A.I.; Small, B.J.; Apte, S.M.; Chon, H.S.; Shahzad, M.M.; Gonzalez, B.D.; et al. Cumulative burden of symptomatology in patients with gynecologic malignancies undergoing chemotherapy. *Health Psychol.* **2022**, *41*, 864–873. [CrossRef]
18. Bulls, H.W.; Hoogland, A.I.; Small, B.J.; Kennedy, B.; James, B.W.; Arboleda, B.L.; Shahzad, M.M.K.; Gonzalez, B.D.; Jim, H.S.L. Lagged relationships among chemotherapy-induced peripheral neuropathy, sleep quality, and physical activity during and after chemotherapy. *Ann. Behav. Med.* **2021**, *55*, 844–852. [CrossRef]
19. Bulls, H.W.; Hoogland, A.I.; Kennedy, B.; James, B.W.; Arboleda, B.L.; Apte, S.; Chon, H.S.; Small, B.J.; Gonzalez, B.D.; Jim, H.S.L. A longitudinal examination of associations between age and chemotherapy-induced peripheral neuropathy in patients with gynecologic cancer. *Gynecol. Oncol.* **2019**, *152*, 310–315. [CrossRef]
20. Hildebrand, M.; Van Hees, V.T.; Hansen, B.H.; Ekelund, U. Age group comparability of raw accelerometer output from wrist- and hip-worn monitors. *Med. Sci. Sports Exerc.* **2014**, *46*, 1816–1824. [CrossRef]

21. Bakrania, K.; Yates, T.; Rowlands, A.V.; Esliger, D.W.; Bunnewell, S.; Sanders, J.; Davies, M.; Khunti, K.; Edwardson, C.L. Intensity thresholds on raw acceleration data: Euclidean norm minus one (ENMO) and mean amplitude deviation (MAD) approaches. *PLoS ONE* **2016**, *11*, e0164045. [CrossRef]
22. Peddle-McIntyre, C.J.; Cavalheri, V.; Boyle, T.; McVeigh, J.A.; Jeffery, E.; Lynch, B.M.; Vallance, J.K. A review of accelerometer-based activity monitoring in cancer survivorship research. *Med. Sci. Sports Exerc.* **2018**, *50*, 1790–1801. [CrossRef]
23. Cole, R.J.; Kripke, D.F.; Gruen, W.; Mullaney, D.J.; Gillin, J.C. Automatic sleep/wake identification from wrist activity. *Sleep* **1992**, *15*, 461–469. [CrossRef]
24. Aadland, E.; Ylvisåker, E. Reliability of the Actigraph GT3X+ accelerometer in adults under free-living conditions. *PLoS ONE* **2015**, *10*, e0134606. [CrossRef] [PubMed]
25. Jones, S.E.; van Hees, V.T.; Mazzotti, D.R.; Marques-Vidal, P.; Sabia, S.; van der Spek, A.; Dashti, H.S.; Engmann, J.; Kocevska, D.; Tyrrell, J.; et al. Genetic studies of accelerometer-based sleep measures yield new insights into human sleep behaviour. *Nat. Commun.* **2019**, *10*, 1585. [CrossRef] [PubMed]
26. Hoffman, L. *Longitudinal Analysis: Modeling Within-Person Fluctuation and Change*; Routledge: New York, NY, USA, 2015.
27. Hoogland, A.I.; Small, B.J.; Oswald, L.B.; Bryant, C.; Rodriguez, Y.; Gonzalez, B.D.; Li, X.; Janelsins, M.C.; Bulls, H.W.; James, B.W.; et al. Relationships among inflammatory biomarkers and self-reported treatment-related symptoms in patients treated with chemotherapy for gynecologic cancer: A controlled comparison. *Cancers* **2023**, *15*, 3407. [CrossRef]
28. Turner, M.D.; Nedjai, B.; Hurst, T.; Pennington, D.J. Cytokines and chemokines: At the crossroads of cell signalling and inflammatory disease. *Biochim. Biophys. Acta* **2014**, *1843*, 2563–2582. [CrossRef] [PubMed]
29. Rockstrom, M.D.; Chen, L.; Taishi, P.; Nguyen, J.T.; Gibbons, C.M.; Veasey, S.C.; Krueger, J.M. Tumor necrosis factor alpha in sleep regulation. *Sleep Med. Rev.* **2018**, *40*, 69–78. [CrossRef]
30. Rodas, L.; Riera-Sampol, A.; Aguilo, A.; Martínez, S.; Tauler, P. Effects of habitual caffeine intake, physical activity levels, and sedentary behavior on the inflammatory status in a healthy population. *Nutrients* **2020**, *12*, 2325. [CrossRef]
31. Taleb-Belkadi, O.; Chaib, H.; Zemour, L.; Fatah, A.; Chafi, B.; Mekki, K. Lipid profile, inflammation, and oxidative status in peri- and postmenopausal women. *Gynecol. Endocrinol.* **2016**, *32*, 982–985. [CrossRef]
32. Islam, H.; Neudorf, H.; Mui, A.L.; Little, J.P. Interpreting 'anti-inflammatory' cytokine responses to exercise: Focus on interleukin-10. *J. Physiol.* **2021**, *599*, 5163–5177. [CrossRef]
33. Collins, Y.; Holcomb, K.; Chapman-Davis, E.; Khabele, D.; Farley, J.H. Gynecologic cancer disparities: A report from the Health Disparities Taskforce of the Society of Gynecologic Oncology. *Gynecol. Oncol.* **2014**, *133*, 353–361. [CrossRef]
34. Salman, A.; Nguyen, C.; Lee, Y.-H.; Cooksey-James, T. A Review of Barriers to Minorities' Participation in Cancer Clinical Trials: Implications for Future Cancer Research. *J. Immigr. Minor. Health* **2016**, *18*, 447–453. [CrossRef]
35. Haß, U.; Herpich, C.; Norman, K. Anti-inflammatory diets and fatigue. *Nutrients* **2019**, *11*, 2315. [CrossRef]
36. Yetkin, D.; Yılmaz, İ.A.; Ayaz, F. Anti-inflammatory activity of bupropion through immunomodulation of the macrophages. *Naunyn Schmiedebergs Arch. Pharmacol.* **2023**, 1–7. [CrossRef]
37. Hajhashemi, V.; Khanjani, P. Analgesic and anti-inflammatory activities of bupropion in animal models. *Res. Pharm. Sci.* **2014**, *9*, 251–257.
38. Zheng, G.; Qiu, P.; Xia, R.; Lin, H.; Ye, B.; Tao, J.; Chen, L. Effect of aerobic exercise on inflammatory markers in healthy middle-aged and older adults: A systematic review and meta-analysis of randomized controlled trials. *Front. Aging Neurosci.* **2019**, *11*, 1–9. [CrossRef] [PubMed]
39. Irwin, M.R.; Olmstead, R.; Carrillo, C.; Sadeghi, N.; Breen, E.C.; Witarama, T.; Yokomizo, M.; Lavretsky, H.; Carroll, J.E.; Motivala, S.J.; et al. Cognitive behavioral therapy vs. Tai Chi for late life insomnia and inflammatory risk: A randomized controlled comparative efficacy trial. *Sleep* **2014**, *37*, 1543–1552. [CrossRef]
40. Khosravi, N.; Stoner, L.; Farajivafa, V.; Hanson, E.D. Exercise training, circulating cytokine levels and immune function in cancer survivors: A meta-analysis. *Brain Behav. Immun.* **2019**, *81*, 92–104. [CrossRef] [PubMed]
41. Savard, J.; Simard, S.; Ivers, H.; Morin, C.M. Randomized study on the efficacy of cognitive-behavioral therapy for insomnia secondary to breast cancer, part II: Immunologic effects. *J. Clin. Oncol.* **2005**, *23*, 6097–6106. [CrossRef] [PubMed]
42. Jang, D.-i.; Lee, A.H.; Shin, H.-Y.; Song, H.-R.; Park, J.-H.; Kang, T.-B.; Lee, S.-R.; Yang, S.-H. The role of tumor necrosis factor alpha (TNF-α) in autoimmune disease and current TNF-α inhibitors in therapeutics. *Int. J. Mol. Sci.* **2021**, *22*, 2719. [CrossRef] [PubMed]

 cancers

Systematic Review

The Effects of Patient-Reported Outcome Screening on the Survival of People with Cancer: A Systematic Review and Meta-Analysis

Caterina Caminiti [1,*], Giuseppe Maglietta [1], Francesca Diodati [1], Matteo Puntoni [1], Barbara Marcomini [1], Silvia Lazzarelli [1], Carmine Pinto [2] and Francesco Perrone [3]

1 Clinical and Epidemiological Research Unit, University Hospital of Parma, 43126 Parma, Italy
2 Medical Oncology, Comprehensive Cancer Center, Azienda USL-IRCCS Reggio Emilia, 42122 Reggio Emilia, Italy
3 Clinical Trials Unit, Istituto Nazionale Tumori IRCCS Fondazione G. Pascale, 80131 Naples, Italy
* Correspondence: ccaminiti@ao.pr.it

Simple Summary: Patient-reported outcomes (PROs) are information collected directly from patients regarding their health status. Emerging evidence has suggested that integrating PRO assessments into oncology clinical practice can have various benefits for patient care and health. This systematic review and meta-analysis investigated the effects of routine PRO monitoring on the overall survival of people with any type of cancer. We included six studies that compared these interventions to the care that is usually provided to cancer patients. The results seemed to indicate that monitoring PROs in cancer care could positively influence overall survival and that benefits could be largest for individuals with advanced lung cancer. Possible explanations for these findings are that PRO surveillance may allow clinicians to respond to problems more rapidly or that better symptom management could improve tolerance to therapy, thus extending its benefits. However, since available studies are few and of suboptimal quality, additional rigorous research is needed to consolidate our results.

Citation: Caminiti, C.; Maglietta, G.; Diodati, F.; Puntoni, M.; Marcomini, B.; Lazzarelli, S.; Pinto, C.; Perrone, F. The Effects of Patient-Reported Outcome Screening on the Survival of People with Cancer: A Systematic Review and Meta-Analysis. *Cancers* **2022**, *14*, 5470. https://doi.org/10.3390/cancers14215470

Academic Editor: António Araújo

Received: 13 October 2022
Accepted: 3 November 2022
Published: 7 November 2022

Publisher's Note: MDPI stays neutral with regard to jurisdictional claims in published maps and institutional affiliations.

Abstract: This study examined the effects of the routine assessment of patient-reported outcomes (PROs) on the overall survival of adult patients with cancer. We included clinical trials and observational studies with a control group that compared PRO monitoring interventions in cancer clinical practice to usual care. The Cochrane risk-of-bias tools were used. In total, six studies were included in the systematic review: two randomized trials, one population-based retrospectively matched cohort study, two pre–post with historical control studies and one non-randomized controlled trial. Half were multicenter, two were conducted in Europe, three were conducted in the USA and was conducted in Canada. Two studies considered any type of cancer, two were restricted to lung cancer and two were restricted to advanced forms of cancer. PRO screening was electronic in four of the six studies. The meta-analysis included all six studies (intervention = 130.094; control = 129.903). The pooled mortality outcome at 1 year was RR = 0.77 (95%CI 0.76–0.78) as determined by the common effect model and RR = 0.82 (95%CI 0.60–1.12; $p = 0.16$) as determined by the random-effects model. Heterogeneity was statistically significant ($I^2 = 73\%$; $p < 0.01$). The overall risk of bias was rated as moderate in five studies and serious in one study. This meta-analysis seemed to indicate the survival benefits of PRO screening. As routine PRO monitoring is often challenging, more robust evidence regarding the effects of PROs on mortality would support systematic applications.

Keywords: patient-reported outcome; overall survival; cancer; symptom monitoring; meta-analysis; systematic review

1. Introduction

Patient-reported outcomes (PROs) are defined by the USA Food and Drug Administration as "any report of the status of a patient's health condition that comes directly from the patient, without interpretation of the patient's response by a clinician or anyone else" [1]. PRO measures (PROMs) are derived from the patient self-assessment of a variety of health and wellbeing indices, including measures for health-related quality of life (HRQoL), symptom reporting, satisfaction with care or treatment, economic impacts and the specific dimensions of patient experience, such as depression and anxiety [2,3]. PRO measures are multidimensional and subjective, grounded on patient perceptions and objectively quantified [4]. As PROs can provide crucial information about unique patient experiences during cancer trajectories, their use in clinical practice is becoming increasingly advocated. Subjective patient perceptions can be particularly relevant as it has been shown that patient experiences do not always coincide with clinician understanding [5,6]. These differences in perspective have inspired, for instance, the development of the Patient-Reported Outcomes version of the Common Terminology Criteria for Adverse Events (PRO-CTCAE) by the USA National Cancer Institute [7], which is an internationally accepted system for the grading and reporting of adverse events by clinician [8]. The tool, originally devised for use in clinical cancer trials, is now frequently employed in clinical practice as well and has been translated and cross-culturally adapted into various languages [9].

Various literature reviews have indicated that PRO collection/symptom monitoring in oncology practice can have numerous advantages for patients, including improved communication with healthcare professionals [2,6,10], higher satisfaction [2,11] and higher levels of health-related quality of life [11,12], as well as economic benefits due to decreased emergency room visits and hospital readmissions [11,12]. The growing body of evidence supporting the impacts of PRO detection on patient survival is even more interesting. In a recent systematic review by Lizan et al. [12], five out of six publications assessing this outcome indicated that patient-reported symptom surveillance led to significantly improved survival compared to usual symptom monitoring. Specifically, the review found that active patient-reported monitoring was associated with increased survival for five months or more compared to usual care. However, that review did not provide an assessment of the study quality using appropriate instruments and did not report any meta-analysis data. To our knowledge, no work has yet been published that fills this gap.

Therefore, we conducted a systematic review and meta-analysis to determine whether the assessment of PROs in clinical practice using validated instruments can influence the overall survival of patients affected by any type or stage of cancer.

Review question: Does the use of PROs in oncology clinical practice have an impact on patient survival?

2. Materials and Methods

Before conducting this work, the PROSPERO database [13] was searched in April 2022 to identify any existing reviews on the subject in order to avoid replication; however, none were found. This review was designed and conducted following the Preferred Reporting Items for Systematic reviews and Meta-Analyses (PRISMA) guidelines [14]. The protocol was registered with PROSPERO (CRD42022328407) on 10 May 2022.

2.1. Search Strategy

Studies were identified by searching the MEDLINE database using the PubMed platform and the Web Of Science Clarivate, with no date or language restrictions. The searches were conducted on 21 June 2022. A "backwards" snowball search was conducted on the references of systematic reviews. The full search strategies and notes on strategy development are provided in the Supplementary Materials.

2.2. Study Eligibility

Clinical trials and observational studies with control groups were considered. Studies had to compare the use of a PROM as an intervention in cancer clinical practice to not using a PROM. Any measure that qualified as a PROM according to the aforementioned FDA definition [1] was eligible, provided that it was detected using a validated screening tool that was administered in any format. Comparison had to be to usual care, i.e., the care that is normally provided at the studied center. Thus, we excluded uncontrolled studies, validation studies and studies using PROMs to evaluate another intervention (e.g., PRO data used to measure treatment benefits or risks in medical product clinical trials), as well as studies comparing PROM intervention modalities. Reviews, editorials, commentaries, methodological articles and case reports, along with duplicates/replicates of studies, were excluded.

2.3. Population Eligibility

The review concerned adult individuals with any type of cancer in any setting and in any phase of their care trajectory (currently receiving cancer treatment or in follow-up). Studies focused on children (<18 years) were not considered.

2.4. Selection Process

Two reviewers independently performed the initial title and abstract screening for relevance to this review using the Rayyan platform [15], which allowed the recording of any discrepancies and the reaching of a consensus. Next, the two reviewers independently examined the full texts of the screened publications and identified eligible papers to include in the review. Any disagreements were resolved by a third independent reviewer.

2.5. Data Extraction

Two reviewers independently extracted data from the selected studies using a Microsoft Excel form and disagreements were resolved through discussions, involving a third reviewer when necessary. The extracted data items included title and first author, country, number of centers, cancer type, number of patients, phase of care (active treatment or follow-up), intervention delivery method, screened PROMs and corresponding instruments, estimates of the effects and measures of variability (standard errors or confidence intervals).

Study investigators were contacted when data confirmation was needed.

2.6. Risk of Bias Assessment

The internal validity (risk of bias) of the included studies was assessed using the two most recommended tools for interventional studies, according to the design in [16], namely, the Cochrane risk of bias tool for randomized trials (RoB 2) [17] and the Risk Of Bias In Non-randomized Studies of Interventions (ROBINS-I) [18].

RoB 2 [17] is structured into five bias domains, which address all important mechanisms by which bias can be introduced into the results of a trial. They cover the randomization process, deviations from intended interventions, missing outcome data, the measurement of the outcomes and the selection of the reported results. Within each domain, users answer one or more signaling questions. These answers lead to a judgment of "low risk of bias", "some concerns" or "high risk of bias". The judgments within each domain lead to an overall judgment for the risk of bias in the results being assessed.

ROBINS-I [18] considers seven domains through which bias can be introduced into non-randomized studies of interventions (NRSIs), covering the confounding and selection of participants into the studies, the classification of the interventions themselves, issues arising after the start of the interventions, biases due to deviations from intended interventions, missing data, the measurement of the outcomes and the selection of the reported results. Responses to "signaling questions" provide the basis for domain-level judgments about the risk of bias, which then provide the basis for an overall risk of bias judgment for particular outcomes using the categories of a "low risk", "moderate risk", "serious risk" and "critical risk" of bias.

To present the results of this assessment in a graphical format, we used a traffic light plot to depict the domain level judgments for each study and a summary bar plot figure to show the proportion of studies with a given risk of bias within each domain, weighted by inverse variance. To provide a combined representation of the judgments obtained using the two selected tools, ROBINS-I was used as a reference (seven domains). Thus, for RCTs, the two domains that were not applicable ("selection bias" and "classification of intervention") were highlighted in gray.

Two reviewers independently applied the tools to each included study and recorded the supporting information and justifications for the judgments of risk of bias for each domain. Doubts were resolved through discussions.

Following the instrument indications, the overall risk of bias was judged according to the following criteria: "low" when all domains were rated as low risk; "some concerns/moderate" when at least one domain was rated as having some concerns/moderate risk but no domain was rated as having a high risk; "high/serious" when at least one domain was rated as having a high/serious risk or if the study was judged to have some concerns for multiple domains in a way that substantially lowered confidence in the results.

2.7. Data Synthesis

To summarize the effects of the interventions, the risk ratio (RR) was estimated using the raw data. We performed random-effects meta-analyses using the Paule and Mandel method for the estimation of between-study variance [19,20]. Due to the great variability between the selected studies in terms of sample size, we assigned weights using an inverse variance matrix. The confidence intervals of the overall effects on survival were adjusted by applying the Hartung–Knapp–Sidik–Jonkman (HKSJ) approach [21–23] to account for the uncertainty in the variance estimates. I^2 statistics tests were calculated to quantify the degree of study heterogeneity [24]. The I^2 value that established significant heterogeneity was 70%. The level of significance was set at $p < 0.050$.

We did not perform formal subgroup analyses due to the insufficient number of included studies. Furthermore, we planned to assess publication bias using funnel plot representation and a Peter's test at a 10% level; however, this was not possible because fewer than 10 studies were considered.

The data were processed using R statistical software (R: a language and environment for statistical computing; the R Foundation for Statistical Computing, Vienna, Austria), v. 4.0.3, with the meta and metasens packages [25].

2.8. Patient and Public Involvement

Patients nor the public were involved in this research.

3. Results

3.1. Study Selection

A total of 3723 articles were retrieved from the two databases and uploaded into the Rayyan platform. After removing duplicates, 2433 records underwent the title and abstract screening. We chose not to apply automation tools to determine ineligibility in order to increase accuracy; thus, all references were screened manually. In total, 14 reports were identified as potentially eligible and underwent a full text review. Of these, six [26–31] were excluded, mainly because the outcomes of interest in this review were not measured or because the types of intervention or study aims were not eligible for our study question (Table S1). Overall, six studies [32–37] were included in our systematic review, to which two follow-up publications [38,39] were added for the meta-analysis.

A flow diagram depicting the selection process is provided in Figure 1.

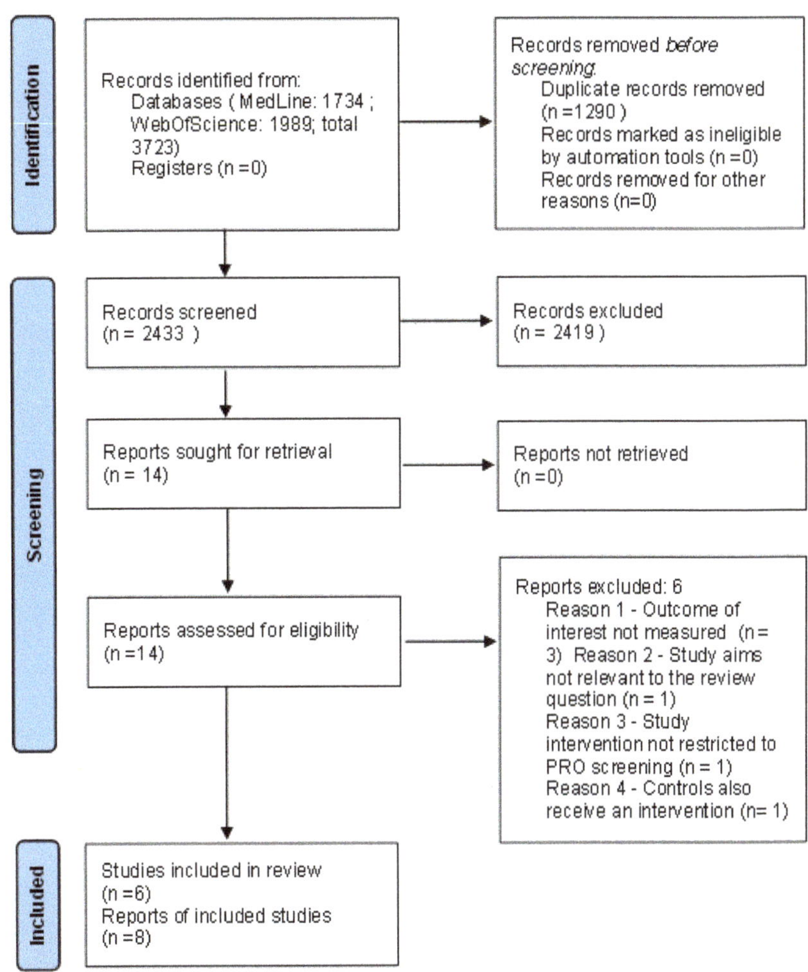

Figure 1. A PRISMA 2020 flow diagram of the process of identifying studies (both included and excluded).

3.2. Study Characteristics

The characteristics of the six studies included in the review are shown in Table 1.

Table 1. Characteristics of included studies.

First Author	Year	Study Design	Country	Single Center or Multicenter	No. of Patients	Age (Years)	Sex (% Female)	Cancer Type	Phase of Care (Active Treatment vs. Follow-Up)	Intervention vs. Control	PRO Instrument Used for Intervention	Mode of Administration	Follow-up Length for Survival	Survival as Primary or Secondary Outcome?
Barbera	2020	Population-based retrospectively matched cohort analysis	Canada	Multicenter	257,786 (128,893 patients with ESAS exposure matched to 128,893 patients without ESAS exposure)	Mean: 64 SD: 13	47.8%	Various	Any	12-month telephonic screening: high-risk patients were called weekly and low-risk patients were called monthly. Historical controls received usual cancer care without standardized symptom screening or management.	Edmonton Symptom Assessment System (ESAS)	Electronic (Touch screen at the center)	5 years (median: 1.4)	Primary
Basch	2016	Randomized controlled trial	USA	Single center	766 (441 intervention vs. 325 control)	Median: 61 Range: 26–91	58%	Advanced solid tumors: metastatic breast, genitourinary, gynecologic or lung cancers	Active treatment	Patients were randomly assigned to either report 12 common symptoms via tablet computers or receive usual care consisting of symptom monitoring at the discretion of clinicians. Those with home computers received weekly email prompts to report symptoms between visits. Treating physicians received symptom printouts at visits and nurses received email alerts when participants reported severe or worsening symptoms.	Symptom Tracking and Reporting (STAR)	Electronic (web-based)	Median follow-up of 7 years (interquartile range 6.5–7.8) $	Secondary
Demedts	2021	Non-randomized controlled study	Belgium	Single center	204 stage IV non-small cell lung cancer patients (89 intervention vs. 115 control)	Median: 66 Range: 32–88	24%	Lung cancer	Active treatment	Patients were invited by email every week to report on the side effects of their treatment. Feedback loops were created with automatically triggered electronic alerts to the care team when a predefined threshold of symptoms was reached. When patients refused, usual care was offered.	Items from the Patient-Reported Outcomes version of the Common Terminology Criteria for Adverse Events (PRO-CTCAE) library	Electronic (email reminders and digital reporting)	4 years	Secondary
Denis	2017	Randomized controlled trial	France	Multicenter	133 patients enrolled: 12 deemed ineligible after randomization and 121 retained in the intent-to-treat analysis (60 intervention vs. 61 control)	Median: 64.5 Range: 35.7–88.1	33%	Advanced-stage lung cancer	Active treatment	Personalized follow-up strategy based on 12 symptoms that were self-scored weekly and transmitted to the oncologist. Clinical follow-ups in both arms included oncology visits at least every 3 months.	e-Follow-up Application (e-FAP)	Electronic (web-based)	2 years *	Primary

Table 1. *Cont.*

First Author	Year	Study Design	Country	Single Center or Multicenter	No. of Patients	Age (Years)	Sex (% Female)	Cancer Type	Phase of Care (Active Treatment vs. Follow-Up)	Intervention vs. Control	PRO Instrument Used for Intervention	Mode of Administration	Follow-up Length for Survival	Survival as Primary or Secondary Outcome?
Patel	2019	Non-randomized study with historical control	USA	Single center	288 (186 intervention vs. 102 control)	Mean: 79 SD: 8	55%	Advanced cancers	Any	12-month telephonic program in which a lay health worker (LHW), supervised by a physician assistant (PA), assessed patient symptoms after diagnosis, with the frequency of symptom screening varying on the basis of patient risk (once a week for high-risk patients and at least once a month for low-risk patients). All participants received usual cancer care.	Edmonton Symptom Assessment System (ESAS)	Telephone	1 year	Secondary
Patel	2020	Non-randomized study with historical control	USA	Multicenter	832 (425 intervention vs. 407 control)	Mean: 79 SD: 8.3	41.5%	Various new diagnoses of solid or hematologic malignant neoplasms	Any	12-month telephonic program in which a lay health worker (LHW), supervised by a physician assistant (PA), assessed patient symptoms after diagnosis, with the frequency of symptom screening varying on the basis of patient risk (once a week for high-risk patients and at least once a month for low-risk patients). All participants received usual cancer care.	For symptoms: Edmonton Symptom Assessment System (ESAS). For depression: the 9-item Patient Health Questionnaire (PHQ-9)	Telephone	1 year	Secondary

$ Results for the 7-year follow-up are reported in the following article on the same study [37] (Basch 2017), * Results for the 2-year follow-up are reported in the following article on the same study [39] (Denis 2019).

Three out of the six studies [32,35,37] were multicenter. Two of the six studies [34,35] were set in Europe, three [33,36,37] were set in the USA and one [32] was set in in Canada. Two studies were RCTs [33,35], one was a population-based retrospectively matched cohort analysis [32], two used a pre–post design with historical controls [36,37] and one was a non-randomized controlled trial in which the controls were patients who refused the intervention [34].

Regarding included the populations, three out of the six studies [33–35] restricted eligibility to patients receiving active treatment, while the others recruited patients in any phase of care. Two studies [34,35] focused exclusively on lung cancer, two [33,36] focused on advanced cancers and two [32,37] included any type and stage of cancer.

For PRO screening instruments, three studies [32,36,37] used ESAS (Edmonton Symptom Assessment System), one study [37] screened for depression using the 9-item Patient Health Questionnaire (PHQ-9), one trial [33] employed STAR (Symptom Tracking and Reporting), one study [34] used items from the NCI's PRO-CTCAE (Patient-Reported Outcomes version of the Common Terminology Criteria for Adverse Events) and one study [35] used the e-FAB (e-follow-up) application. The mode of administration was electronic in all studies except for those by Patel et al. [36,37], in which the interventions were telephonic.

The meta-analysis considered two additional papers that reported follow-up mortality data for the two RCTs [38,39]: Basch 2017 [38] reported the results of a preplanned post-hoc analysis and Denis 2019 [39] described the results of a 2-year follow-up after the study was stopped early. Overall, 130.094 patients in the intervention arms vs. 129.903 in the control arms were considered for the meta-analysis, of whom 124,259 (48%) were women.

3.3. *Impact of Patient-Reported Outcome Monitoring on Overall Survival*

All six studies selected for the systematic review were included in the meta-analysis as they all reported the necessary raw mortality data. The two non-randomized studies with historical controls [36,37] (611 patients in the intervention arms vs. 509 patients in the control arms) reported survival after 1 year as a secondary output, the population-based retrospectively matched cohort study [32] (128,893 patients in the intervention arm vs. 128,893 patients in the control arm) analyzed survival up to 5 years as a primary outcome and the non-randomized controlled study [34] (89 patients in the intervention arm vs. 115 patients in the control arm) analyzed survival up to 4 years as a secondary outcome. Two RCTs [33,35] (overall, 501 patients in the intervention arms vs. 386 patients in the control arms) reported survival results from post-hoc analyses [38,39] as a secondary outcome with a median follow-up of 7 years and the other reported the results as a primary outcome after 2 years.

Figure 2 displays the results of our meta-analysis, together with the risk of bias assessments described in the following section.

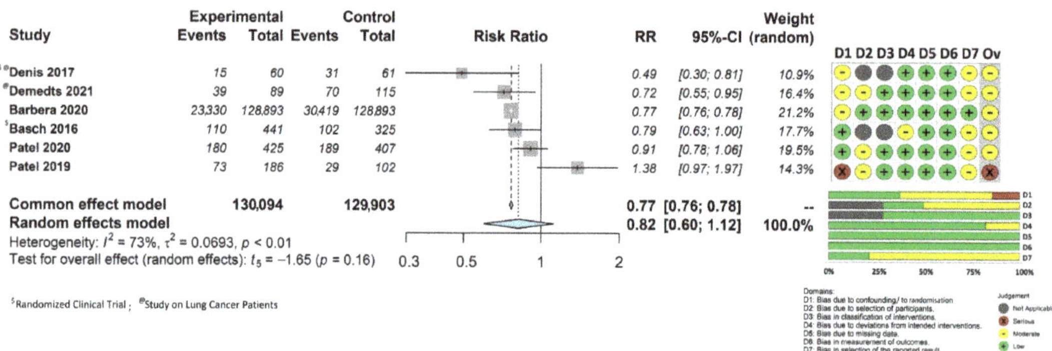

Figure 2. A forest plot of the survival pooled effect and a summary of the risk of bias (traffic light plot and summary bar plot) [32–37].

All studies except for the first study by Patel et al. [36] demonstrated reductions in mortality, which were statistically significant in four studies [32–35] with RRs ranging between 0.49 to 0.79. The pooled mortality outcome after 1 year (the observation timeframe that was common to all studies) was RR = 0.77 (95%CI 0.76–0.78) as determined using the common effect model and RR = 0.82 (95%CI 0.60–1.12; $p = 0.16$) as determined using the random-effects model. Heterogeneity was statistically significant ($I^2 = 73\%$; $p < 0.01$).

3.4. *Quality of Included Studies*

In addition to the forest plot, Figure 2 depicts both a traffic light plot showing the risk of bias judgments for the individual domains for each study and the overall risk of bias and a summary bar plot showing the cumulative risk of bias percentage for each domain. The overall risk of bias was judged to be moderate for all studies, except for one of the Patel studies [36], which was rated as having a serious risk. This was mainly due to selection bias as the intervention group comprised a higher number of patients with baseline stage IV disease than the control group. The most frequent problem, which was present in five out of the six studies (corresponding to a weighted percentage of 78.8%), concerned bias in the selection of the reported results; specifically, findings on survival were not reported in detail or only referred to patient subgroups. The reasons for the judgments for each bias domain are provided in the Supplementary Materials (Table S2).

3.5. *Further Analyses*

The additional outcomes that were planned in the protocol (disease-free survival, progression-free survival and event-free survival) could not be measured because they were not investigated in the included studies.

Since two of the six included studies [34,35] focused on patients with advanced lung cancer, an exploratory subgroup analysis was performed for this cancer type. The results are depicted in the Supplementary Materials (Figure S1).

Finally, we carried out a sensitivity analysis, excluding the study with the serious overall risk of bias [36]. The overall survival determined using the random-effects model was RR = 0.78 (95%CI = 0.63–0.95; $p = 0.02$; $I^2 = 51\%$; $p = 0.08$) (Figure S2).

4. Discussion

To our knowledge, this is the first meta-analysis of the effects of PRO monitoring in oncology practice in terms of overall patient survival. The results seemed to indicate that monitoring patient-reported outcomes in clinical practice could have a positive impact on the overall survival of people with cancer. Specifically, our overall estimate indicated an 18% reduction in the risk of death, although this effect was not statistically significant and adjustment for confounding factors was not possible. Furthermore, we wished to perform a formal subgroup analysis for lung cancer patients that included multiple studies; however, the two authors we contacted could not provide us with the necessary data. The exploratory sub-analysis we were able to perform, which was limited to two studies, suggested that patients with advanced lung cancer might benefit the most from these interventions.

The systematic introduction of PRO monitoring into clinical practice is often difficult due to operational and financial barriers to the implementation of these complex interventions, as well as the uncertainty among physicians regarding their usefulness in actual practice [3,40]. Therefore, it is important to identify the categories of patients who are most likely to benefit and consequently focus efforts on these populations [41].

Survival improvements after PRO surveillance are plausible, although the mechanisms by which this important benefit is achieved remain debatable [12,42]. One hypothesis is that patient-reported surveillance allows doctors to respond to problems earlier, thus preventing complications, unexpected hospitalizations or the discontinuation of chemotherapy. Another possibility is that improvements in symptom management may also allow patients to tolerate their symptoms better and, consequently, benefit from chemotherapy for a longer time than when receiving usual care. Finally, the systematic collection of PROs can support the recognition of problematic symptoms, thus promoting patient empowerment and self-management.

The results of this work must be interpreted with caution, due to both the lack of statistical significance of the pooled estimate and the suboptimal quality of the six included studies (none of which were rated as having a low risk of bias). We also observed an under-representation of women in three of the studies (Demedts et al. [34] (24%), Denis et al. [35] (33%) and Patel et al. [37] (41.5%)), as well as the absence of any formal analyses of gender differences in all studies. Given these limitations, more rigorous research is needed to consolidate the positive signs yielded by this work.

5. Conclusions

In conclusion, the available evidence was insufficient to draw firm conclusions that PRO monitoring could extend survival. Further evidence is expected to emerge from PRO-TECT, a large randomized cluster study that is currently underway in 52 community oncology practices in the United States of America on 1191 subjects with metastatic cancer [43]. The results regarding overall survival as the primary outcome are not yet available. Therefore, we intend to update this meta-analysis after the publication of those results.

In any case, as indicated in the European guidelines, there is substantial evidence to support the benefits and feasibility of implementing PROMs in clinical outpatient cancer care, particularly for patients receiving active therapy or during the observation of therapy with a high risk of recurrence [41]. Furthermore, routine PRO surveillance could help to standardize clinical care in a world in which the volume of patients is increasing and ensure patient engagement during the entirety of their cancer trajectory.

Supplementary Materials: The following are available online at https://www.mdpi.com/article/10.3390/cancers14215470/s1: Full search strategies and notes on strategy development, Table S1: Studies excluded after full text review and corresponding reasons, Table S2: Risk of bias summaries for randomized trials and non-randomized trials of interventions, Figure S1: Subgroup analysis for lung cancer shown as a forest plot of the survival pooled effect, Figure S2: Sensitivity analysis shown as a forest plot of the survival pooled effect excluding one study with the overall risk of bias rated as serious.

Author Contributions: C.C. conceived and designed the systematic review; C.C. coordinated the systematic review; F.D. and C.C. designed the search strategy; C.C., F.D., G.M. and S.L. screened the abstracts and full texts; S.L. and B.M. extracted the data; G.M. and M.P. assessed the data and elaborated on the analysis plan; G.M. performed the statistical analyses; C.C., F.D. and G.M. assessed the risk of bias; F.P. and C.P. contributed to the interpretation of the results; C.C. and F.D. drafted the first version of the manuscript. All authors critically revised the article for important intellectual content and provided final approval for the article. All authors have read and agreed to the published version of the manuscript.

Funding: This research received no external funding.

Acknowledgments: We are very grateful to the members of the Directive of FICOG (Federation of Italian Cooperative Oncology Groups) for their constant commitment to the field of patient-reported outcome research, from which this work was inspired.

Conflicts of Interest: The authors declare no conflict of interest.

References

1. Patient-Reported Outcome Measures: Use in Medical Product Development to Support Labeling Claims. Available online: https://www.fda.gov/regulatory-information/search-fda-guidance-documents/patient-reported-outcome-measures-use-medical-product-development-support-labeling-claims (accessed on 11 October 2022).
2. Chen, J.; Ou, L.; Hollis, S.J. A systematic review of the impact of routine collection of patient reported outcome measures on patients, providers and health organisations in an oncologic setting. *BMC Health Serv. Res.* **2013**, *13*, 211. [CrossRef] [PubMed]

3. Nic Giolla Easpaig, B.; Tran, Y.; Bierbaum, M.; Arnolda, G.; Delaney, G.P.; Liauw, W.; Ward, R.L.; Olver, I.; Currow, D.; Girgis, A.; et al. What are the attitudes of health professionals regarding patient reported outcome measures (PROMs) in oncology practice? A mixed-method synthesis of the qualitative evidence. *BMC Health Serv. Res.* **2020**, *20*, 102. [CrossRef] [PubMed]
4. Silveira, A.; Sequeira, T.; Gonçalves, J.; Lopes Ferreira, P. Patient reported outcomes in oncology: Changing perspectives-a systematic review. *Health Qual. Life Outcomes* **2022**, *20*, 82. [CrossRef] [PubMed]
5. Di Maio, M.; Gallo, C.; Leighl, N.B.; Piccirillo, M.C.; Daniele, G.; Nuzzo, F.; Gridelli, C.; Gebbia, V.; Ciardiello, F.; De Placido, S.; et al. Symptomatic toxicities experienced during anticancer treatment: Agreement between patient and physician reporting in three randomized trials. *J. Clin. Oncol.* **2015**, *33*, 910–915. [CrossRef]
6. Yang, L.Y.; Manhas, D.S.; Howard, A.F.; Olson, R.A. Patient-reported outcome use in oncology: A systematic review of the impact on patient-clinician communication. *Support Care Cancer* **2018**, *26*, 41–60. [CrossRef]
7. Basch, E.; Reeve, B.B.; Mitchell, S.A.; Clauser, S.B.; Minasian, L.M.; Dueck, A.C.; Mendoza, T.R.; Hay, J.; Atkinson, T.M.; Abernethy, A.P.; et al. Development of the National Cancer Institute's patient-reported outcomes version of the common terminology criteria for adverse events (PRO-CTCAE). *J. Natl. Cancer Inst.* **2014**, *106*, dju244. [CrossRef]
8. National Cancer Institute, National Institutes of Health. US Department of Health and Human Services: Common Terminology Criteria for Adverse Events (CTCAE). Available online: https://evs.nci.nih.gov/ftp1/CTCAE/About.html (accessed on 11 October 2022).
9. Caminiti, C.; Bryce, J.; Riva, S.; Ng, D.; Diodati, F.; Iezzi, E.; Sparavigna, L.; Novello, S.; Porta, C.; Del Mastro, L.; et al. Cultural adaptation of the Italian version of the Patient-Reported Outcomes Common Terminology Criteria for Adverse Event (PRO-CTCAE®). *Tumori J.* **2022**, *2022*, 3008916221099558. [CrossRef]
10. Howell, D.; Molloy, S.; Wilkinson, K.; Green, E.; Orchard, K.; Wang, K.; Liberty, J. Patient-reported outcomes in routine cancer clinical practice: A scoping review of use, impact on health outcomes, and implementation factors. *Ann. Oncol.* **2015**, *26*, 1846–1858. [CrossRef]
11. Graupner, C.; Kimman, M.L.; Mul, S.; Slok, A.H.M.; Claessens, D.; Kleijnen, J.; eDirksen, C.D.; Breukink, S.O. Patient outcomes, patient experiences and process indicators associated with the routine use of patient-reported outcome measures (PROMs) in cancer care: A systematic review. *Support Care Cancer* **2021**, *29*, 573–593. [CrossRef]
12. Lizán, L.; Pérez-Carbonell, L.; Comellas, M. Additional Value of Patient-Reported Symptom Monitoring in Cancer Care: A Systematic Review of the Literature. *Cancers* **2021**, *13*, 4615. [CrossRef]
13. International Prospective Register of Systematic Reviews (PROSPERO). PROSPERO. Available online: https://www.crd.york.ac.uk/prospero/ (accessed on 11 October 2022).
14. Page, M.J.; Moher, D.; Bossuyt, P.M.; Boutron, I.; Hoffmann, T.C.; Mulrow, C.D.; Shamseer, L.; Tetzlaff, J.M.; Akl, E.A.; Brennan, S.E.; et al. PRISMA 2020 explanation and elaboration: Updated guidance and exemplars for reporting systematic reviews. *BMJ* **2021**, *372*, n160. [CrossRef] [PubMed]
15. Rayyan–Intelligent Systematic Review. Available online: https://www.rayyan.ai/ (accessed on 11 October 2022).
16. Ma, L.L.; Wang, Y.Y.; Yang, Z.H.; Huang, D.; Weng, H.; Zeng, X.T. Methodological quality (risk of bias) assessment tools for primary and secondary medical studies: What are they and which is better? *Mil. Med Res.* **2020**, *7*, 1–11. [CrossRef] [PubMed]
17. Sterne, J.A.C.; Savović, J.; Page, M.J.; Elbers, R.G.; Blencowe, N.S.; Boutron, I.; Cates, C.J.; Cheng, H.Y.; Corbett, M.S.; Eldridge, S.M.; et al. RoB 2: A revised tool for assessing risk of bias in randomised trials. *BMJ* **2019**, *366*, l4898. [CrossRef] [PubMed]
18. Sterne, J.A.; Hernán, M.A.; Reeves, B.C.; Savović, J.; Berkman, N.D.; Viswanathan, M.; Henry, D.; Altman, D.G.; Ansari, M.T.; Boutron, I.; et al. ROBINS-I: A tool for assessing risk of bias in non-randomised studies of interventions. *BMJ* **2016**, *355*, i4919. [CrossRef]
19. Veroniki, A.A.; Jackson, D.; Viechtbauer, W.; Bender, R.; Bowden, J.; Knapp, G.; Kuss, O.; Higgins, J.P.; Langan, D.; Salanti, G. Methods to estimate the between-study variance and its uncertainty in meta-analysis. *Res. Synth. Methods* **2016**, *7*, 55–79. [CrossRef]
20. Axfors, C.; Janiaud, P.; Schmitt, A.M.; Van't Hooft, J.; Smith, E.R.; Haber, N.A.; Abayomi, A.; Abduljalil, M.; Abdulrahman, A.; Acosta-Ampudia, Y.; et al. Association between convalescent plasma treatment and mortality in COVID-19: A collaborative systematic review and meta-analysis of randomized clinical trials. *BMC Infect. Dis.* **2021**, *21*, 1170. [CrossRef]
21. IntHout, J.; Ioannidis, J.P.; Borm, G.F. The Hartung-Knapp-Sidik-Jonkman method for random effects meta-analysis is straightforward and considerably outperforms the standard DerSimonian-Laird method. *BMC Med. Res. Methodol.* **2014**, *14*, 25. [CrossRef]
22. Brabaharan, S.; Veettil, S.K.; Kaiser, J.E.; Raja Rao, V.R.; Wattanayingcharoenchai, R.; Maharajan, M.; Insin, P.; Talungchit, P.; Anothaisintawee, T.; Thakkinstian, A.; et al. Association of Hormonal Contraceptive Use With Adverse Health Outcomes: An Umbrella Review of Meta-analyses of Randomized Clinical Trials and Cohort Studies. *JAMA Netw. Open* **2022**, *5*, e2143730. [CrossRef]
23. Partlett, C.; Riley, R.D. Random effects meta-analysis: Coverage performance of 95% confidence and prediction intervals following REML estimation. *Stat. Med.* **2017**, *36*, 301–317. [CrossRef]
24. Higgins, J.P.; Thompson, S.G.; Deeks, J.J.; Altman, D.G. Measuring inconsistency in meta-analyses. *BMJ* **2003**, *327*, 557–560. [CrossRef]
25. Schwarzer, G.; Carpenter, J.R.; Rücker, G. *Metaanalysis with R*; Springer: Berlin/Heidelberg, Germany, 2015; Volume 4784.

26. De Rooij, B.H.; Ezendam, N.P.M.; Mols, F.; Vissers, P.A.J.; Thong, M.S.Y.; Vlooswijk, C.C.P.; Oerlemans, S.; Husson, O.; Horevoorts, N.J.E.; van de Poll-Franse, L.V. Cancer survivors not participating in observational patient-reported outcome studies have a lower survival compared to participants: The population-based PROFILES registry. *Qual. Life Res.* **2018**, *27*, 3313–3324. [CrossRef] [PubMed]
27. Geerse, O.P.; Hoekstra-Weebers, J.E.; Stokroos, M.H.; Burgerhof, J.G.; Groen, H.J.; Kerstjens, H.A.; Hiltermann, T.J. Structural distress screening and supportive care for patients with lung cancer on systemic therapy: A randomised controlled trial. *Eur. J. Cancer* **2017**, *72*, 37–45. [CrossRef] [PubMed]
28. Gilbert, S.M.; Dunn, R.L.; Wittmann, D.; Montgomery, J.S.; Hollingsworth, J.M.; Miller, D.C.; Hollenbeck, B.K.; Wei, J.T.; Montie, J.E. Quality of life and satisfaction among prostate cancer patients followed in a dedicated survivorship clinic. *Cancer* **2015**, *121*, 1484–1491. [CrossRef] [PubMed]
29. Hentschel, L.; Richter, S.; Kopp, H.G.; Kasper, B.; Kunitz, A.; Grünwald, V.; Kessler, T.; Chemnitz, J.M.; Pelzer, U.; Schuler, U.; et al. Quality of life and added value of a tailored palliative care intervention in patients with soft tissue sarcoma undergoing treatment with trabectedin: A multicentre, cluster-randomised trial within the German Interdisciplinary Sarcoma Group (GISG). *BMJ Open* **2020**, *10*, e035546. [CrossRef]
30. Oerlemans, S.; Arts, L.P.J.; Kieffer, J.M.; Prins, J.; Hoogendoorn, M.; van der Poel, M.; Koster, A.; Lensen, C.; Stevens, W.B.C.; Issa, D.; et al. Web-Based Return of Individual Patient-Reported Outcome Results Among Patients With Lymphoma: Randomized Controlled Trial. *J. Med. Internet Res.* **2021**, *23*, e27886. [CrossRef]
31. Skovlund, P.C.; Vind Thaysen, H.; Schmidt, H.; Alsner, J.; Hjollund, N.H.; Lomborg, K.; Nielsen, B.K. Effect of patient-reported outcomes as a dialogue-based tool in cancer consultations on patient self-management and health-related quality of life: A clinical, controlled trial. *Acta Oncol.* **2021**, *60*, 1668–1677. [CrossRef]
32. Barbera, L.; Sutradhar, R.; Seow, H.; Mittmann, N.; Howell, D.; Earle, C.C.; Li, Q.; Thiruchelvam, D. The impact of routine Edmonton Symptom Assessment System (ESAS) use on overall survival in cancer patients: Results of a population-based retrospective matched cohort analysis. *Cancer Med.* **2020**, *9*, 7107–7115. [CrossRef]
33. Basch, E.; Deal, A.M.; Kris, M.G.; Scher, H.I.; Hudis, C.A.; Sabbatini, P.; Rogak, L.; Bennett, A.V.; Dueck, A.C.; Atkinson, T.M.; et al. Symptom Monitoring With Patient-Reported Outcomes During Routine Cancer Treatment: A Randomized Controlled Trial. *J. Clin. Oncol.* **2016**, *34*, 557–565. [CrossRef]
34. Demedts, I.; Himpe, U.; Bossuyt, J.; Anthoons, G.; Bode, H.; Bouckaert, B.; Carron, K.; Dobbelaere, S.; Mariën, H.; Van Haecke, P.; et al. Clinical implementation of value based healthcare: Impact on outcomes for lung cancer patients. *Lung Cancer* **2021**, *162*, 90–95. [CrossRef]
35. Denis, F.; Lethrosne, C.; Pourel, N.; Molinier, O.; Pointreau, Y.; Domont, J.; Bourgeois, H.; Senellart, H.; Trémolières, P.; Lizée, T.; et al. Randomized Trial Comparing a Web-Mediated Follow-up With Routine Surveillance in Lung Cancer Patients. *J. Natl. Cancer Inst.* **2017**, *109*, djx029. [CrossRef]
36. Patel, M.I.; Ramirez, D.; Agajanian, R.; Agajanian, H.; Bhattacharya, J.; Bundorf, K.M. Lay Health Worker-Led Cancer Symptom Screening Intervention and the Effect on Patient-Reported Satisfaction, Health Status, Health Care Use, and Total Costs: Results From a Tri-Part Collaboration. *JCO Oncol. Pract.* **2020**, *16*, e19–e28. [CrossRef] [PubMed]
37. Patel, M.I.; Ramirez, D.; Agajanian, R.; Agajanian, H.; Coker, T. Association of a Lay Health Worker Intervention With Symptom Burden, Survival, Health Care Use, and Total Costs Among Medicare Enrollees With Cancer. *JAMA Netw. Open* **2020**, *3*, e201023. [CrossRef] [PubMed]
38. Basch, E.; Deal, A.M.; Dueck, A.C.; Scher, H.I.; Kris, M.G.; Hudis, C.; Schrag, D. Overall Survival Results of a Trial Assessing Patient-Reported Outcomes for Symptom Monitoring During Routine Cancer Treatment. *JAMA* **2017**, *318*, 197–198. [CrossRef]
39. Denis, F.; Basch, E.; Septans, A.L.; Bennouna, J.; Urban, T.; Dueck, A.C.; Letellier, C. Two-Year Survival Comparing Web-Based Symptom Monitoring vs. Routine Surveillance Following Treatment for Lung Cancer. *JAMA* **2019**, *321*, 306–307. [CrossRef]
40. Takvorian, S.U.; Anderson, R.T.; Gabriel, P.E.; Poznyak, D.; Lee, S.; Simon, S.; Barrett, K.; Shulman, L.N. Real-World Adherence to Patient-Reported Outcome Monitoring as a Cancer Care Quality Metric. *JCO Oncol. Pract.* **2022**, *18*, e1454–e1465. [CrossRef]
41. Di Maio, M.; Basch, E.; Denis, F.; Fallowfield, L.J.; Ganz, P.A.; Howell, D.; Kowalski, C.; Perrone, F.; Stover, A.M.; Sundaresan, P.; et al. The role of patient-reported outcome measures in the continuum of cancer clinical care: ESMO Clinical Practice Guideline. *Ann. Oncol.* **2022**, *33*, 878–892. [CrossRef] [PubMed]
42. Hassett, M.J.; Wong, S.; Osarogiagbon, R.U.; Bian, J.; Dizon, D.S.; Jenkins, H.H.; Uno, H.; Cronin, C.; Schrag, D.; SIMPRO Co-Investigators. Implementation of patient-reported outcomes for symptom management in oncology practice through the SIMPRO research consortium: A protocol for a pragmatic type II hybrid effectiveness-implementation multi-center cluster-randomized stepped wedge trial. *Trials* **2022**, *23*, 506. [CrossRef]
43. Basch, E.; Schrag, D.; Henson, S.; Jansen, J.; Ginos, B.; Stover, A.M.; Carr, P.; Spears, P.A.; Jonsson, M.; Deal, A.M.; et al. Effect of Electronic Symptom Monitoring on Patient-Reported Outcomes Among Patients With Metastatic Cancer: A Randomized Clinical Trial. *JAMA* **2022**, *327*, 2413–2422. [CrossRef]

Article

Cancer-Related Symptom Management Intervention for Southwest American Indians

Felicia S. Hodge [1,2,*], Tracy Line-Itty [1] and Rachel H. A. Arbing [1]

1 School of Nursing, University of California, Los Angeles, 700 Tiverton Avenue, Room 5-934A Factor Building, Los Angeles, CA 90095, USA
2 Fielding School of Public Health, University of California, Los Angeles, 640 Charles E Young Drive South, Los Angeles, CA 90095, USA
* Correspondence: fhodge@sonnet.ucla.edu; Tel.: +1-310-210-9887

Simple Summary: Quality of life during, and even after, cancer treatment is greatly affected by cancer symptoms that include pain, fatigue, and changes to mental state and activities of daily living, to name a few. American Indians living in the Southwestern United States have cancer experiences which may be different than the general population and have long been understudied. A randomized controlled trial designed to test the impact of a culturally tailored intervention on the management of individual cancer symptoms was implemented. Outcomes included improvement in pain, depression, fatigue and loss of function management in adult American Indians. Study evaluations at post-test show a significant improvement in scores from pre-test and compared to the control group, demonstrating increased knowledge levels in managing cancer-related symptoms. Study findings guide researchers towards a better understanding of the meaning and impact of cancer symptoms for American Indian cancer survivors, thus their improving care and quality of life.

Abstract: There is limited literature related to culturally embedded meanings of cancer and related symptoms among American Indians. A culturally appropriate intervention to improve management of cancer-related symptoms, including pain, depression, fatigue and loss of function, was tested. Two-hundred and twenty-two adult American Indians with cancer were recruited from eight Southwest sites for a randomized clinical trial. The intervention group received tailored education, a toolkit with a video, and participated in discussion sessions on cancer symptom management; the control group received information on dental care. Pre- and post-test questionnaires were administered to control and intervention groups. Measures included socio-demographics, cancer-related symptom management knowledge and behavior, and quality of life measures. Male cancer survivors reported poorer self-assessed health status and lower scores on quality-of-life indicators as compared to female cancer survivors. Significant improvement was reported in symptom management knowledge scores following the intervention: management of pain ($p = 0.003$), depression ($p = 0.004$), fatigue ($p = 0.0001$), and loss of function ($p = 0.0001$). This study is one of the first to demonstrate a change in physical symptom self-management skills, suggesting culturally appropriate education and interventions can successfully enhance cancer-related symptom management knowledge and practice.

Keywords: American Indian; cancer; survivors; symptom management; quality of life; intervention

Citation: Hodge, F.S.; Line-Itty, T.; Arbing, R.H.A. Cancer-Related Symptom Management Intervention for Southwest American Indians. *Cancers* **2022**, *14*, 4771. https://doi.org/10.3390/cancers14194771

Academic Editor: António Araújo

Received: 12 August 2022
Accepted: 23 September 2022
Published: 29 September 2022

1. Introduction

Cancer is a chronic illness that places additional demands on cancer survivors and their families. American Indians and Alaska Natives are at higher risk for some cancers than the general population. The Centers for Disease Control and Prevention (CDC) reports that American Indians and Alaska Natives are more likely to be diagnosed with, and have higher rates of, certain cancers such as lung, colorectal, liver, stomach, and kidney cancers, than non-Hispanic Whites [1]. Following a cancer diagnosis, treatment can include surgery, chemotherapy and radiation, and American Indian survivors face many additional challenges in managing their healthcare. In addition to long-term surveillance and possible additional treatment, survivors contend with serious cancer-related symptoms, including pain, depression, fatigue, and loss of function. Having adequate skills to respond to cancer symptoms and implement self-management strategies is an essential part of cancer survivorship.

The role of self-management of cancer-related symptoms is broader than simply responding to the physical problems experienced after cancer treatment. Shifting personal perspectives from illness to wellness reinforces holistic cancer care management. Problem solving includes the ability to identify the source of a problem and resources needed, and then acting on the steps needed to improve daily living and thereby quality of life. Maintaining regular medical appointments and on-going surveillance of pain, fatigue and loss of function takes organizational skills and follow-through. Recognizing symptoms and seeking appropriate care is necessary for healthy well-being. Having knowledge of strategies to improve daily living and communication skills is helpful in the self-management process.

Several studies point to improved quality of life [2,3], health [2,4], and psychological and emotional well-being [5,6] as essential goals in the self-management process among cancer survivors. It is also critical that strategies to promote health and wellness are sensitive to survivors' beliefs and respect cultural traditions. Burnette, Roh, Liddell, and Lee [4] conducted a qualitative study in South Dakota identifying American Indian women cancer survivors' needs and preferences, with a particular emphasis on community supports for their cancer experience. Participants identified a need for more community-based support systems and infrastructures to ameliorate the cancer survivor experience. The need for an improved healthcare system that included the integration of spirituality and holistic healing options was emphasized. Recommendations for a community approach to raise awareness, education, and support for American Indian cancer survivors was provided.

Education and supportive interventions by healthcare providers, as well as family and caregivers, can also improve the skills and confidence needed to manage cancer-related symptoms [7–9]. Educational information can be readily obtained at various clinics and non-profit agencies, such as the American Cancer Society. The Indian Health Service also provides pamphlets and videos on cancer screening and cancer care. Encouragement and practical support through needed transportation and daily living tasks can be instrumental in survivors' ability to accept and adopt the needed skills to manage cancer-related symptoms.

Despite cancer rates being twice as high for liver cancer (18.1 vs. 7.1/100,000) and near double for stomach and kidney cancers in American Indians compared to non-Hispanic Whites [10], there has been limited scholarly literature, as well as understanding, related to culturally embedded meanings of cancer and related symptoms among American Indians [11,12]. By listening to individual and family cancer experiences and building on everyday cultural values and strengths of a community, effective cancer-related interventions that are relevant and culturally appropriate may be developed.

This paper reports on a randomized control trial (RCT) designed and coordinated to test a culturally sensitive intervention targeting knowledge and strategies to help American Indian cancer survivors and their families/caregivers to better manage cancer-related symptoms. The aim of the study was to explore the cancer experience and barriers to the management of pain, depression, fatigue and loss of function among American Indians residing in the Southwestern United States. Native language and cultural differences in project educational materials and skill building curriculum, as well as in the study instruments, was respected and incorporated. Improving communication about treatment with healthcare providers, such as asking focused follow-up questions and recording easily forgotten information and instructions at medical visits, was one of many key study intervention strategies emphasized in this research intervention curriculum. This study postulated that good communication between patient and provider would facilitate improved cancer knowledge and improved treatment compliance.

The study was carried out over a seven-year period, from project planning, design and coordination to implementation and evaluation. The study was organized into the following three phases: 1. Interviews, 2. Focus Groups, and 3. Intervention Testing. This paper reports on findings and implications from the intervention (Phase 3).

2. Materials and Methods

Interested tribal councils and health clinics provided written approvals for research in their communities. Institutional Review Board (IRB) approvals were obtained at the beginning of the study from the University of California, Los Angeles and from the Phoenix Area Indian Health Service. Figure 1 illustrates the study design.

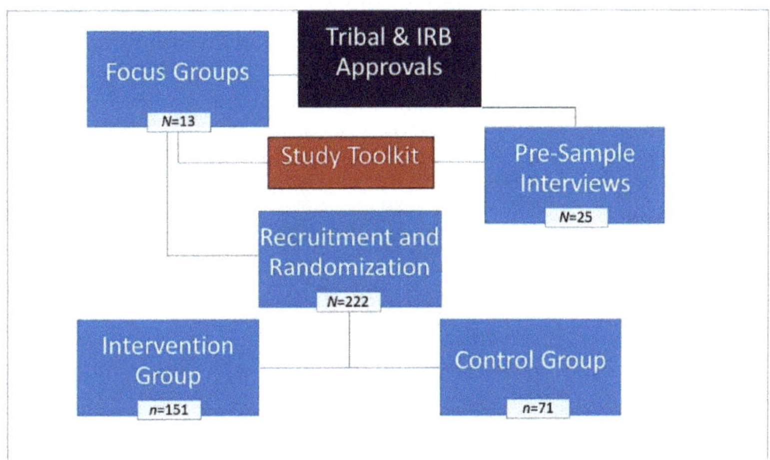

Figure 1. Flow diagram of study design.

2.1. Participant Recruitment

Recruitment efforts consisted of community flyers and notices placed in Tribal/Indian Health Service clinics, and word-of-mouth recruitment lasted two months. The flyers and notices included a description of the project and the contact telephone number and e-mail for those who had questions or who wished to enroll in the study. Participant eligibility criteria (American Indian, age 18 and older, diagnosed with cancer by a medical provider, and resident of the state of Arizona) and the location where the scheduled educational sessions would be held was included. Identified key staff at the clinics/hospitals assisted in distributing the announcements and signing up the interested participants. Project staff made regular visits to sites to register eligible participants into the intervention phase and to read and administer the active consent forms. Three hundred individuals responded to the recruitment efforts and 222 met the study criteria and participated in the intervention phase of the study. Assignment to the Intervention and Control groups were by random assignment using a computerized random assignment program.

The project's educational "toolkit" that served as the study intervention was comprised of culturally sensitive materials. The Toolkit titled, "Weaving Balance Into Life," incorporated American Indian values of health and balance into self-management of the most common and debilitating cancer symptoms [13]. In addition to increasing knowledge about cancer itself and productive strategies for relieving symptoms, educational targets also included approaches for building support and reducing communication barriers among caregivers and "Western" healthcare providers. The tools also included resource materials and culturally appropriate instruments for measuring symptoms that intended to improve American Indian knowledge of, and access to, local cancer symptom management services.

Toolkit development incorporated important themes and findings gleaned from qualitative analysis of interview and focus group transcripts from prior phases of this study. The intervention toolkit, developed to be informative and culturally appropriate, used Southwest Native imagery and themes, including American Indian healing practices and spirituality. The toolkit components included a Cancer Symptom Management educational video, a Cancer Symptom Management Guide (six chapters, skills building exercises, and a glossary), a Cancer Resource Directory (contact information for resources by region), Talking Circle Curriculum Slides and "Fact Sheet" review handouts, along with a journal, pen, post-it pad, and a hand-held back massager. The Cancer Symptom Management video reinforced American Indian survivors' stories about their cancer diagnosis experience, management of cancer-related symptoms, and recommendations to others on health and wellness. American Indian survivors participated in the storytelling phase of the video—cultural advisors from Southwestern tribal groups reviewed and enhanced cultural appropriateness of the study materials. Emphasis for curriculum development was drawn from the needs and desires of the target audience, with goals to improve management of individual symptoms, as well as increase survivor advocacy in healthcare settings [14].

The Intervention was tested at eight locations: four urban sites and four reservation locations. Study participants were randomly assigned to intervention or control groups with the pre- and post-test questionnaires gathered 8 weeks apart.

2.2. Intervention Group

The intervention group received the educational curriculum, cancer information and instruction on how to self-manage cancer symptoms. Participants met weekly at a series of one to one and one-half hour meetings comprised of 15–20 members who met for two months (total of 8 sessions). The meetings were held in a "Talking Circle" format where participants and the facilitator sat in a circle and participants took turns responding to the weekly topic. Seated in a circle, no particular individual is at the head leading the topic, thus all members have equal weight in the discussion of cancer-related matters. At the first session, the trained American Indian facilitator introduced the project and each participant read and signed a consent agreement and completed the 60-minute pre-test questionnaire. Educational materials and a new toolkit chapter were delivered at the subsequent six intervention Talking Circle sessions. The post-test questionnaire was administered at the 8th session. The intervention facilitators discussed with participants a brief lesson following a curriculum guide. Participants then took part in discussions to share and discuss the curriculum information. Participants were encouraged not to use their real names and asked not to discuss personal details about other participants outside of the sessions in order to protect their confidentiality. Each week participants received a new educational component to build their toolkit. All Talking Circles were audiotaped to ensure that the lessons were being taught in a standard manner and to capture important themes that were discussed. A research moderator and an assistant monitored the tape recorder while taking any necessary notes. All moderators received training in focus-group implementation in American Indian populations. Refreshments were offered to participants, as is the custom at American Indian gatherings. Participants received a gift card for travel and other costs associated with participating in the sessions. At the last session, the post-test questionnaire was administered (session 8) and participants received a certificate of completion.

2.3. Control Group

The control groups met for the pre-test questionnaire and received information on dental care at their initial visit. The control groups' post-test questionnaire was administered at week 8. At the end of the project, all participants in the control arm received all toolkit materials.

2.4. *Measures*

2.4.1. Demographic Characteristics

Measures included age, gender, tribal affiliation, degree of Indian blood (reported as 25%, 50%, 75%, or 100%), language (English, Spanish, or tribal language), marital status, number of household members (children and adults), and educational attainment (high school degree and above vs. fewer years of education).

2.4.2. Cancer History

Participants were asked if they have ever been told by their healthcare provider that they have cancer, the type of cancer (e.g., sarcoma or carcinoma), location of where the cancer was found (e.g., breast, colon, etc.), if they were being treated (currently or in the past) and type of treatment (chemotherapy, radiation, surgery, traditional method, other).

2.4.3. Cancer Symptoms

Participants were asked if they experienced pain, depression, fatigue, or loss of function due to their cancer. They were asked to describe the experience (type of pain, depression symptoms, fatigue and limitations in activities of daily living due to their cancer such as mobility, work, social events, and self-care). They were asked what medicines they took for the treatment of symptoms and what they felt worked. They were also asked if they talked to their healthcare providers and/or their family about their symptoms.

2.4.4. Knowledge and Behaviors

Knowledge level of skills needed to manage cancer-related symptoms was measured via a series of true and false questions. Participants were asked if they felt they had or acquired the skills needed to manage cancer-related symptoms.

2.4.5. Quality of Life

Participants were asked about their daily life activities and ability to function, and the impact cancer and cancer treatment had on their lives. Five domains were measured: mobility, self-care, ability to perform usual activities, pain or discomfort, and depression.

3. Results and Discussion

Table 1 reports on the participant baseline characteristics. Two hundred and twenty-two participants enrolled in the study. The study sample was heavily skewed toward females with only 30% representing males—a common occurrence in cancer survivorship studies with Indigenous peoples [11]. The mean age was 43 years. About sixty-three percent of participants had less than a high school education, and the majority were unemployed and not in a relationship.

Table 1. Baseline characteristics ($N = 222$).

Characteristics	Overall ($n = 222$)	Intervention ($n = 151$)	Control ($n = 71$)	p-Value
	(Mean/SE)	(Mean/SE)	(Mean/SE)	
Age	42.58 (15.68)	42.12 (15.88)	43.68 (15.28)	0.47
	%	%	%	
Gender				
Male	29.86	26.49	37.14	0.11
Female	70.14	73.51	62.86	
Education				
<HS degree	63.06	72.19	43.66	<0.0001
≥HS degree	36.94	27.81	56.34	

Table 1. *Cont.*

Characteristics	Overall (n = 222)	Intervention (n = 151)	Control (n = 71)	p-Value
	(Mean/SE)	(Mean/SE)	(Mean/SE)	
Marital Status				
Currently married	34.93	29.08	47.06	0.01
Divorced/Separated/Widowed/Single	65.07	70.92	52.94	
Employed				
Yes	31.53	25.83	43.66	0.008
No	68.47	74.17	56.34	
Cancer Diagnosis and History				
Type of Cancer (n = 50)				
Sarcoma	8.00	96.67	85.00	0.29
Carcinoma	92.00	3.33	15.00	
Currently being treated (n = 187)				
Yes	55.61	56.10	54.69	0.85
No	44.39	43.90	45.31	
Type of treatment				
Chemotherapy	30.18	24.50	42.25	0.007
Radiotherapy	13.96	14.57	12.68	0.70
Surgery	11.26	11.26	11.27	1.00
Traditional method	3.15	1.99	5.63	0.21
Other	7.66	8.61	5.63	0.59

3.1. Cancer Diagnosis and Treatment

The most common type of cancer diagnosed among males was prostate (10.0%), followed by colon/rectal cancer (7.1%), then stomach (2.9%) and lung cancer (2.9). Females reported diagnoses of breast cancer (10.6%), followed by ovarian (2.5%), colorectal (1.9%) and kidney cancer (1.9%). Forty-two percent of males and 53.7% females were currently being treated for their cancer. Thirty percent of participants reported that they were treated with chemotherapy for their cancer and 92% were being treated for carcinomas (Table 1). Although 75.0% of males and 62.3% of females received treatment for cancer in the past, the majority of participants (55.6%) were in the mist of treatment for their cancer; 40% of males reported having had surgery and 41% of females reported radiotherapy as the primary treatment they were undergoing for their cancer at the time of the questionnaire. Only about 3% of the study population reported use of traditional methods (such as healing ceremonies, herbal medicines, and traditional diets) for treatment of their cancers, though which methods were used by participants was not characterized in this study. It is noteworthy that, increasingly, Indigenous cancer survivors have derived perceived cultural, spiritual, and emotional benefits from its use in coping and healing from cancer [15,16].

3.2. Health Status/Quality of Life at Baseline

Information on the health status and quality of life of cancer survivors is important in that it provides useful information about their ability to function in areas of mobility, self-care, daily activities, and during episodes of pain and depression. Pre-test results demonstrate that male cancer survivors report poorer self-assessed health status and lower scores on quality-of-life indicators as compared to female cancer survivors. For instance, more males (31.6%) than females (24.3%) reported poor physical health interfered with their normal social activities with family, friends, neighbors or groups "quite a bit" (during the past 4 weeks). In addition, more males than females (47.4% vs. 16.2%) reported mobility problems and "difficulty performing work or activity," and "accomplished less than you would like" (47.4% male and 40.5% female). Although more females than males reported that they suffered from cancer-related pain (2/3 of females and half of males), in the area of emotional health (depression), male survivors reported their emotional health was currently "much worse" (10.5%), which was almost four times that reported by female survivors (2.7%). Loss of function experienced may have contributed to male survivors' lower scores in emotional health and poor physical health that interfered with normal social activities. Further, it must be acknowledged those who have a holistic view of health and wellness that encompasses a balance of one's physical, spiritual, emotional, as well as mental well-being may perceive their health differently than those who place more emphasis on physical health alone.

3.3. Cancer Symptoms

3.3.1. Pain

Pain is a prevalent symptom in cancer survivors and impacts thinking, concentration, and activities of daily living [17]. In a meta-analysis of 122 studies reporting cancer pain prevalence in adults, rates of pain were 39.3% after curative treatment; 55.0% during anticancer treatment; and 66.4% in advanced, metastatic, or terminal disease; additionally, moderate to severe pain was reported by 38% of all patients [17], indicating a need for improved pain management. This study found that the majority of cancer survivors (67.5% females vs. 49.2% males) suffer from pain due to their cancer experience. The onset of painful episodes occurred before, during, and even after remission, either as a result of the cancer itself or due to the cancer treatment. Although the majority of survivors were prescribed pain medication by their healthcare providers (71.5% females vs. 59.7% males), many survivors did not take pain medication as prescribed due adverse side effects of nausea/drowsiness and fear of becoming addicted to the medication (44.4% females and 39.7% males). Cancer survivors reported that they were generally instructed by their healthcare provider on how to manage or control their pain and that their providers reportedly addressed concerns about the side effects of the pain medication. However, overall, only 49.1% (54.4% females and 37.1% males) had ever been told how to manage pain at time of pre-test. Further, less than one-half (47.5% vs. 42.9% males) reported that they actually knew how to manage their pain. Additionally, at pre-test, 85% females and 68.6% males reported they would like to learn how to manage pain. At pre-test, more than one-half of survivors felt that they would not be able to control (manage) their pain (58.6%). This improved to 29% at post-test ($p = 0.09$). In addition, 67.9% of survivors reported at post-test that they now knew how to manage their pain, a significant increase as compared to the control group ($p = 0.003$). In addition, there was a reduction at post-test of those who reported, "there will always be pain with cancer," as compared to the control group ($p = 0.002$).

3.3.2. Depression

Depression is a common symptom of cancer with pooled prevalence estimates of 8–48% survivors affected that differ based on cancer treatment phase, type of cancer and/or location of tumor, and type of instrument used [18]. The risk for depression far exceeds that found in the general population, with odds being five times higher in cancer survivors [19]. Since it may resemble neurovegetative symptoms, including sleep disturbance, fatigue, and loss of appetite, a depression diagnosis may often be overlooked, especially in the context of busy oncology units where clinicians are not often skilled at diagnosing mental illness and survivors are reluctant to talk about their emotional health [20]. Depression may extend far beyond cancer treatment [18,21,22], making self-management skills even more important for improving quality of life during survivorship. Many participants in this study were hesitant to discuss having depression themselves, although they felt more comfortable calling it "the blues." Acknowledging that depression can exist during various times during cancer treatment, as well as during post-treatment, and understanding that depression is treatable is important for survivors and their families. At pre-test, the majority of participants (69% females and 66.2% males) reported that they feel depressed or "get the blues" "now and then." Sixteen percent of females and 11.6% of males were currently being treated for depression in the form of medication (14.4% females and 10% males) or counseling (6.9% females and 57% males). However, less than half of the survivors (42.5% females and 38.6% males) had been told how to manage their depression or how to manage their life around their medication's side effects (76.7% males and 57.3% females). As a result, 20.5% had concerns about the side effects of medication used to treat depression. A large percentage of participants (78%) reported at pre-test that they would like to learn how to manage their depression (88.2% females and 68% males). Following the intervention, post-test results showed that a significant increase was reported among participants who now knew how to manage their depression ($p = 0.004$), as compared to the control group. This was an increase in knowledge levels from 36.1% at pre-test to 66.4% at post-test.

3.3.3. Fatigue

Feeling fatigue means being so tired that it interferes with daily activities. Cancer-related fatigue is a very common symptom of cancer before, during, and following cancer treatment and it can affect survivors for long periods of time, yet it can go underrecognized [23]. A meta-analysis of 129 studies dating back to the year 1993 estimated prevalence of fatigue to be 49% in patients with cancer, with major differences related to type of cancer, cancer stage, and gender [24]. Understanding that fatigue is a cancer symptom which can be managed is an important message for the survivor, as well as family members and caregivers who often have to adjust their roles in response. Gaining a better understanding that fatigue is a legitimate cancer and treatment symptom and gaining strategies to manage fatigue were learning goals for the project's participants. At pre-test, a majority of participants reported experiencing some fatigue. Less than half of survivors (38.8% females and 30% males) reported knowing how to manage their fatigue and more females than males (41.9% vs. 27.4%) reported that they had been told how to manage fatigue time of pre-test. A large majority (68.1% females and 60% males) reported ever having thought about managing fatigue, and an even larger majority (86.3% females and 68.6% males) reported they would like to learn how to manage their fatigue. At post-test, a statistically significant increase was reported among those participants who now knew how to manage their fatigue (67.2% post-test vs. 32.5% pre-test; $p = 0.0001$). The control group reported no change in knowing how to manage their fatigue.

3.3.4. Loss of Function

Many cancer survivors report loss of physical functions due to the cancer itself and/or due its treatment, with older survivors experiencing greater losses [25]. Loss of function (in all or part of the body) is a common symptom that cancer survivors face and substantially impacts their quality of life. Functional limitations may come about shortly after treatment initiation and resolve at its completion, but others may last for years [26]. Irrespective of onset, functional limitations may affect one or more systems, including cardiovascular, pulmonary, and musculoskeletal, and may include additional symptoms, such as peripheral neuropathy, pain, fatigue, and sleep disturbances [26]. Common musculoskeletal limitations include the ability to walk (partial or full loss), weakness in arms and legs, and inability to easily lift articles, as well as sensory limitations, such as in hearing and sight, and cognition loss in area of memory. There is a known link between functional decline and caregiver dependency, impaired quality of life, comorbidity burden and increased mortality [25]. Early detection and use of evidence-based interventions may partly mitigate risk of functional decline in cancer survivors [25]. Learning how to manage loss of function during daily activities was a learning goal of this project. At pre-test, one-third of participants reported that they knew how to manage loss of function (31.3% females and 32.8% males). In addition, about a quarter of males (28.6% vs. 40% females) had been told how to manage loss of function due to cancer. A majority of participants reported that they had ever thought about managing loss of function, and at pre-test 87.5% females and 71.4% males reported that they would like to learn how to manage loss of function. At post-test, a statistically significant increase was reported among those participants who now knew how to manage their loss of function (61.9% post-test vs. 28.5% pre-test; $p = 0.0001$). The control group reported no change in functional status.

3.4. Knowledge and Symptom Control

Low pre-test scores largely improved at post-test in the intervention group and comparisons with the control group scores showed significant improvement in scores in all targeted categories. This was observed in the survivors' level of knowledge and perceived ability to control their cancer-related pain, knowledge of depression symptoms, and perceived ability to adopt recommended skills in managing depression, fatigue and function. Participant scores improved greatly at post-test among participants who reported that they now knew how to manage their cancer-related symptoms. Statistically significant improvement in scores when intervention groups were compared to the control groups was found in the areas of cancer pain, depression, fatigue and loss of function (see Table 2).

Table 2. Pre-test–Post-test knowledge changes.

Symptom	Measure	*p*-Value
Pain	At post-test 67.9% of participants reported that they now knew how to manage their pain, a significant increase compared to the control group.	0.003
Depression	At post-test, participants knew how to manage depression as compared to the control group. This was an increase in knowledge levels from 35.1% at pre-test to 66.4% at post-test.	0.004
Fatigue	A statistically significant increase was reported at post-test among those participants who now knew how to manage their fatigue (67.2% post-test vs. 32.5% pre-test). The control group reported no change.	0.0001
Loss of Function	At post-test, a statistically significant increase was reported among participants who now knew how to manage their loss of function (61.9% post-test vs. 28.5% pre-test). The control group reported no increase.	0.0001

4. Conclusions

This intervention project, designed to increase the ability of American Indian cancer survivors to better manage cancer-related symptoms such as pain, depression, fatigue and loss of function, improves communication with health care providers, thereby improving cancer survivors' quality of life. The Cancer Symptom Management Toolkit was designed specifically for American Indian cancer survivors and their caregivers, and was shaped by participants in earlier phases of the study. The intervention curriculum promoted culturally guided self-management strategies for cancer survivors and, in addition, provided caregivers with tangible ways to offer support to their loved ones. Study outcomes document the intervention was successful in improving knowledge and perceived skills/strategies in the management of all tested domains of cancer-related symptoms.

When examining pre-test scores, the survivors' responses to pre-test questions present a picture of cancer survivors who had little prior knowledge of cancer-related symptom management, and had limited instruction in pain, depression, fatigue or loss of function management. Survivors reported daily pain and episodes of depression experienced from diagnosis to treatment and beyond. More than one-half to two-thirds of participants suffered from cancer-related pain, yet the majority of survivors who reported that they were prescribed pain medication would take it only during the most painful episodes. They indicated fear of side effects from both of their medications used to treat pain and depression, as well as a fear of addiction to pain medications. Additionally, little was known by participants about pain control and depressive symptoms and the ability to manage or control these symptoms. As to fatigue experiences, most cancer survivors experienced some cancer-related fatigue, however, over one-half had never thought about their ability to manage their fatigue.

Interventions to improve knowledge and cancer symptom management skills and strategies contribute toward improving quality of life during cancer survivorship journeys. Developing relevant, culturally appropriate, effective cancer-related interventions to meet the needs of a diverse set views of health and wellness amongst Indigenous peoples continues to be needed [11,27,28]. A large majority of interventional studies in cancer with Indigenous survivors have shown positive effects on study outcomes, including increasing cancer knowledge, social and spiritual support, cancer service access and communication, yet ours may be the first to demonstrate a change in physical symptom self-management skills [11]. In this study, significant improvement in all targeted domains of cancer symptom management was achieved. This was observed through survivors' level of knowledge and perceived ability to manage their cancer-related pain, knowledge of depressive symptoms and perceived ability to adopt recommended skills in managing depression, fatigue and function.

The findings from this study guides researchers and healthcare providers towards a better understanding of the meaning and impact of cancer symptoms among American Indian cancer survivors. As others have observed [15,28–30], during this project, this study's findings found that American Indians in the Southwest experienced late diagnosis of their cancer all too often, leading to late-stage cancer at diagnosis and therefore poorer prognoses. In addition, once diagnosed, many cancer survivors lacked effective self-management strategies for the commonly experienced symptoms of pain, fatigue, depression and loss of function. Results from this study demonstrate the impact of the culturally appropriate use of the Talking Circle and toolkit intervention on participants' knowledge, attitudes, and ability to manage common cancer symptoms. Moreover, participant satisfaction with materials and experience with the Talking Circles curriculum was very high, with multiple indications for the need to expand the project's reach beyond the study. There are some limitations to this study; for example, the findings of the study may not be generalizable to American Indians living outside of the Southwest, and may not extend to all males since the majority of the study's participants were female. Participant bias may have affected study data. In addition, the toolkit, which was specifically tailored for Southwestern American Indian cultural preferences, would require adaptation in order to be appropriate for other American Indian communities.

Author Contributions: Conceptualization, F.S.H. and T.L.-I.; methodology, F.S.H. and T.L.-I.; validation, T.L.-I. and F.S.H.; formal analysis, T.L.-I. and F.S.H.; investigation, F.S.H. and T.L.-I., writing—original draft preparation, F.S.H.; writing—review and editing, F.S.H., T.L.-I. and R.H.A.A.; supervision, F.S.H.; project administration, T.L.-I.; funding acquisition, F.S.H. All authors have read and agreed to the published version of the manuscript.

Funding: This research was funded by the National Cancer Institute, NIH, grant number R01 CA115358.

Institutional Review Board Statement: The study was conducted in accordance with the Declaration of Helsinki, and approved by the Institutional Review Boards of the University of California Los Angeles (protocol code IRB#11-002460 and date of approval 18 August 2011) and the Phoenix Area Indian Health Service.

Informed Consent Statement: Informed consent was obtained from all subjects involved in the study.

Data Availability Statement: The data presented in this study are available on request from the corresponding author. The data are not publicly available due to confidentiality agreements with Tribes.

Acknowledgments: We would like to express our gratitude to the cancer survivors for participating in our project. This manuscript's contents are solely the responsibility of the authors and do not necessarily represent the official views of the NIH/NCI or the Indian Health Service.

Conflicts of Interest: The authors declare no conflict of interest. The funders had no role in the design of the study; in the collection, analyses, or interpretation of data; in the writing of the manuscript, or in the decision to publish the results.

References

1. CDC—Centers for Disease Control and Prevention. Cancer in American Indians and Alaska Natives in the United States . 2021. Available online: https://www.cdc.gov/cancer/dcpc/research/articles/cancer-AIAN-US.htm (accessed on 19 January 2022).
2. Sawchuk, C.N.; Van Dyke, E.; Omidpanah, A.; Russo, J.E.; Tsosie, U.; Goldberg, J.; Buchwald, D. Barriers to Cancer Care among American Indians and Alaska Natives. *J. Health Care Poor Underserved* **2016**, *27*, 84–96. [CrossRef] [PubMed]
3. Chou, F.-Y. Cancer Illness Perception and Self-Management of Chinese Patients. *Asia-Pac. J. Oncol. Nurs.* **2019**, *6*, 57–63. [CrossRef] [PubMed]
4. Burnette, C.E.; Roh, S.; Liddell, J.; Lee, Y.-S. American Indian Women Cancer Survivor's Needs and Preferences: Community Support for Cancer Experiences. *J. Cancer Educ.* **2019**, *34*, 592–599. [CrossRef] [PubMed]
5. Burnette, C.E.; Roh, S.; Liddell, J.; Lee, Y.-S. American Indian Women Cancer Survivors' Coping with Depressive Symptoms. *J. Psychosoc. Oncol.* **2019**, *37*, 494–508. [CrossRef]
6. Lee, Y.-S.; Burnette, C.E.; Liddell, J.; Roh, S. Understanding the Social and Community Support Networks of Ameri-can Indian Women Cancer Survivors. *J. Evid. Inf. Soc. Work* **2018**, *15*, 481–493. [CrossRef]
7. Hohl, S.; Molina, Y.; Koepl, L.; Lopez, K.; Vinson, E.; Linden, H.; Ramsey, S. Satisfaction with Cancer Care among American Indian and Alaska Natives in Oregon and Washington State: A Qualitative Study of Survivor and Caregiver Perspectives. *Support. Care Cancer* **2016**, *24*, 2437–2444. [CrossRef]
8. Haozous, E.A.; Doorenbos, A.; Alvord, L.A.; Flum, D.R.; Morris, A.M. Cancer Journey for American Indians and Alaska Natives in the Pacific Northwest. *Oncol. Nurs. Forum* **2016**, *43*, 625–635. [CrossRef]
9. McKinley, C.E.; Roh, S.; Lee, Y.-S.; Liddell, J. Family: The Bedrock of Support for American Indian Women Cancer Survivors. *Fam. Community Health* **2020**, *43*, 246–254. [CrossRef]
10. U.S. Cancer Statistics Working Group. *U.S. Cancer Statistics Data Visualizations Tool, Based on 2020 Submission Data (1999–2018)*; U.S. Department of Health and Human Services, Centers for Disease Control and Prevention and National Cancer Institute: Atlanta, GA, USA, June 2021. Available online: www.cdc.gov/cancer/dataviz (accessed on 29 June 2022).
11. Gifford, W.; Rowan, M.; Dick, P.; Modanloo, S.; Benoit, M.; Al Awar, Z.; Wazni, L.; Grandpierre, V.; Thomas, R.; Sikora, L.; et al. Interventions to Improve Cancer Survivorship among Indigenous Peoples and Communities: A Systematic Review with a Narrative Synthesis. *Support. Care Cancer* **2021**, *29*, 7029–7048. [CrossRef]
12. Burhansstipanov, L.; Krebs, L.U.; Seals, B.F.; Bradley, A.A.; Kaur, J.S.; Iron, P.; Dignan, M.B.; Thiel, C.; Gamito, E. Native American Breast Cancer Survivors' Physical Conditions and Quality of Life. *Cancer* **2010**, *116*, 1560–1571. [CrossRef]
13. Hodge, F.S.; Itty, T.L.; Cadogan, M.P.; Martinez, F. "Weaving Balance into Life": Development and Cultural Adaptation of a Cancer Symptom Management Toolkit for Southwest American Indians. *J. Cancer Surviv.* **2012**, *6*, 182–188. [CrossRef] [PubMed]
14. Kern, D.E.; Thomas, P.A.; Bass, E.B.; Howard, D.M. *Curriculum Development for Medical Education: A Six Step Approach*; JHU Press: Baltimore, MD, USA, 1998; ISBN 978-0-8018-5844-4.
15. Gall, A.; Leske, S.; Adams, J.; Matthews, V.; Anderson, K.; Lawler, S.; Garvey, G. Traditional and Complementary Medicine Use Among Indigenous Cancer Patients in Australia, Canada, New Zealand, and the United States: A Systematic Review. *Integr. Cancer Ther.* **2018**, *17*, 568–581. [CrossRef] [PubMed]
16. Gall, A.; Anderson, K.; Adams, J.; Matthews, V.; Garvey, G. An Exploration of Healthcare Providers' Experiences and Perspectives of Traditional and Complementary Medicine Usage and Disclosure by Indigenous Cancer Patients. *BMC Complement. Altern. Med.* **2019**, *19*, 259. [CrossRef] [PubMed]
17. Van den Beuken-van Everdingen, M.H.J.; Hochstenbach, L.M.J.; Joosten, E.A.J.; Tjan-Heijnen, V.C.G.; Janssen, D.J.A. Update on Prevalence of Pain in Patients with Cancer: Systematic Review and Meta-Analysis. *J. Pain Symptom Manag.* **2016**, *51*, 1070–1090.e9. [CrossRef] [PubMed]
18. Bach, A.; Knauer, K.; Graf, J.; Schäffeler, N.; Stengel, A. Psychiatric Comorbidities in Cancer Survivors across Tumor Subtypes: A Systematic Review. *World J. Psychiatry* **2022**, *12*, 623–635. [CrossRef]
19. Hartung, T.J.; Brähler, E.; Faller, H.; Härter, M.; Hinz, A.; Johansen, C.; Keller, M.; Koch, U.; Schulz, H.; Weis, J.; et al. The Risk of Being Depressed Is Significantly Higher in Cancer Patients than in the General Population: Prevalence and Severity of Depressive Symptoms across Major Cancer Types. *Eur. J. Cancer* **2017**, *72*, 46–53. [CrossRef]
20. Krebber, A.M.H.; Buffart, L.M.; Kleijn, G.; Riepma, I.C.; de Bree, R.; Leemans, C.R.; Becker, A.; Brug, J.; van Straten, A.; Cuijpers, P.; et al. Prevalence of Depression in Cancer Patients: A Meta-Analysis of Diagnostic Interviews and Self-Report Instruments. *Psychooncology* **2014**, *23*, 121–130. [CrossRef]

21. Roh, S.; Lee, Y.-S.; Hsieh, Y.-P.; Easton, S.D. Protective Factors against Depressive Symptoms in Female American Indian Cancer Survivors: The Role of Physical and Spiritual Well-Being and Social Support. *Asian Pac. J. Cancer Prev.* **2021**, *22*, 2515–2520. [CrossRef]
22. Bamonti, P.M.; Moye, J.; Naik, A.D. Pain Is Associated with Continuing Depression in Cancer Survivors. *Psychol. Health Med.* **2018**, *23*, 1182–1195. [CrossRef]
23. Jagsi, R.; Griffith, K.A.; Vicini, F.; Boike, T.; Dominello, M.; Gustafson, G.; Hayman, J.A.; Moran, J.M.; Radawski, J.D.; Walker, E.; et al. Identifying Patients Whose Symptoms Are Underrecognized During Treatment with Breast Radiotherapy. *JAMA Oncol.* **2022**, *8*, 887–894. [CrossRef]
24. Al Maqbali, M.; Al Sinani, M.; Al Naamani, Z.; Al Badi, K.; Tanash, M.I. Prevalence of Fatigue in Patients with Cancer: A Systematic Review and Meta-Analysis. *J. Pain Symptom Manag.* **2021**, *61*, 167–189.e14. [CrossRef] [PubMed]
25. Muhandiramge, J.; Orchard, S.G.; Warner, E.T.; van Londen, G.J.; Zalcberg, J.R. Functional Decline in the Cancer Patient: A Review. *Cancers* **2022**, *14*, 1368. [CrossRef] [PubMed]
26. Nekhlyudov, L.; Campbell, G.B.; Schmitz, K.H.; Brooks, G.A.; Kumar, A.J.; Ganz, P.A.; Von Ah, D. Cancer-Related Impairments and Functional Limitations among Long-Term Cancer Survivors: Gaps and Opportunities for Clinical Practice. *Cancer* **2022**, *128*, 222–229. [CrossRef] [PubMed]
27. Cavanagh, B.M.; Wakefield, C.E.; McLoone, J.K.; Garvey, G.; Cohn, R.J. Cancer Survivorship Services for Indigenous Peoples: Where We Stand, Where to Improve? A Systematic Review. *J. Cancer Surviv.* **2016**, *10*, 330–341. [CrossRef]
28. Guadagnolo, B.A.; Petereit, D.G.; Coleman, C.N. Cancer Care Access and Outcomes for American Indian Populations in the US: Challenges and Models for Progress. *Semin. Radiat. Oncol.* **2017**, *27*, 143–149. [CrossRef]
29. Nash, S.H.; Zimpelman, G.L.; Miller, K.N.; Clark, J.H.; Britton, C.L. The Alaska Native Tumour Registry: Fifty Years of Cancer Surveillance Data for Alaska Native People. *Int. J. Circumpolar Health* **2022**, *81*, 2013403. [CrossRef]
30. Decker, K.M.; Kliewer, E.V.; Demers, A.A.; Fradette, K.; Biswanger, N.; Musto, G.; Elias, B.; Turner, D. Cancer Incidence, Mortality, and Stage at Diagnosis in First Nations Living in Manitoba. *Curr. Oncol.* **2016**, *23*, 225–232. [CrossRef]

Article

Outcomes Following Autologous Fat Grafting in Patients with Sequelae of Head and Neck Cancer Treatment

Jorge Masià-Gridilla [1,2,*], Javier Gutiérrez-Santamaría [1], Iago Álvarez-Sáez [1,2], Jorge Pamias-Romero [1,2], Manel Saez-Barba [1,2] and Coro Bescós-Atin [1,2,3]

1 Noves Tecnologies i Microcirurgia Craniofacial, Vall d'Hebron Institut de Recerca (VHIR), Hospital Universitari Vall d'Hebron, Vall d'Hebron Barcelona Hospital Campus, Passeig Vall d'Hebron 119-129, E-08035 Barcelona, Spain
2 Servei de Cirurgia Oral i Maxil·lofacial, Hospital Universitari Vall d'Hebron, Vall d'Hebron Barcelona Hospital Campus, Passeig Vall d'Hebron 119-129, E-08035 Barcelona, Spain
3 Unitat Docent Vall d'Hebron, Facultat de Medicina, Universitat Autònoma de Barcelona, Passeig Vall d'Hebron 119-129, E-08035 Barcelona, Spain
* Correspondence: jmasia@vhebron.net

Simple Summary: In recent years, there have been relevant advances in the use of surgery, radiation therapy, and chemotherapy for the treatment of malignant tumors of the head and neck. Extensive tumor resection and radical radiotherapy frequently result in altered form and function of orofacial structures that can severely impact the patient's quality of life. This study reports the benefits obtained with the injection of autologous fat to correct the deformities and improve functionality in a series of 40 patients who have been treated for head and neck cancer. Esthetic improvement was obtained in 77.5% of patients and functional improvement in 89.2%. In addition, there was a high degree of satisfaction regarding esthetic improvement and 92.5% of patients would recommend the procedure to other patients in the same situation. The injection of autologous fat is an effective procedure for the management of sequelae of head and neck cancer treatment.

Abstract: A single-center retrospective study was designed to assess the outcomes of autologous fat grafting for improving surgery- and radiotherapy-related sequelae in 40 patients with head and neck cancer. All patients underwent surgical resection of primary tumors and radiotherapy (50–70 Gy) and were followed over 12 months after fat grafting. Eligibility for fat grafting procedures included complete remission after at least 3 years of oncological treatment. The cervical and paramandibular regions were the most frequently treated areas. Injected fat volumes ranged between 7.5 and 120 mL (mean: 23 mL). Esthetic improvement was obtained in 77.5% of patients, being significant in 17.5%, and functional improvement in 89.2%, being significant in 29.7% of patients. Minor complications occurred in three patients. There was a high degree of satisfaction regarding esthetic improvement, global satisfaction, and 92.5% of patients would recommend the procedure. This study confirms the benefits of fat grafting as a volumetric correction reconstructive strategy with successful cosmetic and functional outcomes in patients suffering from sequelae after head and neck cancer treatment.

Keywords: autologous fat grafting; head and neck cancer; radiotherapy; reconstruction; sequelae; quality of life

Citation: Masià-Gridilla, J.; Gutiérrez-Santamaría, J.; Álvarez-Sáez, I.; Pamias-Romero, J.; Saez-Barba, M.; Bescós-Atin, C. Outcomes Following Autologous Fat Grafting in Patients with Sequelae of Head and Neck Cancer Treatment. *Cancers* **2023**, *15*, 800. https://doi.org/10.3390/cancers15030800

Academic Editor: António Araújo

Received: 5 January 2023
Revised: 21 January 2023
Accepted: 26 January 2023
Published: 28 January 2023

1. Introduction

Head and neck cancer represents the seventh most common cancer worldwide, with 1.1 million new diagnoses reported annually [1,2]. However, there is a substantial geographical variation in the incidence and anatomical distribution of tumors, predominantly attributed to differences in smoking and alcohol consumption, steady increase in human papillomavirus-related cancer, genetic predisposition, or exposure to ionizing radiation, which are known to play an important role in the pathogenesis of the disease [3,4]. Radiation therapy, surgery or both combined and chemotherapy are currently available standard therapeutic strategies but are often prioritized differently depending on the site of tumor origin, histological diagnosis, tumor burden, quality of life considerations, patient preference, or hospital characteristics with the availability of specialized multidisciplinary care teams [5,6].

Advances in surgery, radiation therapy, and chemoradiotherapy have improved locoregional control and survival, but the outcomes of these treatment modalities have incorporated preservation and restoration of function in the focus of radical ablation and curative efforts [7]. However, despite improvements in the multimodal treatment approach aimed at decreasing cosmetic and functional deficits with resultant psychological, physical, and nutritional detriments [8,9], management of sequelae following treatment of head and neck cancer, particularly in patients with locally advanced tumors, still remains a challenge difficult to solve in daily practice [10,11].

Fat grafting, also referred to as fat transfer or fat injections, dates back to 1893 when Neuber first described the technique and reported successful outcomes after transplanting fat beneath atrophic scars [12]. Structural autologous fat grafts for the enhancement of facial contours were proposed by Coleman in 1997 [13] and the Coleman's lipostructure technique became subsequently recognized as a standard procedure for fat transfer [14,15]. In recent years, autologous fat grafting has been described by different authors as a very useful tool to improve residual esthetic and functional deformities after head and neck cancer treatment, and for its ability to correct volumetric defects and regenerative properties [16–19]. A systematic review and meta-analysis of 52 studies with 1568 patients confirmed that autologous fat transfer is an effective technique in facial reconstruction surgery with a low rate of minor complications [20].

The purpose of this study was to evaluate esthetic and functional outcomes as well as patients' satisfaction and complications associated with autologous fat grafting in the context of integral management of head and neck cancer patients.

2. Materials and Methods

2.1. Design and Study Population

A retrospective study was made of all consecutive patients who required autologous fat grafting procedures between January 2010 and January 2019 at the Service of Oral and Maxillofacial Surgery of Hospital Universitari Vall d'Hebron, in Barcelona, Spain. Fat grafting was indicated for the treatment of sequelae associated with any form of therapy of head and neck cancer. Inclusion criteria were history of head and neck cancer treated with surgery, radiation therapy, or chemotherapy followed by duration of complete clinical remission of at least 3 years; presence of severe or very severe esthetic defects and/or loss of skin flexibility, and severe or very severe dysphonia, dysphagia, alteration in head and neck mobility, and alteration in swallowing or chewing, corresponding to scores 3 or 4 of esthetic and/or functional evaluation of the scoring method described by Pulphin et al. [21]; good health, as confirmed by preoperative work-up studies; and signed informed consent. Patients previously treated with fat infiltration procedures or with insufficient fat tissue deposits for fat transfer were excluded from the study, as were those expected to have poor adherence to follow-up visits scheduled for at least 12 months after the intervention, and ineligibility as judged by the investigators.

The study protocol was approved by the Clinical Research Ethics Committee of Hospital Universitari Vall d'Hebron (code PR (ATR) 57/2016, approval date 26 February 2016) (Barcelona, Spain). Written informed consent was obtained from all participants.

2.2. Fat Grafting and Surgical Procedure

The available fat deposits were evaluated, and the donor site was selected with consent from the patient. Fat harvesting was performed under general anesthesia or local anesthesia with intravenous sedation, and the patient was in the supine position. Ten minutes before liposuction, abdominal infiltration was performed through a 2–3 mm incision puncture at the level of both flanks or in the umbilical region, using a modified Klein solution (500 mL Ringer lactate) with 0.5 mg epinephrine, and adding 1% lidocaine for patients under sedation. Harvesting was performed through the same infiltration incisions using a liposuction cannula (COL-ASP15, 3 mm × 15 cm, Byron Medical Inc., Tucson, AZ, USA or COL-KHU12 Mitmed®, 3 mm × 20 cm, Surgest Medical, Sant Cugat del Vallès, Barcelona, Spain) connected to a 10 mL Luer-Loc syringe, by firm and regular back-and-forth movements under low negative digital pressure until the syringe was filled. Then, fat was purified either by the centrifugation method described by Coleman [22] (3000 rpm for 3 min) (Medigraft-BL® Centrifuge, Surgest Medical) or washing and filtration using the Puregraft system (Cytori Therapeutics, San Diego, CA, USA).

The graft was injected in small amounts, separated between them in order to obtain a better vascularization and therefore longer graft survival, slowly and without overcorrection, from the deep to the superficial cutaneous tissue using an atraumatic cannula (7–9 mm long, 16G, types I-III COL-19, Byron Medical, COL SPA9), creating multiple tunnels in a fan-like fashion following the recommendation of Coleman [22]. An abdominal bandage was applied for 48 h and substituted with an abdominal belt for the following 7 days. Amoxicillin–clavulanic acid (500/250 mg), 1 tablet every 8 h, was administered during the first 7 days after the procedure.

2.3. Evaluation and Follow-Up

Patients were visited postoperatively by the same investigator (J.M.-G.) after 1 week of fat grafting and at 1, 3, 6, and 12 months thereafter. At each visit, patients were questioned and underwent a physical examination to assess the evolution of the graft and the eventual appearance of early or late complications. Twelve months after fat grafting, esthetic and functional results were evaluated using a 4-point scale described by Pulphin et al. [21], including no esthetic or functional problems (score 0), and esthetic defects and/or loss of skin flexibility and functional alterations of dysphonia, dysphagia, neck/head mobility or swallowing or chewing scored as 1 = slight, 2 = moderate, 3 = severe, and 4 = very severe. Improvement was defined in the presence of a postoperative score lower than the preoperative score, and significant improvement was defined if the postoperative score was 2 or more points lower than the preoperative score. The severity of complications was classified according to the Clavien–Dindo classification system [23].

Also, after 12 months of fat grafting, the patient's satisfaction regarding esthetic improvement was evaluated on a scale of 0 to 10 (0 = nothing, 10 = maximum satisfaction) and the overall satisfaction with treatment according to responses to the following four questions: "*What is the degree of satisfaction with the treatment received?*", "*Do you consider that you have received sufficient and clear information?*", "*Did the treatment meet your expectations?*", and "*In case you request an advice, would you recommend this treatment to another patient in the same conditions?*" using a 4-point Likert scale (1 = nothing, 2 = little, 3 = quite a lot, 4 = a lot/very much).

2.4. Statistical Analysis

Descriptive statistics were calculated and presented as frequencies and percentages for categorical variables, and as mean and standard deviation (SD) for continuous variables.

3. Results

The study population included 40 patients, 26 men and 14 women, with a mean age of 60.5 years (range 32–86 years). Complete data at the 12-month follow-up visit were obtained in all participants. Squamous cell carcinoma of the oral cavity was the most frequent primary tumor (n = 25, 62.5%) followed by a malignant tumor of the salivary glands (n = 7, 17.5%). All patients underwent radical surgery of the neoplasms and radiotherapy (50–70 Gy), and 23 of them (57.5%) received chemotherapy. Reconstruction procedures using different types of flaps were performed in 27 (67.5%) patients using microsurgical free fibula flaps in most of them. All patients presented with esthetic sequelae, including scarring, cervicofacial asymmetry, and cutaneous fibrosis. Limitations of neck movements, trismus, and dysphagia were the most frequent functional sequelae.

The abdominal region was the donor site for fat grafting in all patients. The manual low pressure aspiration technique was used to obtain the fat graft in all cases using a 10 mL syringe with Luer-Loc connection and COL-KHU12 Mitmed®, 3 mm × 20 cm liposuction cannula. Regarding the processing method, the centrifugation method following Coleman's recommendations [22] was used in the first series of 16 patients, and the filtration processing system was carried out using the Puregraft device in the remaining 24. In all patients, infiltration was performed following the Coleman technique [22].

Fat grafting was mostly performed under general anesthesia, with sedation and local anesthesia in only three patients. The cervical and paramandibular regions were the most frequently treated areas, with injected fat volumes ranging between 7.5 and 120 mL (mean: 23 mL). The length of surgery varied between 45 and 180 min, with a mean of 89 min. No intraoperative complications were recorded, and all patients were discharged within 24 h after the procedure. Minor complications occurred in three patients (7.5%) with abdominal pain, seroma, and lingual paresthesia in one patient each, and they resolved spontaneously. All these complications were classified as grade I of the Clavien–Dindo classification system [23].

Esthetic improvement was obtained in 31 patients (77.5%), being significant in 7 of them (17.5%). In relation to functional alterations, there were three patients who scored 0 preoperatively. In the remaining 37 patients, functional improvement was found in 33 (89.2%), being significant in 11 of them (29.7%). One of the most widespread findings in the treated patients was clinical improvement in the quality of irradiated skin on the neck or face, with apparent improvement in blood supply, skin smoothness, function, and elasticity. The analysis of graft stability was performed clinically by evaluating the patient and analyzing the photographic documentation, showing a progressive volumetric decrease close to 50% of the injected volume in all patients. Details of treatment characteristics and outcome of the study patients are shown in Table 1. Figures 1–5 show some representative cases. Postoperative photographs of these patients were obtained between 6 and 12 months of follow-up after the fat grafting procedure.

Esthetic improvement evaluated by the patients showed a mean (SD) score of 7.03 (1.83) and a mean satisfaction with treatment of 3.05 (0.68). In addition, 37 patients (92.5%) would recommend treatment with autologous fat grafting to other patients in a similar situation.

After 12 months of follow-up of autologous fat grafting, two patients died; the causes of death were a new lung cancer and heart disease, respectively. Recurrence of the primary head and neck cancer occurred in three patients, but in all cases, the site of recurrence was far from the fat infiltrated area.

Table 1. Clinical characteristics, details of treatment, and outcome in the 40 study patients.

Patient	Histology /Location	Surgery	Chemotherapy	RT Gy	Reconstruction Type	Injection Site	Volume mL	Length of Surgery min	Anesthesia	Esthetic Score Preoperative/ Postoperative	Functional Score Preoperative/ Postoperative
1	ACC/parotid	PT + MD	No	60	Not performed	Laterocervical and parotid area	7.5	50	General	4/3	3/2
2	DFSP/malar	TEOM	No	60	Mustarde cheek flap	Hemifacial	23	45	General	3/2	3/2
3	SCC/tongue	MD + ND	Yes	70	Microsurgical fibula flap	Laterocervical and paramandibular	28	65	General	4/4	4/3
4	SCC/gums	MD + ND	Yes	60	Not performed	Laterocervical and paramandibular	20	120	General	4/2	2/1
5	SCC/gums	MD + ND	No	66	Microsurgical fibula flap	Submaxillary, lower lip, nasolabial and submental fold	24	64	General	3/2	3/2
6	SCC/gums, mouth floor	MD + ND bilateral	Yes	70	Fibula flap + anterolateral thigh flap	Laterocervical and paramandibular	19	90	General	4/4	4/3
7	SCC/gums	MD + ND	Yes	50	Not performed	Laterocervical and paramandibular	24	115	General	4/3	3/2
8	SCC/jugal mucosa	TEOM	Yes	60	Local flap	Jugal region, nasolabial, submental. Laterocervical, tracheocervical	20	100	General	3/2	4/2
9	SCC/gums	TEOM + MD + ND	Yes	60	Microsurgical fibula flap	Laterocervical and paramandibular	37	135	General	4/4	3/2
10	SCC/gums	MD + ND	Yes	70	Fibula flap	Laterocervical and paramandibular	42	120	General	3/3	3/3
11	SCC/tongue, mouth floor	TEOM + MD	No	70	Fibula flap + anterolateral thigh flap	Hemifacial and cervical	20	80	Sedation local	4/3	3/2
12	SCC/retromolar trigone	MD + ND	No	70	Microsurgical fibula flap	Hemifacial and cervical	27	180	General	4/3	3/2
13	SCC/jugal mucosa	TEOM + ND	No	60	Radial flap	Laterocervical, paramandibular, jugal	23	110	General	3/2	2/1
14	SCC/mouth floor	MD + ND	Yes	70	Microsurgical fibula flap	Laterocervical, paramandibular, tracheal	70	100	General	4/3	4/2

Table 1. *Cont.*

Patient	Histology /Location	Surgery	Chemotherapy	RT Gy	Reconstruction Type	Injection Site	Volume mL	Length of Surgery min	Anesthesia	Esthetic Score Preoperative/ Postoperative	Functional Score Preoperative/ Postoperative
15	Myoepithelial carcinoma/ minor salivary gland, maxillary	Maxillectomy	No	60	Temporal muscle flap	Temporal	13	150	General	3/1	0/0
16	Myoepithelial carci-noma/parotid gland	Radical Parotidectomy + ND	Yes	66	Not performed	Laterocervical and parotid region	8.5	58	General	3/2	2/1
17	SCC/gums	MD + ND	Yes	70	Microsurgical fibula flap	Hemifacial and cervical	15	110	General	3/2	3/1
18	SCC/gums	Maxillectomy + ND	Yes	64	Microsurgical fibula flap	Hemifacial and cervical	23	70	General	3/2	3/2
19	Adenocarcinoma parotid gland	Total parotidectomy	No	50	Not performed	Hemifacial	13	87	General	2/0	2/0
20	SCC/retromolar trigone	MD + ND	Yes	70	Not performed	Nasolabial fold, upper and lower lip, jugal and laterocervical	120	170	General	4/3	4/2
21	SCC/cervical unknown origin	ND	No	60	Not performed	Laterocervical	15	52	Sedation local	3/4	4/4
22	SCC/tongue, mouth floor	Glossectomy + MD + ND	No	66	Microsurgical fibula flap	Upper and lower lips, paramandibular, superior laterocervical, bilateral submandibular, bilateral nasolabial folds	20	117	General	4/2	3/1
23	SCC/mouth floor	MD + ND bilateral	No	70	Microsurgical fibula flap	Lower lip, paramandibular, laterocervical	12	97	General	4/2	3/2
24	Ductal carci-noma/parotid gland	Superficial parotidectomy	No	60	Not performed	Laterocervical and parotid region	15	71	General	2/1	2/1

Table 1. *Cont.*

Patient	Histology /Location	Surgery	Chemotherapy	RT Gy	Reconstruction Type	Injection Site	Volume mL	Length of Surgery min	Anesthesia	Esthetic Score Preoperative/ Postoperative	Functional Score Preoperative/ Postoperative
25	SCC/retromolar trigone	MD + ND	Yes	70	Microsurgical fibula flap	Paramandibular, submaxillary, upper and lower lip	23	63	General	4/3	4/2
26	SCC/jugal mucosa	TEOM + ND	Yes	66	Local flap	Laterocervical, paramandibular, jugal	23	79	General	3/2	4/2
27	SCC/mandibular intraosseous	MD + ND bilateral	Yes	63	Microsurgical fibula flap	Paramandibular, laterocervical, nasolabial fold, lower lip	23	103	General	4/2	4/2
28	Undifferentiated parotid carcinoma	Total parotidectomy	No	60	Not performed	Paramandibular and parotid region	25	83	General	4/2	3/1
29	SCC/cervical unknown origin	ND	Yes	60	Not performed	Laterocervical	14	47	General	3/2	4/3
30	SCC/tongue	Glossectomy + ND	Yes	54	Nor performed	Laterocervical, lingual	18	115	General	3/3	4/4
31	ACC/minor salivary gland maxillary	Maxillectomy	No	66	Temporal muscle flap	Malar bilateral, left nasolabial fold, upper lip, left jugal	20	47	General	3/2	0/0
32	SCC/gums	MD + ND	Yes	70	Pectoral flap	Paramandibular and jugal	10	48	General	4/2	2/1
33	SCC/retromolar trigone	MD + ND	Yes	70	Pectoral flap + fibula flap	Hemifacial, cervical, labial, tracheal	15	114	General	4/2	3/2
34	SCC/lip	TEOM + ND	No	55	Anterolateral thigh flap	Jugal and labial	8	54	Sedation local	2/2	0/0
35	ACC/ oropharynx-tongue	Glossectomy + ND	No	66	Anterolateral thigh flap	Laterocervical	20	104	General	3/3	3/3
36	SCC/gums, mouth floor	MD + ND	Yes	69	Not performed	Right laterocervical, paramandibular	15	55	General	4/4	4/2
37	Angiofibroma/ nasal	TEOM	No	50	Not performed	Temporal	20	95	General	3/2	1/0
38	SCC/mandibular symphysis	MD + ND bilateral	Yes	70	Microsurgical fibula flap	Laterocervical and submental	10	48	General	4/3	3/2

Table 1. *Cont.*

Patient	Histology /Location	Surgery	Chemotherapy	RT Gy	Reconstruction Type	Injection Site	Volume mL	Length of Surgery min	Anesthesia	Esthetic Score Preoperative/ Postoperative	Functional Score Preoperative/ Postoperative
39	Osteosarcoma mandibular	MD + maxillectomy + ND	Yes	60	Anterolateral thigh flap	Hemifacial and cervical	10	70	General	4/3	3/2
40	ACC/upper maxilla	Maxillectomy + ND	Yes	72	Temporal muscle flap + microsurgical fibula flap	Hemifacial, cervical and temporal	27	76	General	3/2	3/1

ACC: adenoid cystic carcinoma, DFSP: dermatofibrosarcoma protuberans, MD: mandibulectomy, ND: neck dissection, PT: parotidectomy, RT: radiotherapy, SCC: squamous cell carcinoma, TEOM: tumoral exeresis with oncological margins.

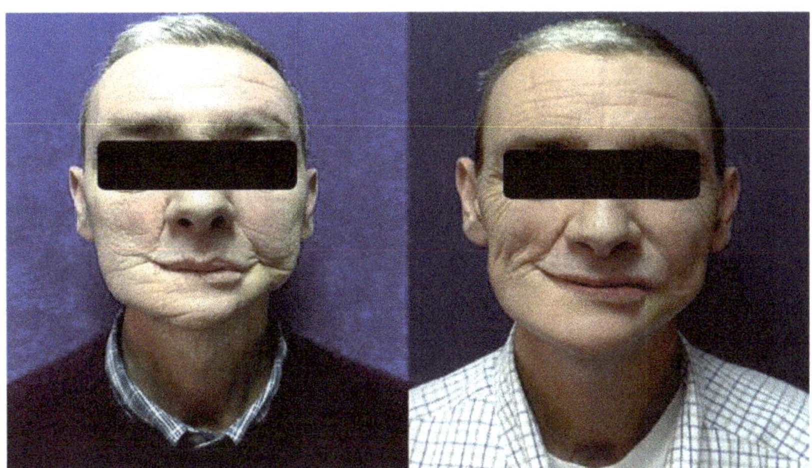

Figure 1. A 59-year-old male patient treated with left segmental mandibulectomy and microsurgical reconstruction with a microsurgical fibula flap, ipsilateral cervical lymph node dissection, and postoperative radiotherapy at a dose of 66 Gy for a squamous cell carcinoma of the left alveolar crest. He was treated with autologous fat grafting in the paramandibular and cervical regions with a total of 24 mL of fat (**left**). The postoperative photograph at follow-up shows the improvement of the paramandibular and cervical deformity (**right**).

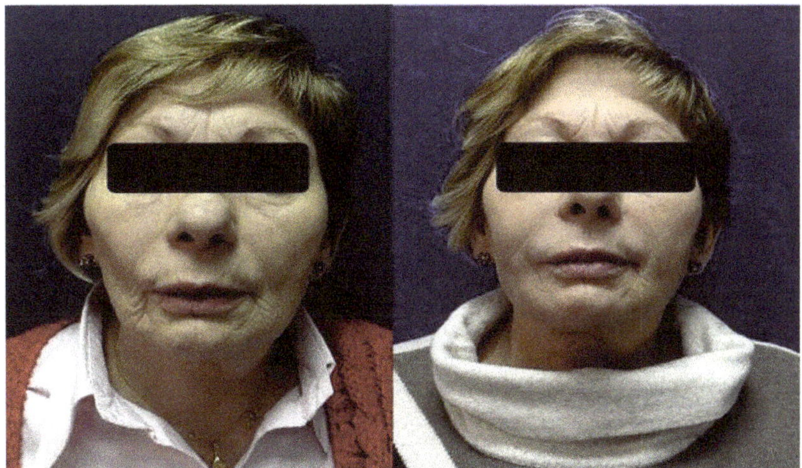

Figure 2. A 71-year-old female patient treated with right buccal mucosa excision, and postoperative radiotherapy at a dose of 66 Gy for an oral squamous cell carcinoma. She was treated with autologous fat grafting in cheek and cervical regions with a total of 23 mL of fat (**left**). The postoperative photograph at follow-up shows the improvement of the facial and cervical deformity (**right**).

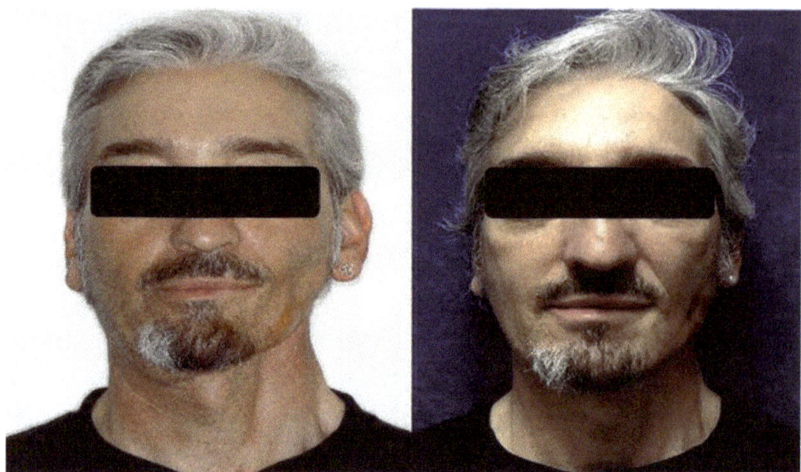

Figure 3. A 54-year-old male patient treated for left segmental mandibulectomy and microsurgical reconstruction with a microsurgical fibula flap, ipsilateral cervical lymph node dissection, and postoperative radiotherapy at a dose of 64 Gy for a squamous cell carcinoma of the left alveolar crest. He was treated with autologous fat grafting in the paramandibular and cervical regions with a total of 23 mL of fat (**left**). The postoperative photograph at follow-up shows the improvement of the paramandibular and cervical deformity (**right**).

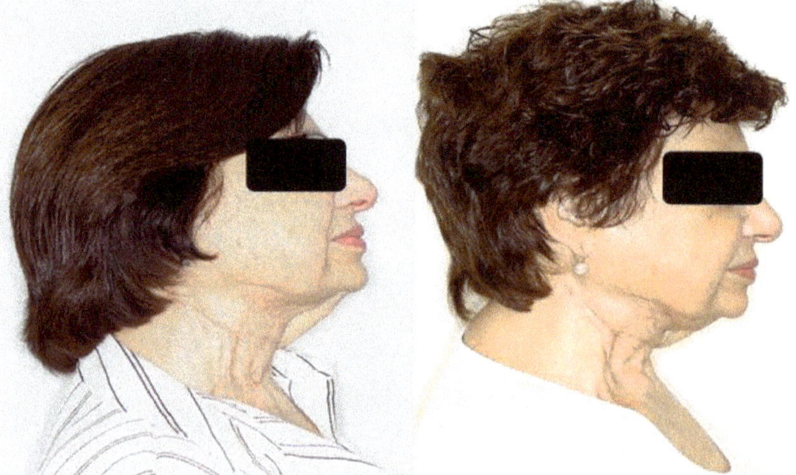

Figure 4. A 71-year-old female patient treated for right buccal squamous cell carcinoma with local resection and ipsilateral radical cervical lymph node dissection and postoperative radiotherapy at a dose of 66 Gy for a squamous cell carcinoma. She was treated with autologous fat grafting in the right cervical region with a total of 23 mL of fat (**left**). The postoperative photograph at follow-up shows the improvement of the paramandibular and cervical deformity (**right**).

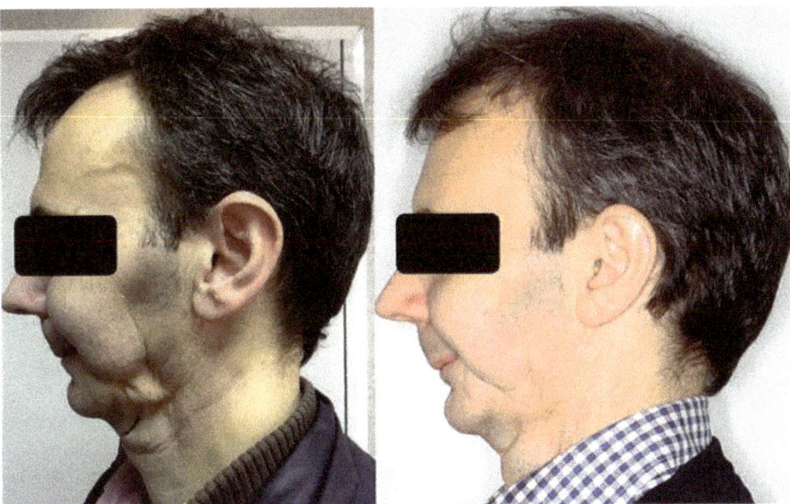

Figure 5. A 43-year-old male patient treated for an intraosseous carcinoma of the left mandible with segmental mandibulectomy and microsurgical reconstruction with a fibula flap, ipsilateral cervical lymph node dissection, and postoperative radiotherapy at a dose of 63 Gy. He was treated with autologous fat grafting in the paramandibular and cervical regions with a total of 24 mL of fat (**left**). The postoperative photograph at follow-up shows the improvement of the paramandibular and cervical deformity (**right**).

4. Discussion

Autologous fat grafting is a feasible and valuable technique for patients with sequelae following surgery and radiotherapy of primary head and neck cancer tumors [17,24]. However, the experience with the use of fat grafting for esthetic and functional improvement in these patients is still limited [17–22,25]. The present clinical series is the largest published of head and neck cancer patients treated with combined surgery and radiation therapy of at least 50 Gy, undergoing autologous fat grafting for the correction of esthetic and functional sequelae of treatment. In all cases, fat grafting was performed after a disease-free interval of 3 years, a time period with the highest risk of tumor recurrence. In other studies, fat grafting has been performed after a minimum follow-up of 1 year [19,26].

All patients were operated on following the technique described by Coleman [13,22], although in 60% of cases ($n = 14$), purification was performed using the Puregraft system as it was considered that this method better preserved the sterility of the circuit and eliminated the exposure of fat to air, thus avoiding rapid desiccation and preserving the survival of adipocytes. Zhu et al. [27] have compared three preparation methods for fat grafts in twenty-two donors: gravity separation, Coleman centrifugation, and simultaneous washing with filtration using the Puregraft system. Grafts prepared by washing with filtration exhibited significantly reduced blood cell and free lipid content, with significantly greater adipose tissue viability than other methods.

In our study, esthetic and functional outcomes were evaluated at 12 months after fat grafting using the 4-point scale described by Pulphin et al. [21]. Esthetic improvement was obtained in 77.5% of patients, being significant in 17.5%, and functional improvement in 89.2%, being significant in 29.7% of patients (significant improvement defined as postoperative score of 2 or more points lower than preoperative score). The rate of improvement is difficult to compare to other previously published studies because of differences in the scoring system for the assessment of results, except for similar findings in a preliminary feasibility study of 12 patients reported by our group [17], and the clinical series of 11 patients reported by Pulphin et al. [21] who were the authors that described the evaluation score system. In this study, significant esthetic improvement was obtained in nine patients (81.8%) and functional improvement in seven (63.6%). The total injected volume ranged between 10 and 119 mL, with an average of 48.5 mL, which is a somewhat greater volume than the 23 mL used in our study. No complications were recorded. Patients were followed for a mean of 39.9 months (range 2–88 months), but the resorption of engrafted fat was observed for all patients and was estimated to be approximately 20% to 40%. Because of the importance of the defects, reinjection was performed in six patients. In addition, histological examination of biopsies taken from the treated areas of six patients showed reduction in irradiated morphology patterns, with normal histological structure, high vascular network density, and reduction in fibrosis. In our study, biopsies from fat grafting areas were not obtained.

In 2003, Ducic et al. [19] reported data of a retrospective series of 23 patients undergoing lipotransfer as part of their craniofacial reconstructive procedure. In this study, six patients underwent a total of eight fat transfer procedures (two procedures in two patients), with good results in five and inadequate results in one. No intraoperative or postoperative complications were observed. Vitagliano et al. [26] described 10 patients with squamous and basal cell carcinomas of the lower or upper lips treated with resection and nasolabial flaps. After 6 months of the primary surgery, 5 of these 10 patients underwent fat grafting to improve persistent depressions and deformities. All treated patients showed favorable cosmetic and favorable results in terms of improvement of their clinical appearance, oral competence, sensitivity, and lip movements. In the study of Karmali et al. [18], 116 patients with head and neck cancer (or benign locally aggressive tumors), with history of radiotherapy in 69% of cases, underwent 190 fat grafting procedures. However, the esthetic outcomes were evaluated in only 17 patients after a mean follow-up of over 2 years, with significant improvements in all of them according to a 5-point Likert scale as evaluated by 10 plastic surgeons and 10 laypersons. Procedural-related complications were observed in 5.1% of cases (infection, oil cysts, fat necrosis) and all four locoregional recurrences were in areas outside of where the fat was grafted. Griffin et al. [28] reported a retrospective analysis of 38 patients who underwent fat grafting, with a history of head and neck malignancy, multimodal treatment including at least surgery or radiotherapy, and at least 2-year disease-free survival. Esthetic and functional improvements in their radiation-induced skin fibrosis, and volumetric defects at a follow-up of 32 months were shown in 37 (97%) patients. Lipotransfer was also associated with psychological and quality of life improvement. In this study, recurrence was detected in two patients (5.3%) after a mean follow-up of 10 years.

Patients' satisfaction was also evaluated in our study, showing a high degree of satisfaction in terms of esthetic improvement, global satisfaction with treatment, and percentage of patients who would recommend fat grafting to other patients in similar conditions.

However, despite the refinement of technical aspects of lipotransfer and encouraging results for improving esthetic and functional sequelae of surgery and radiotherapy in head and neck cancer patients, the variability of fat absorption rates has been recognized as a limitation of the procedure. Although the restoration of altered contour can be achieved reproducibly intraoperatively and in the early postoperative period, the long-term durability of results remains to be established. Moreover, methods for quantifying the stability of grafted fat have not been standardized. Hörl et al. [29] reported an average volume decline of 55% at 6 months, evaluated by resonance magnetic imaging (RMI) studies in a group of 53 patients with facial defects repaired using autogenous fat tissue. Meier et al. [30] provided three-dimensional volumetric measurements demonstrating an average graft survival of 32% at 16 months after autologous fat grafting for midfacial rejuvenation. However, Coleman [25] indicates that the volume of the graft stabilizes at 3–4 months, and a subtle volumetric decrease may occur up to 1 year after infiltration; beyond that, he states that the volume remains constant for 8–12 years. Quantifiable data of graft survival are rarely reported. In our series, clinical examination and comparison of photographs over the follow-up period showed a progressive volumetric decrease, close to 50% of the injected volume.

Although recurrences in our study, like others [18,28], occurred in areas far from the treatment site, the use of autologous fat grafting in a bed with a history of cancer involvement is a matter of concern. Further clinical studies with longer follow-up periods are needed to confirm these findings. Finally, patients should be informed regarding the possibility of having to repeat fat grafting in order to achieve more stable and visible results.

5. Conclusions

Autologous fat grafting is a valuable technique for improving esthetic and functional sequelae of extensive surgical resections and radiation therapy in patients with malignant head and neck tumors. The technique is a minimally invasive procedure for which a sufficient volume of abdominal fat can be easily obtained. The results of the present study confirm the benefits of fat grafting as a volumetric correction reconstructive strategy, with successful cosmetic and functional outcomes, a high degree of patient satisfaction, low complication rate, and no evidence of being associated with cancer recurrence.

Author Contributions: Conceptualization, J.M.-G., J.P.-R. and C.B.-A.; methodology, J.M.-G., J.G.-S., I.Á.-S., J.P.-R. and C.B.-A.; validation, J.M.-G.; formal analysis, J.M.-G. and J.P.-R.; resources, J.P.-R. and C.B.-A.; data curation, J.M.-G.; writing—original draft preparation, J.M.-G.; writing—review and editing, J.M.-G., J.G.-S., I.Á.-S., M.S.-B., J.P.-R. and C.B.-A.; supervision, J.P.-R. and C.B.-A. All authors have read and agreed to the published version of the manuscript.

Funding: This research received no external funding.

Institutional Review Board Statement: The study was conducted in accordance with the Declaration of Helsinki, and approved by the Institutional Review Board of Hospital Universitari Vall d'Hebron (code PR (ATR) 57/2016, date of approval 26 February 2016) (Barcelona, Spain).

Informed Consent Statement: Informed consent was obtained from all subjects involved in the study.

Data Availability Statement: Clinical data are not shared.

Acknowledgments: The authors would like to thank Rosa Pujol Pina, Research Project Manager, for logistic support and Marta Pulido, for editing the manuscript and editorial assistance.

Conflicts of Interest: The authors declare no conflict of interest.

References

1. Chow, L.Q.M. Head and neck cancer. *N. Engl. J. Med.* **2020**, *382*, 60–72. [CrossRef] [PubMed]
2. Mody, M.D.; Rocco, J.W.; Yom, S.S.; Haddad, R.I.; Saba, N.F. Head and neck cancer. *Lancet* **2021**, *398*, 2289–2299. [CrossRef]
3. Vigneswaran, N.; Williams, M.D. Epidemiologic trends in head and neck cancer and aids in diagnosis. *Oral Maxillofac. Surg. Clin. N. Am.* **2014**, *26*, 123–141. [CrossRef] [PubMed]
4. Lewis, A.; Kang, R.; Levine, A.; Maghami, E. The new face of head and neck cancer: The HPV epidemic. *Oncology* **2015**, *29*, 616–626. [PubMed]
5. Chen, M.M.; Roman, S.A.; Yarbrough, W.G.; Burtness, B.A.; Sosa, J.A.; Judson, B.L. Trends and variations in the use of adjuvant therapy for patients with head and neck cancer. *Cancer* **2014**, *120*, 3353–3360. [CrossRef] [PubMed]
6. Miserocchi, G.; Spadazzi, C.; Calpona, S.; De Rosa, F.; Usai, A.; De Vita, A.; Liverani, C.; Cocchi, C.; Vanni, S.; Calabrese, C.; et al. Precision medicine in head and neck cancers: Genomic and preclinical approaches. *J. Pers. Med.* **2022**, *12*, 854. [CrossRef]
7. Cognetti, D.M.; Weber, R.S.; Lai, S.Y. Head and neck cancer: An evolving treatment paradigm. *Cancer* **2008**, *113* (Suppl. S7), 1911–1932. [CrossRef]
8. Stepnick, D.; Gilpin, D. Head and neck cancer: An overview. *Semin. Plast. Surg.* **2010**, *24*, 107–116. [CrossRef]
9. Wehage, I.C.; Fansa, H. Complex reconstructions in head and neck cancer surgery: Decision making. *Head Neck Oncol.* **2011**, *3*, 14. [CrossRef]
10. Hanasono, M.M. Reconstructive surgery for head and neck cancer patients. *Adv. Med.* **2014**, *2014*, 795483. [CrossRef] [PubMed]
11. Riechelmann, H.; Dejaco, D.; Steinbichler, T.B.; Lettenbichler-Haug, A.; Anegg, M.; Ganswindt, M.; Ganswindt, U.; Gamerith, G.; Riedl, D. Functional outcomes in head and neck cancer patients. *Cancers* **2022**, *14*, 2135. [CrossRef] [PubMed]
12. Zielins, E.R.; Brett, E.A.; Longaker, M.T.; Wan, D.C. Autologous fat grafting: The science behind the surgery. *Aesthet. Surg. J.* **2016**, *36*, 488–496. [CrossRef] [PubMed]
13. Coleman, S.R. Facial recontouring with lipostructure. *Clin. Plast Surg.* **1997**, *24*, 347–367. [CrossRef]

14. Paolini, G.; Amoroso, M.; Longo, B.; Sorotos, M.; Karypidis, D.; Santanelli di Pompeo, F. Simplified lipostructure: A technical note. *Aesthet. Plast. Surg.* **2014**, *38*, 78–82. [CrossRef]
15. Burnouf, M.; Buffet, M.; Schwarzinger, M.; Roman, P.; Bui, P.; Prévot, M.; Deleuze, J.; Morini, J.P.; Franck, N.; Gorin, I.; et al. Evaluation of Coleman lipostructure for treatment of facial lipoatrophy in patients with human immunodeficiency virus and parameters associated with the efficiency of this technique. *Arch. Dermatol.* **2005**, *141*, 1220–1224. [CrossRef]
16. Drochioi, C.I.; Sulea, D.; Timofte, D.; Mocanu, V.; Popescu, E.; Costan, V.V. Autologous fat grafting for craniofacial reconstruction in oncologic patients. *Medicina* **2019**, *55*, 655. [CrossRef] [PubMed]
17. Gutiérrez Santamaría, J.; Masiá Gridilla, J.; Pamias Romero, J.; Giralt López-de-Sagredo, J.; Bescós Atín, M.S. Fat grafting is a feasible technique for the sequelae of head and neck cancer treatment. *J. Craniomaxillofac. Surg.* **2017**, *45*, 93–98. [CrossRef] [PubMed]
18. Karmali, R.J.; Hanson, S.E.; Nguyen, A.T.; Skoracki, R.J.; Hanasono, M.M. Outcomes following autologous fat grafting for oncologic head and neck reconstruction. *Plast. Reconstr. Surg.* **2018**, *142*, 771–780. [CrossRef]
19. Ducic, Y.; Pontius, A.T.; Smith, J.E. Lipotransfer as an adjunct in head and neck reconstruction. *Laryngoscope* **2003**, *113*, 1600–1604. [CrossRef]
20. Krastev, T.K.; Beugels, J.; Hommes, J.; Piatkowski, A.; Mathijssen, I.; van der Hulst, R. Efficacy and safety of autologous fat transfer in facial reconstructive surgery: A systematic review and meta-analysis. *JAMA Facial Plast. Surg.* **2018**, *20*, 351–360. [CrossRef]
21. Phulpin, B.; Gangloff, P.; Tran, N.; Bravetti, P.; Merlin, J.L.; Dolivet, G. Rehabilitation of irradiated head and neck tissues by autologous fat transplantation. *Plast. Reconstr. Surg.* **2009**, *123*, 1187–1197. [CrossRef] [PubMed]
22. Coleman, S.R. Structural fat grafting. *Aesthet. Surg. J.* **1998**, *18*, 386–388. [CrossRef] [PubMed]
23. Clavien, P.A.; Barkun, J.; de Oliveira, M.L.; Vauthey, J.N.; Dindo, D.; Schulick, R.D.; de Santibañes, E.; Pekolj, J.; Slankamenac, K.; Bassi, C.; et al. The Clavien-Dindo classification of surgical complications: Five-year experience. *Ann. Surg.* **2009**, *250*, 187–196. [CrossRef] [PubMed]
24. Mojallal, A.; Foyatier, J.L. Historical review of the use of adipose tissue transfer in plastic and reconstructive surgery. *Ann. Chir Plast Esthet.* **2004**, *49*, 419–425. [CrossRef] [PubMed]
25. Coleman, S.R. Structural fat grafting: More than a permanent filler. *Plast Reconstr Surg.* **2006**, *118* (Suppl. 3), 108S–120S. [CrossRef] [PubMed]
26. Vitagliano, T.; Curto, L.S.; Greto Ciriaco, A.; Gareri, P.; Ribuffo, D.; Greco, M. Two-thirds lip defects: A new combined reconstructive technique for patients with epithelial cancer. *J. Craniofac. Surg.* **2016**, *27*, 1995–2000. [CrossRef] [PubMed]
27. Zhu, M.; Cohen, S.R.; Hicok, K.C.; Shanahan, R.K.; Strem, B.M.; Yu, J.C.; Arm, D.M.; Fraser, J.K. Comparison of three different fat graft preparation methods: Gravity separation, centrifugation, and simultaneous washing with filtration in a closed system. *Plast. Reconstr. Surg.* **2013**, *131*, 873–880. [CrossRef]
28. Griffin, M.F.; Drago, J.; Almadori, A.; Kalavrezos, N.; Butler, P.E. Evaluation of the efficacy of lipotransfer to manage radiation-induced fibrosis and volume defects in head and neck oncology. *Head Neck* **2019**, *41*, 3647–3655. [CrossRef]
29. Hörl, H.W.; Feller, A.M.; Biemer, E. Technique for liposuction fat reimplantation and long-term volume evaluation by magnetic resonance imaging. *Ann. Plast. Surg.* **1991**, *26*, 248–258. [CrossRef]
30. Meier, J.D.; Glasgold, R.A.; Glasgold, M.J. Autologous fat grafting: Long-term evidence of its efficacy in midfacial rejuvenation. *Arch. Facial Plast. Surg.* **2009**, *11*, 24–28. [CrossRef]

Article

Unraveling Desmoid-Type Fibromatosis-Specific Health-Related Quality of Life: Who Is at Risk for Poor Outcomes

Anne-Rose W. Schut [1,2], Emma Lidington [3], Milea J. M. Timbergen [1,2], Eugenie Younger [3], Winette T. A. van der Graaf [1,4], Winan J. van Houdt [5], Johannes J. Bonenkamp [6], Robin L. Jones [3,7], Dirk. J. Grünhagen [2], Stefan Sleijfer [1], Cornelis Verhoef [2], Spyridon Gennatas [3,8] and Olga Husson [2,4,7,*]

1 Department of Medical Oncology, Erasmus MC Cancer Institute, 3015 GD Rotterdam, The Netherlands; a.schut@erasmusmc.nl (A.-R.W.S.); m.timbergen@erasmusmc.nl (M.J.M.T.); w.vd.graaf@nki.nl (W.T.A.v.d.G.); s.sleijfer@erasmusmc.nl (S.S.)
2 Department of Surgical Oncology, Erasmus MC Cancer Institute, 3015 GD Rotterdam, The Netherlands; d.grunhagen@erasmusmc.nl (D.J.G.); c.verhoef@erasmusmc.nl (C.V.)
3 Sarcoma Unit, Royal Marsden NHS Foundation Trust, London SW3 6JJ, UK; emma.lidington@kcl.ac.uk (E.L.); eugenie.younger@doctors.org.uk (E.Y.); robin.jones4@nhs.net (R.L.J.); spyridon.gennatas@nhs.net (S.G.)
4 Department of Medical Oncology, Netherlands Cancer Institute, 1066 CX Amsterdam, The Netherlands
5 Department of Surgical Oncology, Netherlands Cancer Institute, 1066 CX Amsterdam, The Netherlands; w.v.houdt@nki.nl
6 Department of Surgical Oncology, Radboud University Medical Center, 6525 GA Nijmegen, The Netherlands; han.bonenkamp@radboudumc.nl
7 Division of Clinical Studies, Institute of Cancer Research, Royal Marsden NHS Foundation Trust, London SM2 5NG, UK
8 Department of Medical Oncology, Guy's and St Thomas' NHS Foundation Trust, London SE1 9RT, UK
* Correspondence: olga.husson@icr.ac.uk; Tel.: +44-2034-376-342

Citation: Schut, A.-R.W.; Lidington, E.; Timbergen, M.J.M.; Younger, E.; van der Graaf, W.T.A.; van Houdt, W.J.; Bonenkamp, J.J.; Jones, R.L.; Grünhagen, D.J.; Sleijfer, S.; et al. Unraveling Desmoid-Type Fibromatosis-Specific Health-Related Quality of Life: Who Is at Risk for Poor Outcomes. *Cancers* **2022**, *14*, 2979. https://doi.org/10.3390/cancers14122979

Academic Editor: António Araújo

Received: 13 May 2022
Accepted: 12 June 2022
Published: 16 June 2022

Simple Summary: Desmoid-type fibromatosis (DTF) is an uncommon, non-metastasising soft-tissue tumour. Patients can experience a wide variety of disease-specific issues related to the unpredictable clinical course and aggressiveness of DTF, negatively impacting their health-related quality of life (HRQoL). Little is known about which DTF patients are particularly affected by an impaired HRQoL. In the current study, HRQoL was evaluated among different groups of DTF patients, using the EORTC QLQ-C30 and the DTF-QoL, a DTF-specific HRQoL questionnaire. Age, sex, presence of comorbidities, and type of treatment were found to be most strongly associated with DTF-specific HRQoL outcomes. In general, socio-demographic factors had the greatest impact on generic HRQoL, whereas the influence of clinical factors was mainly seen on the DTF-QoL, underlining the importance of a disease-specific questionnaire. Knowledge of the differences in DTF-specific HRQoL between subgroups can be used to individualize the HRQoL-measurement strategy for research and clinical practice.

Abstract: Desmoid-type fibromatosis (DTF) is a rare, soft-tissue tumour. These tumours do not metastasize, but their local aggressive tumour growth and unpredictable behaviour can have a significant impact on health-related quality of life (HRQoL). Little is known about which DTF patients are particularly affected by an impaired HRQoL. The objectives of this study were to assess HRQoL among different groups of DTF patients and to investigate which socio-demographic and clinical characteristics were associated with DTF-specific HRQoL. A cross-sectional study was conducted among DTF patients from the United Kingdom and the Netherlands. HRQoL was assessed using the European Organization for Research and Treatment of Cancer Quality of Life Questionnaire Core 30 (EORTC QLQ-C30), accompanied by the DTF-QoL to assess DTF-specific HRQoL. The scores were compared amongst subgroups, based on the socio-demographic and clinical characteristics of DTF patients. Multiple linear regression analyses with a backward elimination were conducted to identify the factors associated with DTF-specific HRQoL. A total of 235 DTF patients completed the questionnaires. Female patients, patients with more than two comorbidities, or patients who received treatment other than only active surveillance (AS) or surgery scored significantly worse on the subscales of both the EORTC QLQ-C30 and DTF-QoL. Patients that were $\geq$ 40 years scored significantly worse on the physical functioning scale of the EORTC QLQ-C30, while younger patients

(18–39 years) scored significantly worse on several DTF-QoL subscales. Differences in the DTF-QoL subscales were found for tumour location, time since diagnosis and the presence of recurrent disease. Furthermore, treatments other than AS or surgery only, female sex, younger age and the presence of comorbidities were most frequently associated with worse scores on the DTF-QoL subscales. This study showed that (DTF-specific) HRQoL differs between groups of DTF patients. Awareness of these HRQoL differences could help to provide better, personalised care that is tailored to the needs of a specific subgroup.

Keywords: desmoid-type fibromatosis; rare diseases; health-related quality of life; patient-reported outcomes; disease-specific measures

1. Introduction

Desmoid-type fibromatosis (DTF) is a rare, intermediate-grade, soft-tissue tumour [1]. The estimated incidence in the population is 5–6 patients per million people per year. It usually affects young adult patients and tumours can be located in nearly any part of the body, most commonly, in the extremities and abdominal wall [2–4]. DTF does not metastasize, but it can display locally aggressive tumour growth, causing significant morbidity [1]. The biological behaviour of DTF is unpredictable and variable, and includes phases of progressive growth or growth stabilisation and spontaneous regression in 28% of tumours [5–7]. Regardless of the tumour's behaviour or size, patients may experience a variety of symptoms, from no symptoms at all to extreme pain or functional limitations.

The most recent global consensus guideline recommends active surveillance (AS) as a frontline approach for asymptomatic and mildly symptomatic patients, independent of the tumour's location or size [8]. After initial AS, the majority of DTF patients do not need active treatment, minimising overtreatment and potential treatment-related morbidity [7,9]. In the case of radiological or clinically significant progression or increasing symptoms, active treatment, including systemic therapies, surgical resection and local therapies, such as radiotherapy, can be considered [8]. With high local recurrence rates for DTF at anatomic sites other than the abdominal wall and treatment-related toxicities, these interventions do not guarantee tumour reduction or clinical benefit [3,8,10,11]. For a substantial proportion of patients, DTF is a chronic condition and the primary goal in treating DTF patients is to maintain an acceptable health-related quality of life (HRQoL) [12,13].

HRQoL is a multidimensional concept that includes the patient's perception of the impact of their disease and treatment on their physical, psychological, and social functioning [14]. There are a limited number of studies focusing on HRQoL in DTF patients. These studies have shown that the diagnosis of DTF, its treatment, or both can have a significant impact on different domains of their HRQoL. From qualitative interview studies, it is known that DTF patients experience a variety of disease-specific issues associated with the rarity of DTF, the unpredictable clinical course and the variable treatment efficacies. Additionally, DTF patients report pain and physical symptoms caused by the tumour itself, or as a side effect of treatment [13,15,16]. These DTF-specific HRQoL issues are not captured by generic or cancer-generic HRQoL questionnaires, such as the European Organization for Research and Treatment of Cancer Quality of Life Questionnaire Core 30 (EORTC QLQ-C30), which are predominantly used in DTF studies and in clinical care [17]. Therefore, a DTF-specific HRQoL questionnaire, the DTF-QoL, was recently developed by our group, which can be used to evaluate the prevalence of HRQoL issues in DTF patients [18,19]. Furthermore, the small number of previous studies focused on the population of DTF patients as a whole because of small sample sizes. Consequently, little is known about the differences between subgroups of DTF patients, for example, about the differences in HRQoL between patients receiving different types of treatment or with tumours in different anatomic locations. The objectives of this study are to evaluate the HRQoL in different groups of DTF patients using the DTF-QoL and the EORTC QLQ-C30, and to investigate which socio-demographic and clinical characteristics are associated with DTF-specific HRQoL. The results of this study will provide important insights into the problems and needs of specific groups of DTF patients, which will help to identify patients at risk of a poor HRQoL and to better provide personalised care.

2. Materials and Methods

2.1. Study Sample and Data Collection

The sample included DTF patients from the United Kingdom (UK) and the Netherlands (NL), who participated in the QUALIFIED study (The evaluation of health-related quality of life issues experienced by patients with desmoid-type fibromatosis, registered at clinicaltrials.gov (accessed on 12 May 2022): NCT04289077) [18]. The QUALIFIED study is an international, multicentre, cross-sectional, observational study among adult (≥18 years) patients with sporadic DTF who were treated in one of the participating centres (one centre in the UK, three centres in the NL). After obtaining their informed consent, the patients completed a set of questionnaires, including the EORTC QLQ-C30 and DTF-QoL. Questionnaire data were collected via the PROFILES management system—an established international registry for the collection of cancer patient-reported outcomes [20]. Ethical and institutional approval was obtained in each participating centre in the UK and the NL. Further details of the protocol are described elsewhere [18].

2.2. Study Measures

2.2.1. Socio-Demographic and Clinical Characteristics

Socio-demographic and clinical data were extracted from the questionnaire (patient-reported) and from the patient medical records. The questionnaire included single items on age, sex, race, marital status, family composition, educational level, employment status, tumour location, details regarding the diagnosis, received treatments and tumour recurrence. Comorbidities were assessed using an adapted self-administered comorbidity questionnaire (SCQ) [21], which included one question about the presence of comorbidities in the previous twelve months. Additional medical data were obtained from the electronic patient records to ensure correct and detailed reporting [18]. To compare the HRQoL between the different types of treatment, DTF patients were assigned to one of the following three treatment groups: "only AS", "only surgery" and "other treatment". Receiving treatment with non-steroidal anti-inflammatory drugs (NSAIDs) or other analgesics was not considered an active treatment [8]. The other treatment group included patients who received systemic therapy (i.e., chemotherapy, hormonal therapy, targeted medical therapy), local therapy (i.e., radiotherapy, isolated limb perfusion, high-intensity-focused ultrasound, cryoablation) or a combination of any form of active treatments. In addition, patients who received "only systemic therapy", "only local therapy" or "combination of active treatments" were assessed as separate groups.

2.2.2. Questionnaires

The EORTC QLQ-C30 was used to measure HRQoL [17]. This 30-item HRQoL questionnaire consists of five functional scales, a global quality of life scale, three symptom scales and a number of single items that assess common symptoms and the perceived financial impact of the disease. The timeframe of the questions is during the past week. Each item is scored on a Likert scale ranging from 1, "not at all" to 4, "very much", with the exception of the global QoL scale, which is scored on a seven-point response scale ranging from 1, "very poor" to 7 "excellent". Scores of all scales and single items are linearly transformed to a score between 0 and 100, according to the guidelines of the EORTC quality of life group [22]. A higher score on the functional scales and global quality of life means better functioning and HRQoL, whereas a higher score on the symptom scales means a higher symptom burden.

The DTF-specific HRQoL was measured by the DTF-QoL [19]. The DTF-QoL was developed according to the guidelines of the EORTC Quality of Life Group to supplement the EORTC QLQ-C30 and to assess the disease-specific issues that DTF patients experience [19,23]. The questionnaire consists of 96 items, which are divided into 3 symptom scales, 11 disease impact scales, and 6 single items. The timeframe of the symptom scales is the past week; the disease impact scales and single items have a timeframe of since diagnosis, except for the question on sexual interest, which has a timeframe of four weeks. Items are scored on a Likert scale, with a range of 1, "not at all" to 4 "very much", with an additional "not applicable" option for certain questions. Scores of the DTF-QoL scales are calculated according to the EORTC QLQ-C30 scoring manual for symptom scales/items [22]. First, a raw score is obtained by estimating the average of the items that contribute to a scale. After a linear transformation of the raw scores of all scales and single items, scores range from 0 to 100. A higher score indicates a higher level of symptoms or problems.

2.3. Statistical Analyses

Patient characteristics were summarised using descriptive statistics. Continuous variables were presented as a mean and standard deviation (SD) or median and interquartile range (IQR) where skewed. The categorical variables were described as numbers and percentages. The differences in mean scores of the DTF-QoL and EORTC QLQ-C30 subscales between the subgroups of DTF patients were analysed using the Mann–Whitney *U* test in the case of two groups. In the case of more than two groups, an analysis of variance (ANOVA) with post hoc Bonferroni analysis was used. The clinically relevant differences in DTF-QoL scores between the treatment groups were determined with Norman's "rule of thumb", using the value of 0.5 SD as the default value for a clinically relevant difference [24]. A series of multiple linear regression analyses were conducted to investigate the association between clinical (comorbidity, time since diagnosis, treatment received, recurrence and tumour location) and socio-demographic characteristics (age, sex, relationship status, education level and current employment status) and the DTF-QoL scores. The categorical variables education level, comorbidity, treatment received and tumour anatomic location, had >2 categories and were transformed into dummy variables, with, respectively, low, none, only AS and abdominal wall as the reference groups. A manual backward elimination method was applied to determine the inclusion of variables in the final model, whereby, only those variables with a $p < 0.05$ were retained [25]. If any of the dummy variables had a *p*-value of <0.05, the entire categorical variable was retained. If one of the dummy variables had the largest *p*-value and none of the dummy variables had a *p*-value of <0.05, the entire categorical variable was eliminated. Given the large number of subscales, we decided not to give an extensive description in the text of the differences in scale scores and between which groups these differences were observed, but to refer to the tables as much as possible instead. All analyses were performed using SPSS software, version 25.0 (SPSS Inc., Chicago, IL, USA) and the figures were generated with GraphPad Prism, version 5.0 (GraphPad Software, La Jolla, CA, USA). For all analyses, *p*-values of <0.05 were considered statistically significant.

3. Results

3.1. Patient Characteristics

Two hundred and thirty-five DTF patients completed the DTF-QoL and EORTC QLQ-C30 questionnaires (response rate 46%). No statistically significant differences in sex, age at the time of diagnosis, and age at the time of the questionnaire were observed between the responders and non-responders. The socio-demographic and clinical characteristics of the study sample are described in Table 1. Most patients were female (n = 173, 73.6%) with a median age of 39.3 years (IQR 31.4–50.6) at the time of diagnosis. The median time since diagnosis for all patients was 4.7 years (IQR 2.3–7.8). The most common tumour locations were the abdominal wall (n = 58, 24.7%) and trunk (n = 54, 23.0%). Eighty-seven patients (37.0%) were treated with AS only and 64 patients (27.2%) with surgery only. The other active treatment types are specified in Table S1. Sixteen patients (6.8%) were undergoing active treatment at the time they completed the questionnaire. Back pain (n = 46, 19.6%), depression/anxiety (n = 41, 17.4%), joint condition (n = 26, 11.1%) and high blood pressure (n = 26, 11.1%) were the most common self-reported comorbidities.

Table 1. Desmoid-type fibromatosis patient characteristics (N = 235).

		n (%)
Nationality	United Kingdom	79 (33.6)
	The Netherlands	156 (66.4)
Sex	Male	62 (26.4)
	Female	173 (73.6)
Age in years at time of diagnosis (in years)—Mean (SD)		41.7 (14.4)
Age in years at time of questionnaire (in years)—Mean (SD)		47.2 (14.0)
Time since diagnosis (in years)—Mean (SD)		5.7 (4.5)
Tumour localization	Head/neck	13 (5.5)
	Upper extremity/shoulder	29 (12.3)
	Trunk [1]	54 (23.0)
	Abdominal wall	58 (24.7)
	Intra-abdominal	39 (16.6)
	Hip/pelvis/gluteal region	20 (8.5)
	Lower extremity	22 (9.4)
Recurrent disease after surgery (n = 98, 41.7%)	Yes	41 (41.8)
	No	57 (58.2)
Treatment received [2]	Only active surveillance	87 (37.0)
	Only surgery	64 (27.2)
	Only systemic therapy	32 (13.6)
	Only local therapy	8 (3.4)
	Combination of active treatments	44 (18.7)
Comorbidity (self-report)	None	90 (38.3)
	1	74 (31.5)
	≥2	71 (30.2)
Relationship status	Partnered	181 (77.0)
	Not partnered	53 (22.6)
	Missing	1 (0.4)
Education level	Low (primary/secondary)	36 (15.3)
	Medium (vocation/college/diploma)	126 (53.6)
	High (university/post-graduate)	73 (31.1)
Current employment status	Working	155 (66.0)
	Not working	80 (33.9)

[1]. Including thoracic wall, breast and back. [2]. Active surveillance, surgery, systemic therapy or local therapy only: including patients who received analgesics; Systemic therapy includes: chemotherapy, hormonal therapy and targeted medical therapy (tyrosine kinase and gamma-secretase inhibitors); Local therapy includes: radiotherapy, isolated limb perfusion, high-intensity-focused ultrasound, cryoablation; Combination of active treatments: including patients who received different combinations of surgery, systemic therapy or local therapy.

3.2. Comparison of DTF-Specific HRQoL between Different Groups of DTF Patients

The mean HRQoL scores for the total sample and all subgroups of DTF patients on the DTF-QoL subscales and single items are presented in Tables 2 and S2. Several differences were found for socio-demographic factors. Younger patients (18–39 years) experienced significantly more problems in six subscales, with the largest difference in the subscale parenting and fertility, previously described as the "parents and fertility" subscale. Female patients had significantly higher scores, indicating more problems, on four subscales. Unemployed patients experienced more problems in three subscales, with the highest score on the impact scale related to job and education.

Significant differences in the subscales of the DTF-QoL were also seen for clinical factors (Table 2). Having multiple comorbidities resulted in significantly worse scores on eight subscales. A longer time since diagnosis (≥5 years) resulted in significantly higher scores on eight subscales. Patients with recurrent disease experienced more problems in six subscales. Compared to tumours in some other anatomic locations, patients with tumours in the upper and lower extremities, hip/pelvis/gluteal region, and head and neck, scored significantly worse on several subscales. The lower extremity and hip/pelvis/gluteal group experienced significantly more symptoms that were related to physical consequences. Patients with tumours in the upper extremities or hip/pelvis/gluteal region scored higher on pain and discomfort. Tumours in the head and neck region resulted in more problems with employment and education.

With the exception of the subscales doctor-patient relationship and supportive care, and the single item wasting the time of cancer specialists, significant differences between the three treatment groups were seen for all DTF-QoL subscales and single items, with the other treatment group scoring higher than the group of patients who received AS or surgery only (Tables 2 and S2). Figure 1 presents the mean DTF-QoL scores per treatment type and the clinically relevant differences between the treatment groups, considering systemic therapy and local therapy as separate groups.

3.3. *Comparison of HRQoL between Different Groups of DTF Patients*

The mean HRQoL scores for the EORTC QLQ-C30 are presented in Table 3 for the total sample and all the subgroups of DTF patients. Patients that were ≥40 years scored significantly lower on physical functioning and had significantly more problems with dyspnoea and sleep. Female patients had significantly worse scores on six subscales. Unemployed patients scored significantly lower on all functioning scales and on global health and had higher scores on the single items fatigue, dyspnoea, sleep and financial difficulties. Having multiple comorbidities resulted in lower scores on all subscales. No differences were seen in the time since diagnosis. There were significant differences between the three treatment groups in physical, role, emotional and social functioning, in global health and in fatigue, pain, sleep, diarrhoea and financial difficulties symptom items and scales. For most of these scales and symptoms, patients who received other treatments experienced more problems or symptoms than those patients receiving AS or surgery only. The presence of recurrent disease resulted in significantly worse scores in two subscales. Patients with tumours located in the hip/pelvis/gluteal/ region and the lower and upper extremities scored significantly higher on the pain items.

3.4. *Factors Associated with DTF-Specific HRQoL*

Multiple linear regression analyses with backward elimination were conducted to identify the socio-demographic and clinical characteristics associated with DTF-specific HRQoL (Table 4). An older age (≥40 years) was negatively associated with physical symptoms, while a younger age (18–39 years) was negatively associated with the impact of DTF on concerns about condition, relationships, parenting and fertility, body image concerns about treatment and its consequences, and the unpredictable disease course. Female sex was associated with more physical symptoms and problems related to job and education, physical limitations, parenting and fertility, and body image. Having one or more comorbidities was negatively associated with all the subscales, except for job and education, diagnostic and treatment trajectory, and parenting and fertility. Time since diagnosis was associated with only two scales, with fewer years since diagnosis being negatively associated with pain and discomfort, and a longer diagnosis with problems related to supportive care. Treatment other than AS or surgery only was associated with more problems on all DTF-QoL subscales, except for doctor-patient relationship and supportive care.

Table 2. Mean DTF-QoL scores (±SD) in relation to socio-demographic and clinical characteristics.

	DTF-QoL													
	Symptom Scales [+]			Impact Scales [+]										
	W1 Emotional	**W2 Physical**	**W3 Pain**	**1 Concerns Condition**	**2 Job & Education**	**3 Doctor-Patient**	**4 Relationships**	**5 Physical Consequences**	**6 Diagnostic**	**7 Parenting**	**8 Body Image**	**9 Support**	**10 Treatment Concerns**	**11 Behaviour DTF**
Study population	15.3 (18.7)	11.6 (16.5)	19.6 (25.6)	41.3 (24.1)	29.2 (31.6)	26.7 (19.7)	24.8 (21.8)	18.7 (18.9)	28.1 (19.1)	21.1 (23.1)	29.0 (25.9)	36.0 (14.1)	22.1 (21.4)	38.8 (22.4)
Age (years)														
18–39	16.1 (19.8)	10.9 (16.4)	22.0 (28.5)	44.6 (24.4)	31.8 (31.9)	27.8 (20.7)	28.3 (22.4)	20.8 (19.6)	29.6 (20.0)	28.2 (25.4)	33.2 (26.8)	36.3 (14.6)	25.9 (23.2)	42.5 (23.9)
≥40	14.6 (17.5)	12.2 (16.7)	17.1 (22.1)	38.0 (23.5)	25.6 (30.9)	25.7 (18.6)	21.3 (20.7)	16.7 (18.1)	26.5 (18.2)	8.6 (10.1)	24.8 (24.5)	35.7 (13.7)	18.2 (18.8)	35.0 (20.3)
p-value	0.905	0.159	0.344	**0.024**	0.088	0.518	**0.004**	0.124	0.229	**<0.001**	**0.007**	0.943	**0.019**	**0.032**
Sex														
Male	12.4 (15.2)	7.5 (11.5)	15.1 (22.6)	39.3 (24.3)	25.4 (28.9)	22.8 (16.3)	20.5 (21.5)	14.0 (15.1)	25.3 (18.1)	8.7 (13.3)	21.5 (19.7)	34.5 (13.2)	22.4 (19.1)	37.9 (21.9)
Female	16.4 (19.7)	13.1 (17.8)	21.2 (26.4)	42.0 (24.1)	30.7 (32.5)	28.1 (20.6)	26.4 (21.8)	20.5 (19.9)	29.0 (19.4)	25.6 (24.3)	31.7 (27.4)	36.5 (14.4)	21.9 (22.2)	39.2 (22.7)
p-value	0.264	0.055	0.097	0.443	0.476	0.099	**0.013**	**0.044**	0.158	**<0.001**	**0.017**	0.271	0.502	0.719
Relationship status														
Partnered	14.0 (18.3)	10.6 (15.4)	19.2 (24.9)	41.0 (24.5)	27.2 (30.6)	26.8 (19.6)	23.2 (21.1)	17.0 (18.0)	28.0 (19.3)	22.1 (23.4)	26.7 (24.9)	35.9 (14.4)	21.5 (21.5)	38.7 (22.6)
Not partnered	20.0 (19.2)	15.2 (19.6)	21.0 (28.0)	42.0 (23.2)	36.5 (34.8)	26.8 (20.1)	30.2 (23.9)	24.5 (21.3)	27.7 (18.4)	15.2 (20.6)	37.6 (27.8)	36.1 (13.2)	23.7 (21.3)	39.4 (22.4)
p-value	**0.017**	0.122	0.779	0.755	0.095	0.914	0.050	**0.020**	0.999	0.185	**0.005**	0.551	0.430	0.837
Education level														
Low	16.1 (17.1)	14.3 (21.0)	18.5 (23.8)	34.8 (22.5)	21.1 (28.4)	25.7 (20.8)	21.2 (19.2)	16.1 (17.6)	23.5 (16.8)	15.3 (22.7)	28.8 (25.2)	33.3 (13.5)	16.9 (16.4)	36.4 (21.9)
Medium	16.0 (19.6)	11.4 (15.7)	19.5 (25.5)	43.3 (23.7)	35.2 (33.3)	27.1 (18.8)	25.2 (21.6)	20.8 (19.3)	29.5 (19.7)	19.5 (22.1)	27.8 (25.2)	35.1 (12.4)	21.7 (20.2)	39.3 (22.3)
High	13.8 (17.9)	10.7 (15.5)	20.1 (26.8)	41.0 (25.3)	21.7 (27.7)	26.6 (20.8)	25.8 (23.5)	16.5 (18.8)	27.9 (19.1)	25.0 (24.6)	31.3 (27.7)	38.8 (16.7)	25.2 (25.2)	39.3 (23.1)
p-value [#]	0.712	0.554	0.956	0.181	**0.011** [a]	0.928	0.548	0.212	0.259	0.369	0.657	0.097	0.192	0.781

Table 2. *Cont.*

	DTF-QoL													
	Symptom Scales +			Impact Scales +										
	W1 Emotional	W2 Physical	W3 Pain	1 Concerns Condition	2 Job & Education	3 Doctor-Patient	4 Relationships	5 Physical Consequences	6 Diagnostic	7 Parenting	8 Body Image	9 Support	10 Treatment Concerns	11 Behaviour DTF
Employment status														
Working	13.6 (17.4)	9.5 (15.1)	18.4 (24.8)	41.5 (24.1)	22.9 (25.9)	26.7 (20.2)	23.4 (19.9)	16.3 (16.9)	28.5 (19.1)	20.2 (23.7)	28.0 (24.9)	36.0 (14.9)	19.8 (19.9)	37.6 (21.7)
Not working	18.6 (20.5)	15.6 (18.4)	21.9 (27.0)	40.7 (24.2)	50.5 (39.4)	26.8 (18.8)	27.6 (25.0)	23.5 (21.7)	27.1 (19.3)	24.3 (20.8)	31.0 (27.9)	36.0 (14.9)	26.5 (23.6)	41.3 (23.9)
p-value	0.053	**0.002**	0.307	0.752	**<0.001**	0.876	0.488	**0.017**	0.597	0.210	0.529	0.331	0.063	0.298
Comorbidity														
None	11.3 (15.8)	8.9 (14.7)	16.2 (23.7)	37.3 (22.0)	24.9 (29.4)	23.3 (21.4)	18.6 (18.7)	14.7 (17.3)	27.4 (21.4)	17.7 (21.1)	24.7 (23.2)	33.5 (13.1)	19.8 (19.7)	34.0 (20.7)
1	16.9 (19.6)	9.9 (14.6)	17.8 (26.4)	41.5 (25.0)	26.1 (30.4)	26.2 (16.8)	26.5 (21.6)	17.5 (17.3)	26.1 (17.7)	21.5 (25.1)	29.4 (27.6)	36.5 (12.1)	19.7 (21.7)	37.5 (22.3)
≥2	18.7 (20.2)	16.8 (19.3)	25.6 (26.3)	46.1 (25.2)	38.8 (34.3)	31.6 (19.4)	31.0 (23.9)	25.1 (20.9)	30.9 (17.4)	26.4 (24.7)	34.1 (26.9)	38.7 (16.7)	27.6 (22.6)	46.3 (23.0)
p-value #	**0.028** [b]	**0.005** [b,c]	0.050	0.074	**0.026** [b]	**0.027** [b]	**0.001** [b]	**0.002** [b,c]	0.295	0.271	0.074	0.067	**0.048** *	**0.002** [b]
Time since diagnosis														
<5 years	14.2 (17.2)	9.5 (14.3)	20.5 (26.3)	38.8 (24.3)	25.8 (30.5)	24.1 (19.0)	22.2 (21.2)	16.7 (17.1)	27.9 (18.2)	14.4 (18.2)	24.4 (23.6)	33.6 (13.1)	18.7 (18.5)	37.1 (23.2)
≥5 years	16.6 (20.2)	14.2 (18.6)	18.3 (24.7)	44.3 (23.6)	33.1 (32.4)	30.0 (20.1)	28.1 (22.3)	21.2 (20.8)	28.3 (20.2)	26.1 (25.1)	34.7 (27.6)	38.9 (14.8)	25.8 (23.7)	40.9 (21.4)
p-value	0.670	**0.041**	0.488	0.057	**0.040**	**0.016**	**0.021**	0.163	0.892	**0.013**	**0.003**	**0.002**	**0.043**	0.102
Treatment received [1]														
Only active surveillance	9.1 (14.0)	6.5 (10.9)	13.9 (21.4)	31.1 (18.1)	10.3 (16.8)	25.7 (16.7)	15.7 (15.7)	8.8 (11.6)	23.5 (15.4)	17.4 (23.0)	15.6 (15.6)	34.0 (11.3)	10.3 (14.5)	29.2 (18.6)
Only surgery	12.9 (16.7)	11.1 (18.5)	15.0 (22.8)	37.2 (25.3)	28.5 (28.4)	29.0 (26.3)	22.9 (19.8)	18.6 (18.1)	27.0 (20.7)	12.0 (19.4)	28.3 (24.5)	39.1 (14.5)	22.2 (19.9)	34.7 (21.3)
Other treatment	23.5 (21.4)	17.2 (18.0)	28.8 (28.9)	55.0 (22.5)	49.2 (33.4)	26.1 (16.4)	35.8 (24.1)	29.3 (20.2)	33.6 (20.2)	32.6 (21.8)	43.4 (28.1)	35.7 (16.1)	32.8 (22.3)	51.9 (20.8)
p-value #	**<0.001** [d,e]	**<0.001** [d]	**<0.001** [d,e]	**<0.001** [d,e]	**<0.001** [d,e,f]	0.563	**<0.001** [d,e]	**<0.001** [d,e,f]	**0.002** [d]	**<0.001** [d,e]	**<0.001** [d,e,f]	0.081	**<0.001** [d,e,f]	**<0.001** [d,e]

Table 2. *Cont.*

	DTF-QoL													
	Symptom Scales +			Impact Scales +										
	W1 Emotional	W2 Physical	W3 Pain	1 Concerns Condition	2 Job & Education	3 Doctor-Patient	4 Relationships	5 Physical Consequences	6 Diagnostic	7 Parenting	8 Body Image	9 Support	10 Treatment Concerns	11 Behaviour DTF
Recurrent disease														
Yes	18.7 (20.2)	16.7 (20.6)	20.9 (25.5)	52.7 (22.9)	44.0 (32.7)	27.6 (18.2)	29.3 (23.2)	26.0 (20.9)	29.7 (21.4)	28.7 (26.4)	39.1 (28.1)	36.3 (16.1)	34.1 (21.4)	49.3 (21.7)
No	14.6 (18.3)	10.5 (15.3)	19.3 (25.6)	38.8 (23.7)	25.9 (30.5)	26.6 (20.0)	23.9 (21.4)	17.2 (18.2)	27.7 (18.7)	19.5 (22.2)	26.9 (25.0)	35.9 (13.7)	19.3 (20.5)	36.6 (22.0)
p-value	0.114	0.068	0.645	**0.001**	0.001	0.623	0.098	**0.005**	0.671	0.091	**0.006**	0.671	**<0.001**	**0.001**
Recurrent disease after surgery (n = 98)														
Yes	18.7 (20.2)	16.7 (20.6)	20.9 (25.5)	52.7 (22.9)	44.0 (32.7)	27.6 (18.2)	29.3 (23.2)	26.0 (20.9)	29.7 (21.4)	28.7 (26.4)	39.1 (28.1)	36.3 (16.1)	34.1 (21.4)	49.3 (21.7)
No	11.9 (15.8)	9.6 (14.8)	13.4 (21.3)	35.6 (24.8)	32.1 (29.5)	29.8 (26.4)	23.2 (20.2)	19.5 (17.7)	28.7 (20.7)	12.6 (15.7)	32.0 (27.6)	39.7 (15.2)	18.4 (18.1)	32.9 (19.0)
p-value	**0.040**	0.063	0.162	**0.001**	0.078	0.971	0.150	0.132	0.820	**0.018**	0.190	0.362	**<0.001**	**<0.001**
Tumour location														
Abdominal wall	14.4 (17.2)	8.8 (17.4)	15.2 (24.1)	35.0 (23.8)	19.9 (28.0)	27.2 (20.2)	24.2 (23.5)	16.2 (20.9)	27.3 (19.3)	18.2 (20.9)	25.0 (26.1)	35.2 (12.0)	17.1 (20.8)	34.0 (23.6)
Intra-abdominal	11.8 (19.0)	8.0 (16.0)	8.0 (16.5)	41.7 (26.1)	28.0 (26.5)	24.2 (18.4)	20.7 (19.6)	16.6 (16.0)	23.7 (16.9)	13.5 (18.7)	16.5 (18.8)	37.9 (11.9)	15.4 (16.4)	34.5 (22.9)
Upper extremity	19.1 (17.6)	9.9 (10.9)	31.4 (25.0)	42.7 (23.1)	29.5 (32.6)	25.1 (13.3)	26.0 (23.1)	16.8 (18.3)	28.7 (20.0)	16.1 (14.4)	34.4 (24.5)	37.5 (17.2)	33.8 (23.0)	43.3 (22.0)
Lower extremity	17.8 (20.8)	25.8 (21.3)	21.2 (22.7)	42.0 (21.1)	43.5 (37.0)	24.0 (18.5)	30.0 (18.8)	29.3 (21.6)	31.8 (17.2)	31.2 (30.4)	40.3 (25.6)	40.2 (11.4)	37.8 (23.6)	45.7 (20.9)
Head/neck	18.9 (23.8)	11.3 (14.8)	22.2 (36.3)	46.2 (25.5)	60.0 (32.5)	26.9 (21.7)	27.6 (24.8)	25.0 (20.0)	31.9 (23.9)	35.6 (36.7)	37.8 (32.9)	29.1 (10.7)	20.4 (16.5)	42.7 (21.7)
Trunk	13.8 (17.7)	8.8 (11.4)	17.7 (22.2)	40.1 (23.4)	21.4 (29.3)	28.5 (22.7)	23.1 (20.8)	15.0 (15.3)	29.6 (20.6)	22.3 (28.5)	29.6 (26.7)	35.4 (17.1)	17.9 (18.7)	38.1 (21.0)

Table 2. *Cont.*

	DTF-QoL													
	Symptom Scales +			Impact Scales +										
	W1 Emotional	W2 Physical	W3 Pain	1 Concerns Condition	2 Job & Education	3 Doctor-Patient	4 Relationships	5 Physical Consequences	6 Diagnostic	7 Parenting	8 Body Image	9 Support	10 Treatment Concerns	11 Behaviour DTF
Hip/pelvis/gluteal region	18.3 (20.6)	21.3 (18.4)	38.9 (33.3)	55.8 (23.1)	40.4 (29.7)	30.8 (21.2)	30.2 (23.5)	27.2 (19.7)	26.9 (16.9)	35.4 (21.2)	37.5 (23.8)	33.9 (14.2)	25.6 (22.2)	47.1 (22.2)
p-value #	0.613	**<0.001** g,h,i,j,k,l	**<0.001** h,j,m,n	0.062	**0.001** o,p	0.862	0.607	**0.012** *	0.687	0.153	**0.003** i	0.341	**<0.001** g,i,l,m,q,r	0.109

DTF-QoL scales: W1: emotional and psychological consequences; W2: physical consequences; W3: pain and discomfort; 1: concerns about condition, 2: job and education; 3: doctor-patient relationship, communication and information; 4: effect of desmoid-type fibromatosis (DTF) on relationships; 5: physical limitations and consequences; 6: diagnostic and treatment trajectory of DTF; 7: parenting and fertility; 8: body image and sensations; 9: supportive care; 10: concerns around treatment and its consequences; 11: unpredictable course and nature of DTF. Abbreviations: conseq: consequences. + Higher scores indicate a higher level of symptomatology/problems. [1] Active surveillance only and surgery only: including patients who received analgesics. Other treatment, including patients who received only systemic therapy (i.e., chemotherapy, hormonal therapy, targeted medical therapy) or local therapy (i.e., radiotherapy, isolated limb perfusion, high-intensity-focused ultrasound, cryoablation) or a combination of any form of active treatments. # *p*-value of ANOVA for differences between the subgroups. Bold values indicate significant variables ($p < 0.05$). * No statistically significant differences in Bonferroni post hoc analysis. [a, b, c, d, e, f, g, h, i, j, k, l, m, o, p, q, r] Shows which groups are significantly different according to the Bonferroni post hoc analysis ($p < 0.05$): Medium education level versus: [a] high; ≥ 2 comorbidities versus: [b] none, [c] 1; Other treatment versus: [d] surveillance only, [e] surgery only; Surveillance only versus: [f] surgery only; Abdominal wall versus: [g] lower extremity, [h] hip/pelvis/gluteal region, [o] head and neck, [q] upper extremity; Intra-abdominal versus: [i] lower extremity, [j] hip/pelvis gluteal region, [m] upper extremity; Lower extremity versus: [k] upper extremity; Trunk versus: [l] lower extremity, [n] hip/pelvis/gluteal region, [p] head and neck, [r] upper extremity.

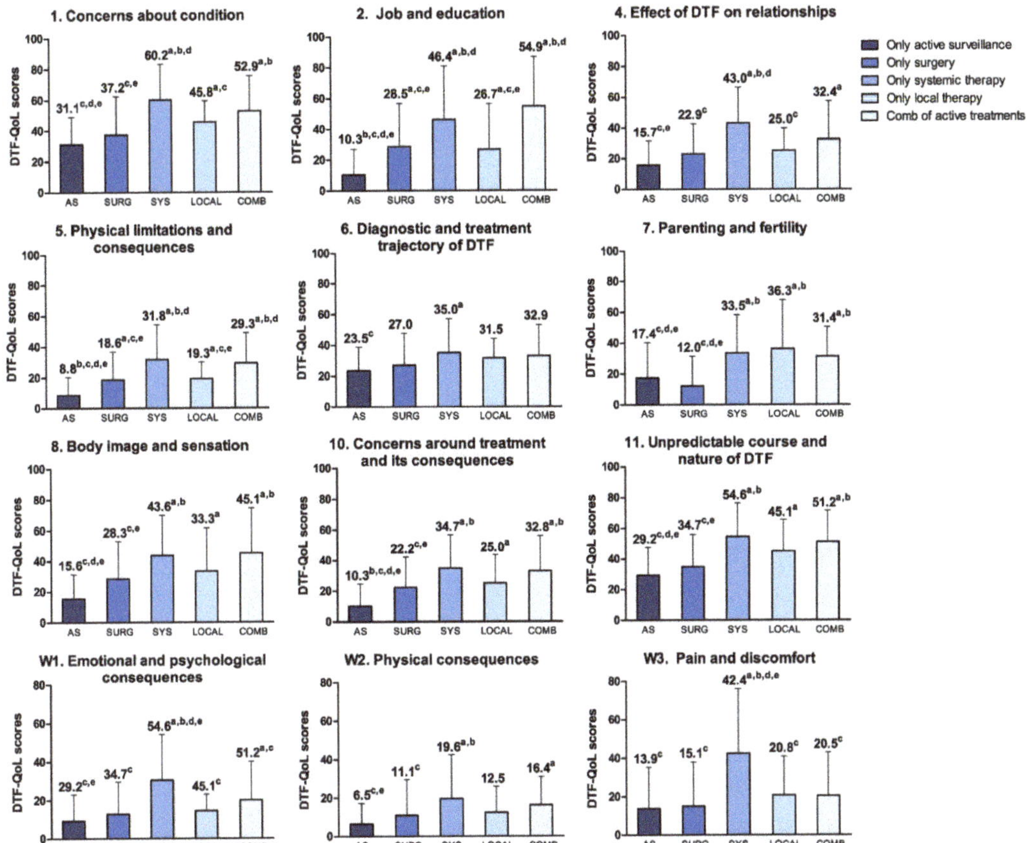

Figure 1. Mean DTF-QoL scores per treatment type. Differences in mean scores of DTF-QoL scales between treatment groups. Higher scores indicate a higher level of symptomatology/problems. Scale 3 (doctor-patient relationship, communication and information) and 9 (supportive care) are not shown because no significant differences were found between the treatment groups for these scales. Active surveillance, surgery, systemic therapy or local therapy only: including patients who received analgesics. Systemic therapy includes: chemotherapy, hormonal therapy and targeted medical therapy (tyrosine kinase and gamma-secretase inhibitors). Local therapy includes: radiotherapy, isolated limb perfusion, high-intensity-focused ultrasound, cryoablation. Combination of active treatments: including patients who received different combinations of surgery, systemic therapy or local therapy. [a,b,c,d,e] Corresponds to whether the score of the respective treatment group is clinically relevant different (difference $\geq$ 0.5 SD) compared to: [a] only active surveillance, [b] only surgery, [c] only systemic therapy, [d] only local therapy, [e] combination of active treatments. Abbreviations: AS, only active surveillance; Surg, only surgery; Sys, only systemic therapy; Comb, combination of active treatments.

Table 3. Mean EORTC QLQ-C30 scores (±SD) in relation to socio-demographic and clinical characteristics.

	EORTC QLQ-C30														
	Functional Scales ++						Symptom Scales/Items +								
	PF	**RF**	**EF**	**CF**	**SF**	**Global QoL**	**Fatigue**	**Nausea**	**Pain**	**Dysp-Noea**	**Sleep**	**Appetite Loss**	**Constipation**	**Diarrhoea**	**FD**
Study population	86.1 (18.6)	82.1 (27.0)	79.3 (21.3)	84.8 (21.2)	83.8 (26.0)	76.3 (19.6)	23.0 (24.9)	3.2 (8.9)	22.5 (26.5)	8.8 (19.2)	25.4 (31.6)	6.0 (16.4)	10.9 (21.3)	8.9 (21.1)	10.5 (25.3)
Age (years)															
18–39	87.9 (17.9)	81.3 (28.8)	79.0 (23.7)	86.2 (22.1)	84.3 (26.2)	77.6 (16.7)	22.7 (26.1)	3.4 (9.9)	24.3 (27.9)	4.8 (12.6)	21.6 (32.0)	5.7 (16.0)	9.7 (21.0)	7.7 (19.8)	12.0 (26.8)
≥40	84.3 (19.2)	82.9 (25.3)	79.7 (18.9)	83.3 (20.2)	83.3 (26.0)	75.0 (22.1)	23.3 (23.6)	3.0 (7.7)	20.6 (25.1)	12.7 (23.4)	29.1 (31.0)	6.2 (16.9)	12.1 (21.6)	10.2 (22.4)	9.0 (23.7)
p-value	**0.043**	0.924	0.552	0.087	0.603	0.717	0.511	0.969	0.318	**0.006**	**0.017**	0.857	0.292	0.332	0.305
Sex															
Male	93.4 (10.2)	90.1 (22.3)	83.3 (17.9)	86.8 (18.9)	87.4 (22.1)	80.0 (15.5)	16.5 (20.9)	2.2 (7.1)	14.5 (21.0)	4.8 (13.3)	17.2 (23.2)	2.7 (11.0)	6.5 (16.9)	9.1 (21.9)	5.4 (17.3)
Female	83.5 (20.2)	79.3 (28.1)	77.9 (22.3)	84.0 (22.0)	82.6 (27.2)	75.0 (20.8)	25.3 (25.8)	3.6 (9.4)	25.3 (27.8)	10.2 (20.8)	28.3 (33.7)	7.1 (17.8)	12.5 (22.5)	8.9 (20.9)	12.3 (27.4)
p-value	**<0.001**	**0.002**	0.147	0.486	0.318	0.195	**0.022**	0.224	**0.005**	0.075	0.052	**0.046**	**0.040**	0.888	0.094
Relationship status															
Partnered	87.1 (16.7)	83.7 (24.9)	80.6 (20.9)	85.6 (21.1)	85.5 (24.6)	77.1 (18.7)	22.7 (24.4)	3.3 (9.2)	22.8 (25.8)	8.5 (18.3)	25.4 (31.3)	5.3 (15.0)	10.9 (21.6)	9.2 (21.4)	8.5 (22.8)
Not partnered	82.3 (24.0)	76.4 (33.1)	74.8 (22.6)	81.4 (21.6)	78.6 (29.8)	73.6 (22.6)	23.9 (26.7)	2.8 (7.8)	21.7 (29.2)	10.1 (22.3)	25.2 (33.3)	8.2 (20.6)	11.3 (20.6)	6.9 (18.9)	17.6 (31.7)
p-value	0.351	0.265	0.075	0.132	0.159	0.421	0.857	0.769	0.399	0.877	0.788	0.452	0.732	0.470	0.022
Education level															
Low	81.5 (24.1)	79.6 (28.5)	80.1 (21.0)	80.6 (23.7)	83.3 (26.7)	72.2 (23.1)	25.6 (27.5)	7.4 (14.6)	22.2 (26.1)	13.0 (24.3)	26.9 (32.7)	13.9 (24.4)	13.9 (21.6)	13.0 (25.5)	13.0 (27.9)
Medium	84.6 (18.5)	79.0 (28.9)	77.4 (22.6)	83.7 (21.1)	80.2 (28.1)	73.1 (19.7)	25.2 (25.2)	2.9 (8.2)	23.8 (27.5)	9.5 (18.8)	25.7 (31.9)	5.3 (16.0)	12.2 (23.7)	10.6 (23.3)	12.4 (27.2)
High	90.6 (14.6)	88.8 (21.5)	82.3 (19.0)	88.6 (19.6)	90.4 (20.4)	83.9 (15.4)	17.8 (22.3)	1.6 (4.9)	20.3 (25.2)	5.5 (16.7)	24.2 (31.1)	3.2 (9.9)	7.3 (16.0)	4.1 (12.4)	5.9 (19.5)
p-value #	**0.018** a	**0.038** b	0.285	0.129	**0.027** b	**<0.001** a,b	0.100	**0.005** a,c	0.671	0.132	0.910	**0.004** a,c	0.200	0.052	0.178
Employment status															
Working	89.8 (14.4)	86.8 (22.8)	82.3 (19.5)	88.1 (17.5)	87.7 (21.3)	78.9 (17.6)	19.7 (22.5)	2.3 (6.3)	19.7 (24.1)	7.1 (17.8)	21.1 (29.2)	4.7 (13.9)	9.2 (19.2)	8.2 (20.9)	6.7 (22.0)
Not working	78.8 (23.2)	73.1 (32.0)	73.6 (23.7)	78.3 (25.9)	76.3 (32.1)	71.3 (22.4)	29.3 (28.0)	5.0 (12.3)	27.9 (30.2)	12.1 (21.4)	33.8 (34.6)	8.3 (20.2)	14.2 (24.7)	10.4 (21.6)	17.9 (29.5)

Table 3. *Cont.*

	EORTC QLQ-C30														
	Functional Scales [++]					Symptom Scales/Items [+]									
	PF	RF	EF	CF	SF	Global QoL	Fatigue	Nausea	Pain	Dysp-Noea	Sleep	Appetite Loss	Constipation	Diarrhoea	FD
p-value	**<0.001**	**<0.001**	**0.006**	**0.004**	**0.012**	**0.016**	**0.011**	0.139	0.073	**0.027**	**0.004**	0.231	0.129	0.292	**<0.001**
Comorbidity															
None	91.5 (15.2)	86.5 (24.6)	85.0 (19.5)	90.7 (15.6)	90.2 (22.3)	84.6 (14.4)	12.8 (20.1)	0.9 (4.6)	17.0 (25.7)	3.7 (12.7)	17.0 (28.8)	2.2 (9.7)	5.9 (14.6)	3.3 (14.2)	5.9 (19.7)
1	88.3 (15.0)	84.9 (25.8)	79.4 (18.1)	84.7 (22.0)	86.9 (21.4)	76.2 (17.1)	23.3 (24.1)	4.1 (10.6)	20.5 (22.8)	5.9 (16.9)	23.9 (30.0)	5.4 (14.6)	10.4 (19.8)	6.8 (15.6)	9.5 (23.1)
≥2	76.9 (22.4)	73.7 (29.6)	72.1 (24.5)	77.2 (24.1)	72.5 (31.0)	65.8 (22.8)	35.5 (25.6)	5.2 (10.4)	31.5 (29.1)	18.3 (24.4)	37.6 (33.3)	11.3 (22.5)	17.8 (27.5)	18.3 (29.2)	17.4 (31.8)
p-value [#]	**<0.001** [d,e]	**0.006** [d,e]	**0.001** [d]	**<0.001** [d]	**<0.001** [d,e]	**<0.001** [d,e,f]	**<0.001** [d,e,f]	**0.006** [d]	**0.002** [d,e]	**<0.001** [d,e]	**<0.001** [d,e]	**0.002** [d]	**0.002** [d]	**<0.001** [d,e]	**0.015** [d]
Time since diagnosis															
<5 years	87.1 (17.7)	81.1 (27.7)	79.9 (20.4)	84.2 (21.6)	85.3 (23.4)	76.8 (20.6)	23.2 (25.3)	3.5 (9.2)	23.8 (26.1)	7.0 (17.0)	24.3 (32.7)	4.9 (13.2)	9.8 (19.3)	9.8 (22.6)	7.2 (20.4)
≥5 years	84.9 (19.8)	83.3 (26.2)	78.6 (22.5)	85.4 (20.8)	82.1 (28.9)	75.7 (18.5)	22.7 (24.4)	2.8 (8.4)	20.9 (27.1)	11.0 (21.4)	26.7 (30.3)	7.2 (19.5)	12.3 (23.6)	7.9 (19.3)	14.5 (29.8)
p-value	0.411	0.515	0.790	0.770	0.587	0.473	0.969	0.413	0.237	0.134	0.289	0.610	0.604	0.690	0.062
Treatment received [1]															
Only active surveillance	90.2 (16.2)	88.5 (19.2)	86.3 (18.3)	88.9 (20.0)	92.0 (17.4)	80.0 (18.4)	17.9 (21.7)	3.6 (10.9)	19.5 (22.8)	8.4 (19.2)	20.3 (28.0)	5.7 (14.6)	10.7 (21.3)	4.6 (13.6)	3.1 (12.1)
Only surgery	86.4 (19.3)	88.8 (22.6)	81.4 (20.1)	83.9 (23.0)	87.2 (21.2)	78.3 (19.8)	21.5 (24.5)	3.1 (7.9)	18.8 (27.9)	6.8 (17.0)	21.4 (31.1)	5.7 (16.3)	14.1 (25.1)	12.5 (24.8)	8.9 (23.2)
Other treatment	81.6 (19.6)	70.4 (32.8)	70.5 (22.3)	81.2 (20.5)	72.8 (32.5)	71.0 (19.8)	29.4 (27.0)	2.8 (7.2)	28.4 (28.3)	10.7 (20.8)	33.7 (34.1)	6.4 (18.3)	8.7 (18.0)	10.7 (23.8)	19.4 (33.2)
p-value [#]	**0.010** [g]	**<0.001** [g,h]	**<0.001** [g,h]	0.053	**<0.001** [g,h]	**0.007** [g]	**0.008** [g]	0.817	**0.039** *	0.455	**0.010** [g]	0.964	0.321	**0.047** *	**<0.001** [g,h]
Recurrent disease															
Yes	84.5 (19.2)	80.5 (25.5)	78.0 (22.3)	82.9 (24.3)	77.2 (29.5)	78.9 (16.9)	23.0 (24.8)	4.5 (9.9)	26.8 (29.1)	4.1 (11.0)	26.0 (30.3)	6.5 (18.6)	11.4 (21.9)	16.3 (29.0)	19.5 (31.6)
No	86.4 (18.5)	82.5 (27.4)	79.6 (21.2)	85.1 (20.5)	85.2 (25.1)	75.8 (20.2)	23.0 (24.9)	2.9 (8.7)	21.6 (26.0)	9.8 (20.4)	25.3 (32.0)	5.8 (15.9)	10.8 (21.3)	7.4 (18.8)	8.6 (23.4)

Table 3. *Cont.*

	EORTC QLQ-C30														
	Functional Scales ++						Symptom Scales/Items +								
	PF	RF	EF	CF	SF	Global QoL	Fatigue	Nausea	Pain	Dysp-Noea	Sleep	Appetite Loss	Constipation	Diarrhoea	FD
p-value	0.766	0.451	0.782	0.799	0.074	0.433	0.947	0.284	0.304	0.124	0.738	0.912	0.929	**0.039**	**0.003**
Recurrent disease after surgery ($n = 98$)															
Yes	84.6 (19.2)	80.5 (25.5)	78.0 (22.3)	82.9 (24.3)	77.2 (29.5)	78.9 (16.9)	23.0 (24.8)	4.5 (9.9)	26.8 (29.1)	4.1 (11.0)	26.0 (30.3)	6.5 (18.6)	11.4 (21.9)	16.3 (29.0)	19.5 (31.6)
No	87.4 (18.4)	87.1 (26.4)	80.8 (19.5)	85.4 (18.9)	87.7 (20.3)	76.3 (20.7)	22.0 (25.2)	2.3 (6.6)	16.7 (27.3)	7.6 (17.8)	22.8 (33.4)	4.7 (16.0)	12.3 (24.1)	7.6 (18.9)	7.6 (20.9)
p-value	0.626	0.059	0.641	0.937	0.086	0.554	0.747	0.295	**0.036**	0.420	0.387	0.404	0.954	0.128	**0.019**
Tumour location															
Abdominal wall	87.1 (20.8)	82.8 (29.8)	80.2 (23.9)	83.0 (25.5)	87.4 (25.0)	76.1 (21.4)	23.0 (28.5)	4.9 (12.9)	19.3 (25.3)	9.8 (24.2)	22.4 (33.3)	8.6 (21.2)	13.8 (27.2)	7.5 (19.8)	7.5 (20.7)
Intra-abdominal	90.9 (15.0)	91.5 (16.6)	87.0 (16.4)	89.3 (18.5)	87.2 (21.8)	78.6 (22.0)	19.4 (18.1)	3.4 (8.7)	8.1 (13.2)	9.4 (20.2)	16.2 (21.5)	5.1 (16.3)	7.7 (16.2)	17.1 (30.5)	6.0 (18.5)
Upper extremity	87.4 (15.3)	81.0 (26.6)	76.1 (21.1)	81.0 (19.3)	77.0 (28.3)	75.6 (18.2)	21.1 (25.3)	1.7 (5.2)	33.9 (30.0)	6.9 (16.4)	29.9 (31.3)	2.3 (8.6)	11.5 (20.5)	3.4 (10.3)	10.3 (23.7)
Lower extremity	80.9 (21.1)	76.5 (28.0)	76.9 (23.6)	81.8 (22.9)	76.5 (31.1)	75.0 (16.1)	25.3 (28.5)	2.3 (5.9)	31.8 (30.8)	6.1 (13.2)	34.8 (37.8)	4.5 (11.7)	10.6 (21.5)	6.1 (13.2)	22.7 (37.6)
Head/neck	83.1 (18.4)	67.9 (35.7)	75.6 (24.9)	87.2 (16.9)	83.3 (28.9)	78.8 (17.9)	24.8 (28.4)	1.3 (4.6)	28.2 (35.6)	10.3 (21.0)	35.9 (37.2)	2.6 (9.2)	10.3 (21.0)	12.8 (16.9)	17.9 (37.6)
Trunk	86.4 (17.4)	84.9 (24.1)	78.5 (19.1)	86.7 (16.9)	86.1 (22.4)	76.5 (19.2)	23.7 (21.6)	3.4 (8.2)	21.3 (22.3)	9.3 (17.6)	22.2 (29.7)	6.2 (17.2)	11.1 (19.4)	9.3 (22.8)	8.6 (23.5)
Hip/pelvis/ gluteal region	78.7 (21.9)	71.7 (31.1)	73.8 (22.3)	82.5 (26.2)	79.2 (33.3)	72.5 (18.9)	27.2 (28.3)	1.7 (5.1)	32.5 (30.8)	8.3 (14.8)	36.7 (34.0)	8.3 (14.8)	8.3 (18.3)	5.0 (16.3)	15.0 (27.5)
p-value #	0.225	**0.044** *	0.250	0.632	0.372	0.949	0.934	0.629	**<0.001** i,j,k	0.985	0.091	0.650	0.895	0.141	0.157

PF: physical functioning; RF: role functioning; EF: emotional functioning; CF: cognitive functioning; SF: social functioning; Global QoL: global quality of life/health status; Sleep: sleep/insomnia; FD: financial difficulties. ++ Higher scores indicate better functioning; + Higher scores indicate a higher level of symptomatology/problems. [1] Active surveillance only and surgery only: including patients who received analgesics. Other treatment, including patients who received only systemic therapy (i.e., chemotherapy, hormonal therapy, targeted medical therapy) or local therapy (i.e., radiotherapy, isolated limb perfusion, high-intensity-focused ultrasound, cryoablation) or a combination of any form of active treatments. # *p*-value of ANOVA for differences between the subgroups. Bold values indicate significant variables ($p < 0.05$). * No statistically significant differences in Bonferroni post hoc analysis. [a, b, c, d, e, f, g, h, i, j, k] Shows which groups are significantly different according to the Bonferroni post hoc analysis ($p < 0.05$): High education level versus: [a] low, [b] medium; Low education level versus: [c] medium; ≥2 comorbidities versus: [d] none, [e] 1; 1 comorbidity versus: [f] none. Other treatment versus: [g] surveillance only, [h] surgery only; Intra-abdominal versus: [I] upper extremity, [j] lower extremity, [k] hip/pelvis gluteal region.

Table 4. Standardised betas of multiple linear regression models evaluating the association of independent variables ($p < 0.05$) with the DTF-QoL, using backward elimination.

	DTF-QoL													
	Symptom Scales +			Impact Scales +										
	W1 Emotional	W2 Physical	W3 Pain	1 Concerns Condition	2 Job & Education	3 Doctor-Patient	4 Relationships	5 Physical Consequences	6 Diagnostic	7 Parenting	8 Body Image	9 Support	10 Treatment Concerns	11 Behaviour DTF
Age	-	0.152 *	-	−0.123 *	-	-	−0.185 **	-	-	−0.324 **	−0.132 *	-	−0.137 *	−0.167 **
Sex	-	−0.138 *	-	-	−0.169 **	-	-	−0.187 **	-	−0.275 **	−0.191 **	-	-	-
Relationship status	-	-	-	-	-	-	-	-	-	-	-	-	-	-
Education level														
Low	-	-	-	-	-	-	-	-	-	-	-	Ref	-	-
Medium	-	-	-	-	-	-	-	-	-	-	-	0.057	-	-
High	-	-	-	-	-	-	-	-	-	-	-	0.215 *	-	-
Employment status	-	-	-	-	−0.260 **	-	-	-	-	-	-	-	-	-
Comorbidity														
None	Ref	Ref	Ref	Ref	-	Ref	Ref	Ref	-	-	Ref	Ref	Ref	Ref
1	0.157 *	0.030	0.086	0.133 *	-	0.069	0.227 **	0.085	-	-	0.127 *	0.117	0.080	0.133 *
≥2	0.156 *	0.190 **	0.203 **	0.167 *	-	0.195 **	0.268 **	0.201 **	-	-	0.137 *	0.193 *	0.191 **	0.264 **
Time since diagnosis	-	-	−0.146 *	-	-	-	-	-	-	-	-	0.157 *	-	-
Treatment received [1]														
Only active surveillance	Ref	Ref	Ref	Ref	Ref	-	Ref	Ref	Ref	Ref	Ref	-	Ref	Ref
Only surgery	0.094	0.124	0.072	0.061	0.206 **	-	0.145 *	0.247 **	0.081	−0.094	0.241 **	-	0.254 **	0.050
Other treatment	0.368 **	0.297 **	0.256 **	0.415 **	0.502 **	-	0.417 **	0.529 **	0.252 **	0.335 **	0.522 **	-	0.443 **	0.414 **
Recurrence	-	-	-	0.138 *	0.141 *	-	-	-	-	-	-	-	-	0.136 *
Tumour location														
Abdominal wall	-	Ref	Ref	-	-	-	-	-	-	-	-	-	Ref	-
Intra-abdominal	-	−0.068	−0.148 *	-	-	-	-	-	-	-	-	-	−0.099	-

Table 4. *Cont.*

	DTF-QoL													
	Symptom Scales +			Impact Scales +										
	W1 Emotional	W2 Physical	W3 Pain	1 Concerns Condition	2 Job & Education	3 Doctor-Patient	4 Relationships	5 Physical Consequences	6 Diagnostic	7 Parenting	8 Body Image	9 Support	10 Treatment Concerns	11 Behaviour DTF
Upper extremity	-	−0.054	0.131	-	-	-	-	-	-	-	-	-	0.144 *	-
Lower extremity	-	0.259 **	0.037	-	-	-	-	-	-	-	-	-	0.198 **	-
Head/neck	-	−0.048	−0.011	-	-	-	-	-	-	-	-	-	−0.056	-
Trunk	-	−0.065	−0.026	-	-	-	-	-	-	-	-	-	−0.022	-
Hip/pelvis/gluteal region	-	0.190 **	0.223 **	-	-	-	-	-	-	-	-	-	0.023	-

Age: 18–39 vs. ≥40 years; sex: female vs. male; relationship status: non-partnered vs. partnered; current employment status: not working vs. working; comorbidity: none vs. 1; none vs. ≥2; time since diagnosis: <5 years vs. ≥5 years; recurrence: no recurrence vs. recurrence. DTF-QoL scales: W1: emotional and psychological consequences; W2: physical consequences; W3: pain and discomfort; 1: concerns about condition; 2: job and education; 3: doctor-patient relationship, communication and information; 4: effect of desmoid-type fibromatosis (DTF) on relationships; 5: physical limitations and consequences; 6: diagnostic and treatment trajectory of DTF; 7: parenting and fertility; 8: body image and sensations; 9: supportive care; 10: concerns around treatment and its consequences; 11: unpredictable course and nature of DTF. [1] Active surveillance only and surgery only: including patients who received analgesics. Other treatment, including patients who received only systemic therapy (i.e., chemotherapy, hormonal therapy, targeted medical therapy) or local therapy (i.e., radiotherapy, isolated limb perfusion, high-intensity-focused ultrasound, cryoablation) or a combination of any form of active treatments. * $p < 0.05$; ** $p < 0.01$. + Higher scores indicate a higher level of symptomatology/problems.

4. Discussion

This international, cross-sectional study evaluating HRQoL in DTF patients, showed that both generic and disease-specific HRQoL differ between subgroups based on socio-demographic and clinical characteristics of DTF patients. In multivariate analyses, younger age, female sex, presence of comorbidities, and treatment other than AS or surgery only, were most strongly associated with worse DTF-specific HRQoL outcomes.

The type of treatment a patient received was found to be one of the most important factors associated with both the generic and DTF-specific HRQoL. The group of patients who received systemic therapy or a combination of active treatments scored significantly worse than patients who received AS or surgery alone, with the differences in the HRQoL scores being clinically relevant. These results may be explained by the fact that patients who require systematic therapy or multiple treatments are those with more complicated DTF tumours, with a more aggressive disease course and/or in whom an eventual resection would be mutilating. The greater impact of these types of treatment may therefore be partly caused by a higher tumour burden. The variable response to systemic and local therapies in DTF may exacerbate the differences between those who need active treatment and those who do not. The treatment itself, or its side effects, could also affect HRQoL. For example, DTF patients undergoing systemic therapy reported comparable hair and skin problems to soft-tissue sarcoma patients who received chemotherapy, which can have a negative impact on the patient's self-image [26]. In addition, a failure of (multiple) treatments can lead to uncertainties about the disease and treatment efficacy [13,15,16]. In general, HRQoL outcomes of patients who received AS or surgery only were comparable. Compared to AS alone, surgery was negatively associated with concerns about treatment and subscales with items related to the physical consequences of a surgical resection, such as body image and sensations, and physical limitations in daily life or work. It has been reported that AS is associated with increased anxiety and uncertainties [27]. In the current study, patients receiving only AS did not experience greater negative physical or psychological effects than patients undergoing active treatment. Our results clearly demonstrate that the type of treatment DTF patients received, which is related to the complexity of the tumour, can have a severe impact on their HRQoL. The potential risks and benefits of treatments should therefore be considered carefully, and patients should be informed about the possible side effects associated with treatments. Since this was a cross-sectional study, it did not assess the magnitude of the impact of treatment on patients' HRQoL over time. In future (longitudinal) studies, and clinical follow-up, the HRQoL outcome measures should be included alongside the objective outcome measures to evaluate treatment efficacy and also to facilitate shared decision making, e.g., between AS and surgery.

Differences in the time since diagnosis were only found for several subscales of the DTF-QoL and not the EORTC QLQ-C30, with significantly worse scores for patients who were $\geq$5 years after diagnosis. These differences were particularly seen on the impact scales, possibly reflecting the chronic character of DTF, since these items cover a timeframe since diagnosis. Another possible explanation may be that active treatments were more common in the past, and that these worse HRQoL scores are a result of these active treatments. This could explain why time since diagnosis affected only two scales after adjusting for the treatment type. A longer time since diagnosis was associated with higher scores on the supportive care subscale, indicating that these patients experienced more lack of support in the past. Therefore, these results suggest that recognition and awareness of HRQoL issues, using the DTF-QoL, is important, even long after the time of diagnosis.

Differences between tumour locations were mainly seen on the subscales of the DTF-QoL and not of the EORTC QLQ-C30, except for the pain items. These results are in line with a study of sarcoma patients by van Eck et al., who assessed HRQoL between different sarcoma locations using the EORTC QLQ-C30 and additional treatment-specific items from the EORTC Item Library [26]. They found no significant differences in the HRQoL domains of the EORTC QLQ-C30 between different tumour locations, however, they did find treatment-specific HRQoL issues that differed per sarcoma location, underlining the importance of using a disease-specific HRQoL-measurement strategy. In our study, worse scores on the DTF-specific questionnaire were observed for DTF patients with tumours in the upper and lower extremities and hip/pelvis/gluteal region on the subscales about physical limitations, pain and concerns around treatment and its consequences. These subscales, consisting of site-specific items, such as "Have you had any trouble walking?" or "Have you been afraid of needing a limb amputation?" are, therefore, particularly useful for these specific tumour sites.

The presence of comorbidities generally has a negative impact on HRQoL [28,29]. DTF patients with two or more comorbidities reported significantly worse scores on all scales and items of the EORTC QLQ-C30, which is in agreement with the previous studies conducted among patients with different types of cancer [30,31]. In addition, the results of our study indicate that the presence of comorbidities significantly affects DTF-specific HRQoL as well. Given the cross-sectional study design, it is unclear whether the self-reported comorbidities were present before a DTF diagnosis or if they developed thereafter. Moreover, comorbidities may interfere with treatment effects [29,30]. It is important to be aware of the impact of comorbidities on HRQoL, not only to assess a true treatment efficacy, but also to provide the necessary support in clinical care.

The socio-demographic factors sex, age, relationship status, education level and employment status are known to be associated with generic HRQoL [32–35]. The results of this study indicate that the female sex is not only associated with worse generic HRQoL scores, but also with DTF-specific HRQoL. It is generally assumed that HRQoL decreases with increasing age [33,36]. However, our results show that, while a higher age was negatively associated with physical symptoms, patients aged between 18 and 39 years scored significantly worse on several of the DTF-QoL impact scales. Younger DTF patients reported similar concerns to adolescent and young adult (AYA) cancer patients, e.g., concerns about their ability to have children [37,38]. The greater impact of DTF on younger patients can be explained by the fact that these patients define their identity in this period of their lives, face important life choices and often have high expectations of themselves at work and in their social lives [38]. A study by Drabbe et al. also found that AYA-sarcoma patients (aged 18–39 years) had significantly lower scores on the emotional, cognitive and social functioning scales of the EORTC QLQ-C30 compared to older patients [36]. Interestingly, in our study, a significant difference was only seen on the physical functioning scale of the EORTC QLQ-C30, with older patients scoring worse. This shows that by only using a generic questionnaire, the impact of DTF on younger patients could be missed, emphasising the importance of AYA-specific and disease-specific questionnaires [39]. It is noteworthy that, in general, socio-demographic factors had the greatest impact on generic HRQoL, whereas the influence of clinical factors was mainly seen on the DTF-QoL, indicating that the DTF-QoL provides relevant additional information about the HRQoL of these specific subgroups.

To the best of our knowledge, this is the first study exploring the heterogeneity in both the generic and disease-specific HRQoL in DTF patients. The strengths of this study are the large study population and the use of generic and disease-specific HRQoL questionnaires. Given the limited data available on HRQoL for DTF patients and the heterogeneous characteristics of DTF, the subgroup analyses are a valuable contribution to providing further insight into which patients are at risk of a poor HRQoL. Furthermore, knowledge of the differences between subgroups of DTF patients can be used to develop an individualised measurement strategy by not using all items of the DTF-QoL, but only the specific scales in which problems can be expected for that particular subgroup. For example, the parenting and fertility impact scale of the DTF-QoL could be used for patients aged 18–39 years and the physical consequences symptom scale could be used for patients with DTF located in the lower extremities or hip/pelvis/gluteal region.

The present study also has some limitations. First, there may be selection bias, as it is unknown whether DTF patients did not respond or participate, due to either the absence of symptoms or poor health [40]. The non-responder analysis did not reveal any differences, however, clinical characteristics were unavailable for these patients. Secondly, as there is no accurate national registration system in both countries, it is not possible to say with certainty which DTF patients attended the participating centres. It is assumed that at least the more complex patients were treated in the participating centres, as these were tertiary referral centres. However, it is unknown how many more complex cases have remained in the peripheral hospitals, which may also have led to selection bias. Thirdly, although we were able to analyse the clinically relevant subgroups of DTF patients, differences may also exist within these groups. Due to small numbers, we did not assess these differences in HRQoL scores. The future use of the DTF-QoL in large international cohorts will provide more data to investigate these differences within subgroups. Fourthly, the cross-sectional study design limits the possibility of drawing conclusions about causal associations. In addition, tumour behaviour was not included in our analyses. Since some DTF patients were discharged at the time of the questionnaire, the information regarding their current disease status was unavailable. Furthermore, tumour behaviour can vary during follow-up due to the unpredictable biological behaviour, making it difficult to classify patients into one particular group and to draw any conclusions about the association between the tumour's behaviour and HRQoL. A longitudinal assessment of HRQoL data will help to determine the impact of socio-demographic and clinical factors on HRQoL over time.

5. Conclusions

DTF can result in a wide variety of disease-specific issues and the impact of DTF on HRQoL differs between subgroups. The use of the DTF-QoL, alongside generic HRQoL instruments, is essential to gain insight into the patient's specific problems and needs. Together, these insights will help clinicians to provide better and more personalised care to patients with DTF.

Supplementary Materials: The following are available online at https://www.mdpi.com/article/10.3390/cancers14122979/s1, Table S1: Specification of active treatment types (n = 148); Table S2: Mean DTF-QoL single item scores (±SD) in relation to socio-demographic and clinical characteristics.

Author Contributions: Conceptualisation, A.-R.W.S., E.L., W.T.A.v.d.G., C.V., D.J.G., S.G. and O.H.; methodology, A.-R.W.S., E.L. and O.H.; validation, A.-R.W.S. and O.H.; formal analysis, A.-R.W.S. and O.H.; resources, A.-R.W.S., E.Y., E.L., R.L.J., M.J.M.T., D.J.G., W.T.A.v.d.G., W.J.v.H., S.S., J.J.B., S.G., C.V. and O.H.; data curation, A.-R.W.S., E.Y., E.L., S.G. and O.H.; writing—original draft preparation, A.-R.W.S.; writing—review and editing, all authors; visualisation, A.-R.W.S., E.L. and O.H.; supervision, O.H.; project administration: A.-R.W.S., E.L. and M.J.M.T.; funding acquisition, M.J.M.T., C.V. and O.H. All authors have read and agreed to the published version of the manuscript.

Funding: This research was funded by Stichting Coolsingel, Rotterdam, the Netherlands (grant number 566), the National Institute for Health Research (NIHR) Biomedical Research Centre at the Royal Marsden NHS Foundation Trust and the Institute of Cancer Research, London (B038). O.H. is supported by a grant from the Netherlands Organisation for Scientific Research (VIDI198.007).

Institutional Review Board Statement: The study was conducted according to the guidelines of the Declaration of Helsinki, and approved by the Royal Marsden and Institute of Clinical Research Joint Committee for Clinical Research for ethical review (SE806) in the United Kingdom, and in the Netherlands, the Institutional Review Board (or Ethics Committee) at each participating centre (Erasmus Medical Centre: MEC-2019-0816, Radboudumc: file number 2020-6235, Netherlands Cancer Institute: IRBd20-088).

Informed Consent Statement: Informed consent was obtained from all subjects involved in the study.

Data Availability Statement: The datasets used and/or analysed during the current study are available from the corresponding author on reasonable request.

Acknowledgments: The authors would like to thank all patients who participated in the QUALIFIED study.

Conflicts of Interest: R.L.J. has received research grants from MSD and GSK and is a consultant for Adaptimmune, Athenex, Bayer, Boehringer Ingelheim, Blueprint, Clinigen, Eisai, Epizyme,

Daichii, Decipheara, Immunedesign, Lilly, Merck, Pharmacar, Springworks, Tracon and UptoDate. W.T.A.v.d.G. has received research grants from Novartis and Lilly, and advisory board fees from Bayer, and is a consultant to Springworks and GSK. All fees were given to the institute. The funders had no role in the design of the study; in the collection, analyses, or interpretation of data; in the writing of the manuscript, or in the decision to publish the results.

References

1. WHO Classification of Tumours Editorial Board. *Soft Tissue and Bone Tumours, WHO Classification of Tumours*, 5th ed.; WHO Press: Geneva, Switzerland, 2020.
2. Penel, N.; Coindre, J.M.; Bonvalot, S.; Italiano, A.; Neuville, A.; Le Cesne, A.; Terrier, P.; Ray-Coquard, I.; Ranchere-Vince, D.; Robin, Y.M.; et al. Management of desmoid tumours: A nationwide survey of labelled reference centre networks in France. *Eur. J. Cancer* **2016**, *58*, 90–96. [CrossRef]
3. Penel, N.; Le Cesne, A.; Bonvalot, S.; Giraud, A.; Bompas, E.; Rios, M.; Salas, S.; Isambert, N.; Boudou-Rouquette, P.; Honore, C.; et al. Surgical versus non-surgical approach in primary desmoid-type fibromatosis patients: A nationwide prospective cohort from the French Sarcoma Group. *Eur. J. Cancer* **2017**, *83*, 125–131. [CrossRef]
4. van Broekhoven, D.L.; Grunhagen, D.J.; den Bakker, M.A.; van Dalen, T.; Verhoef, C. Time trends in the incidence and treatment of extra-abdominal and abdominal aggressive fibromatosis: A population-based study. *Ann. Surg. Oncol.* **2015**, *22*, 2817–2823. [CrossRef]
5. Bonvalot, S.; Ternes, N.; Fiore, M.; Bitsakou, G.; Colombo, C.; Honore, C.; Marrari, A.; Le Cesne, A.; Perrone, F.; Dunant, A.; et al. Spontaneous regression of primary abdominal wall desmoid tumors: More common than previously thought. *Ann. Surg. Oncol.* **2013**, *20*, 4096–4102. [CrossRef]
6. van Houdt, W.J.; Husson, O.; Patel, A.; Jones, R.L.; Smith, M.J.F.; Miah, A.B.; Messiou, C.; Moskovic, E.; Al-Muderis, O.; Benson, C.; et al. Outcome of Primary Desmoid Tumors at All Anatomic Locations Initially Managed with Active Surveillance. *Ann. Surg. Oncol.* **2019**, *26*, 4699–4706. [CrossRef]
7. Schut, A.W.; Timbergen, M.J.M.; van Broekhoven, D.L.M.; van Dalen, T.; van Houdt, W.J.; Bonenkamp, J.J.; Sleijfer, S.; Grunhagen, D.J.; Verhoef, C. A Nationwide Prospective Clinical Trial on Active Surveillance in Patients with Non-Intra-Abdominal Desmoid-Type Fibromatosis: The GRAFITI Trial. *Ann. Surg.* **2022**. [CrossRef]
8. Desmoid Tumor Working Group. The management of desmoid tumours: A joint global consensus-based guideline approach for adult and paediatric patients. *Eur. J. Cancer* **2020**, *127*, 96–107. [CrossRef]
9. Colombo, C.; Fiore, M.; Grignani, G.; Tolomeo, F.; Merlini, A.; Palassini, E.; Collini, P.; Stacchiotti, S.; Casali, P.G.; Perrone, F.; et al. A prospective observational study of Active surveillance in primary desmoid fibromatosis. *Clin. Cancer Res.* **2022**. [CrossRef]
10. Bonvalot, S.; Eldweny, H.; Haddad, V.; Rimareix, F.; Missenard, G.; Oberlin, O.; Vanel, D.; Terrier, P.; Blay, J.Y.; Le Cesne, A.; et al. Extra-abdominal primary fibromatosis: Aggressive management could be avoided in a subgroup of patients. *Eur. J. Surg. Oncol.* **2008**, *34*, 462–468. [CrossRef]
11. van Broekhoven, D.L.M.; Verschoor, A.J.; van Dalen, T.; Grunhagen, D.J.; den Bakker, M.A.; Gelderblom, H.; Bovee, J.; Haas, R.L.M.; Bonenkamp, H.J.; van Coevorden, F.; et al. Outcome of Nonsurgical Management of Extra-Abdominal, Trunk, and Abdominal Wall Desmoid-Type Fibromatosis: A Population-Based Study in the Netherlands. *Sarcoma* **2018**, *2018*, 5982575. [CrossRef]
12. Kasper, B.; Raut, C.P.; Gronchi, A. Desmoid tumors: To treat or not to treat, that is the question. *Cancer* **2020**, *126*, 5213–5221. [CrossRef] [PubMed]
13. Timbergen, M.J.M.; van der Graaf, W.T.A.; Grunhagen, D.J.; Younger, E.; Sleijfer, S.; Dunlop, A.; Dean, L.; Verhoef, C.; van de Poll-Franse, L.V.; Husson, O. Assessing the Desmoid-Type Fibromatosis Patients' Voice: Comparison of Health-Related Quality of Life Experiences from Patients of Two Countries. *Sarcoma* **2020**, *2020*, 2141939. [CrossRef] [PubMed]
14. Kluetz, P.G.; Slagle, A.; Papadopoulos, E.J.; Johnson, L.L.; Donoghue, M.; Kwitkowski, V.E.; Chen, W.H.; Sridhara, R.; Farrell, A.T.; Keegan, P.; et al. Focusing on Core Patient-Reported Outcomes in Cancer Clinical Trials: Symptomatic Adverse Events, Physical Function, and Disease-Related Symptoms. *Clin. Cancer Res.* **2016**, *22*, 1553–1558. [CrossRef] [PubMed]
15. Husson, O.; Younger, E.; Dunlop, A.; Dean, L.; Strauss, D.C.; Benson, C.; Hayes, A.J.; Miah, A.; van Houdt, W.; Zaidi, S.; et al. Desmoid fibromatosis through the patients' eyes: Time to change the focus and organisation of care? *Support Care Cancer* **2019**, *27*, 965–980. [CrossRef]
16. Timbergen, M.J.M.; van de Poll-Franse, L.V.; Grunhagen, D.J.; van der Graaf, W.T.; Sleijfer, S.; Verhoef, C.; Husson, O. Identification and assessment of health-related quality of life issues in patients with sporadic desmoid-type fibromatosis: A literature review and focus group study. *Qual. Life Res.* **2018**, *27*, 3097–3111. [CrossRef]
17. Aaronson, N.K.; Ahmedzai, S.; Bergman, B.; Bullinger, M.; Cull, A.; Duez, N.J.; Filiberti, A.; Flechtner, H.; Fleishman, S.B.; de Haes, J.C.; et al. The European Organization for Research and Treatment of Cancer QLQ-C30: A quality-of-life instrument for use in international clinical trials in oncology. *J. Natl. Cancer Inst.* **1993**, *85*, 365–376. [CrossRef]
18. Schut, A.W.; Timbergen, M.J.M.; Lidington, E.; Grunhagen, D.J.; van der Graaf, W.T.A.; Sleijfer, S.; van Houdt, W.J.; Bonenkamp, J.J.; Younger, E.; Dunlop, A.; et al. The Evaluation of Health-Related Quality of Life Issues Experienced by Patients with Desmoid-Type Fibromatosis (The QUALIFIED Study)-A Protocol for an International Cohort Study. *Cancers* **2021**, *13*, 3068. [CrossRef]

19. Schut, A.W.; Lidington, E.; Timbergen, M.J.M.; Younger, E.; van der Graaf, W.T.A.; van Houdt, W.J.; Bonenkamp, J.J.; Jones, R.L.; Grunhagen, D.J.; Sleijfer, S.; et al. Development of a Disease-Specific Health-Related Quality of Life Questionnaire (DTF-QoL) for Patients with Desmoid-Type Fibromatosis. *Cancers* **2022**, *14*, 709. [CrossRef]
20. van de Poll-Franse, L.V.; Horevoorts, N.; van Eenbergen, M.; Denollet, J.; Roukema, J.A.; Aaronson, N.K.; Vingerhoets, A.; Coebergh, J.W.; de Vries, J.; Essink-Bot, M.L.; et al. The Patient Reported Outcomes Following Initial treatment and Long term Evaluation of Survivorship registry: Scope, rationale and design of an infrastructure for the study of physical and psychosocial outcomes in cancer survivorship cohorts. *Eur. J. Cancer* **2011**, *47*, 2188–2194. [CrossRef]
21. Sangha, O.; Stucki, G.; Liang, M.H.; Fossel, A.H.; Katz, J.N. The Self-Administered Comorbidity Questionnaire: A new method to assess comorbidity for clinical and health services research. *Arthritis Rheum.* **2003**, *49*, 156–163. [CrossRef]
22. Fayers, P.; Aaronson, N.K.; Bjordal, K.; Groenvold, M.; Curran, D.; Bottomley, A. *EORTC QLQ-C30 Scoring Manual*, 3rd ed.; European Organisation for Research and Treatment of Cancer: Brussels, Belgium, 2001.
23. Johnson, C.; Aaronson, N.; Blazeby, J.M.; Bottomley, A.; Fayers, P.; Koller, M.; Kuliś, D.; Ramage, J.; Sprangers, M.; Velikova, G.; et al. *Guidelines for Developing Questionnaire Modules*, 4th ed.; EORTC Quality of Life Group: Brussels, Belgium, 2011. Available online: http://www.eortc.org/app/uploads/sites/2/2018/02/guidelines_for_developing_questionnaire-_final.pdf (accessed on 15 October 2018).
24. Norman, G.R.; Sloan, J.A.; Wyrwich, K.W. Interpretation of changes in health-related quality of life: The remarkable universality of half a standard deviation. *Med. Care* **2003**, *41*, 582–592. [CrossRef] [PubMed]
25. Field, A.P. *Discovering Statistics Using SPSS: (And Sex and Drugs and Rock 'n' Roll)*; SAGE Publications: Los Angeles, CA, USA; Thousand Oaks, CA, USA; London, UK, 2009.
26. van Eck, I.; den Hollander, D.; Desar, I.M.E.; Soomers, V.; van de Sande, M.A.J.; de Haan, J.J.; Verhoef, C.; Vriens, I.J.H.; Bonenkamp, J.J.; van der Graaf, W.T.A.; et al. Unraveling the Heterogeneity of Sarcoma Survivors' Health-Related Quality of Life Regarding Primary Sarcoma Location: Results from the SURVSARC Study. *Cancers* **2020**, *12*, 3083. [CrossRef] [PubMed]
27. Rittenmeyer, L.; Huffman, D.; Alagna, M.; Moore, E. The experience of adults who choose watchful waiting or active surveillance as an approach to medical treatment: A qualitative systematic review. *JBI Database Syst.Rev. Implement Rep.* **2016**, *14*, 174–255. [CrossRef] [PubMed]
28. Gijsen, R.; Hoeymans, N.; Schellevis, F.G.; Ruwaard, D.; Satariano, W.A.; van den Bos, G.A. Causes and consequences of comorbidity: A review. *J. Clin. Epidemiol.* **2001**, *54*, 661–674. [CrossRef]
29. Xuan, J.; Kirchdoerfer, L.J.; Boyer, J.G.; Norwood, G.J. Effects of comorbidity on health-related quality-of-life scores: An analysis of clinical trial data. *Clin. Ther.* **1999**, *21*, 383–403. [CrossRef]
30. Vissers, P.A.; Thong, M.S.; Pouwer, F.; Zanders, M.M.; Coebergh, J.W.; van de Poll-Franse, L.V. The impact of comorbidity on Health-Related Quality of Life among cancer survivors: Analyses of data from the PROFILES registry. *J. Cancer Surviv.* **2013**, *7*, 602–613. [CrossRef]
31. Schoormans, D.; Czene, K.; Hall, P.; Brandberg, Y. The impact of co-morbidity on health-related quality of life in breast cancer survivors and controls. *Acta Oncol.* **2015**, *54*, 727–734. [CrossRef]
32. Eichler, M.; Hentschel, L.; Richter, S.; Hohenberger, P.; Kasper, B.; Andreou, D.; Pink, D.; Jakob, J.; Singer, S.; Grutzmann, R.; et al. The Health-Related Quality of Life of Sarcoma Patients and Survivors in Germany-Cross-Sectional Results of a Nationwide Observational Study (PROSa). *Cancers* **2020**, *12*, 3590. [CrossRef]
33. Schwarz, R.; Hinz, A. Reference data for the quality of life questionnaire EORTC QLQ-C30 in the general German population. *Eur. J. Cancer* **2001**, *37*, 1345–1351. [CrossRef]
34. Michelson, H.; Bolund, C.; Nilsson, B.; Brandberg, Y. Health-related quality of life measured by the EORTC QLQ-C30–reference values from a large sample of Swedish population. *Acta Oncol.* **2000**, *39*, 477–484. [CrossRef]
35. Husson, O.; Haak, H.R.; Buffart, L.M.; Nieuwlaat, W.A.; Oranje, W.A.; Mols, F.; Kuijpens, J.L.; Coebergh, J.W.; van de Poll-Franse, L.V. Health-related quality of life and disease specific symptoms in long-term thyroid cancer survivors: A study from the population-based PROFILES registry. *Acta Oncol.* **2013**, *52*, 249–258. [CrossRef] [PubMed]
36. Drabbe, C.; Van der Graaf, W.T.A.; De Rooij, B.H.; Grunhagen, D.J.; Soomers, V.; Van de Sande, M.A.J.; Been, L.B.; Keymeulen, K.; van der Geest, I.C.M.; Van Houdt, W.J.; et al. The age-related impact of surviving sarcoma on health-related quality of life: Data from the SURVSARC study. *ESMO Open* **2021**, *6*, 100047. [CrossRef] [PubMed]
37. Kaal, S.E.J.; Lidington, E.K.; Prins, J.B.; Jansen, R.; Manten-Horst, E.; Servaes, P.; van der Graaf, W.T.A.; Husson, O. Health-Related Quality of Life Issues in Adolescents and Young Adults with Cancer: Discrepancies with the Perceptions of Health Care Professionals. *J. Clin. Med.* **2021**, *10*, 1833. [CrossRef] [PubMed]
38. Zebrack, B.J. Psychological, social, and behavioral issues for young adults with cancer. *Cancer* **2011**, *117*, 2289–2294. [CrossRef]
39. Husson, O.; Sodergren, S.C.; Darlington, A.S. The importance of a collaborative health-related quality of life measurement strategy for adolescents and young adults with cancer. *Cancer* **2021**, *127*, 1712–1713. [CrossRef]
40. de Rooij, B.H.; Ezendam, N.P.M.; Mols, F.; Vissers, P.A.J.; Thong, M.S.Y.; Vlooswijk, C.C.P.; Oerlemans, S.; Husson, O.; Horevoorts, N.J.E.; van de Poll-Franse, L.V. Cancer survivors not participating in observational patient-reported outcome studies have a lower survival compared to participants: The population-based PROFILES registry. *Qual. Life Res.* **2018**, *27*, 3313–3324. [CrossRef]

Article

Fear of Cancer Recurrence in Patients with Sarcoma in the United Kingdom

Anika Petrella [1], Lesley Storey [2], Nicholas J. Hulbert-Williams [3], Lorna A. Fern [1], Maria Lawal [4], Craig Gerrand [5], Rachael Windsor [6], Julie Woodford [5], Jennie Bradley [7], Hatty O'Sullivan [7], Mary Wells [8] and Rachel M. Taylor [9,*]

1. Cancer Clinical Trials Unit, University College London Hospitals NHS Foundation Trust, London NW1 2PG, UK
2. Department of Psychology, Anglia Ruskin University, Cambridge CB1 1PT, UK
3. Department of Psychology, Edge Hill University, Ormskirk L39 4QP, UK
4. Patient Representative, University College London Hospitals NHS Foundation Trust, London NW1 2PG, UK
5. Sarcoma Unit, The Royal National Orthopaedic Hospital, Stanmore HA7 4LP, UK
6. Paediatric Directorate, University College London Hospitals NHS Foundation Trust, London NW1 2PG, UK
7. Iqvia, Ltd., Reading RG1 3JH, UK
8. Nursing Directorate, Imperial College Healthcare NHS Foundation Trust, London W2 1NY, UK
9. Centre for Nurse, Midwife and Allied Health Profession Research (CNMAR), University College London Hospitals NHS Foundation Trust, London NW1 2PG, UK

* Correspondence: rtaylor13@nhs.net

Simple Summary: After a cancer diagnosis, the fear that it could come back is one of the most difficult negative emotions to manage. Sarcoma is a rare cancer of connective tissue affecting soft tissue and bone that has a high rate of recurrence and metastases. It can present itself in any age group from childhood to older adulthood. The experience of fear of cancer recurrence has not yet been explored in-depth among those with sarcoma. We, therefore, conducted an online survey to identify the prevalence of fear of cancer recurrence and factors that may be associated with it. A total of 229 people with sarcoma submitted responses, and the majority expressed interest in receiving support for fear of cancer recurrence. Overall, fear of cancer recurrence levels was found to be higher than those reported by patients with most other types of cancer. Emotional distress and being able to manage emotions were associated with fear of cancer recurrence.

Abstract: Fear of cancer recurrence (FCR) is a persistent concern among those living with cancer and is associated with a variety of negative psychosocial outcomes. However, people with sarcoma have been underrepresented within this area of research. We aimed to determine the prevalence of FCR experienced by people with sarcoma in the United Kingdom and explore factors that may predict FCR, such as the perceived impact of cancer and psychological flexibility. Participants (n = 229) with soft tissue (n = 167), bone (n = 25), and gastrointestinal stromal tumours (n = 33) completed an online survey including the self-reported measures of FCR, the perceived physical and psychological impact of cancer and psychological flexibility, and demographic information. Data were analysed using ANOVA and multiple regression modelling. Mean FCR scores (M = 91.4; SD = 26.5) were higher than those reported in meta-analytic data inclusive of all cancer types (M = 65.2; SD = 28.2). Interest in receiving support for FCR was also high (70%). Significant factors associated with FCR included cognitive and emotional distress and psychological flexibility, but not perceptions of the physical impact of cancer (R^2 = 0.56). The negative association between psychological flexibility and FCR suggests the potential benefit of intervention approaches which foster psychological flexibility, such as acceptance and commitment therapy.

Keywords: sarcoma; fear of recurrence; psychological flexibility; distress; psychological impact

Citation: Petrella, A.; Storey, L.; Hulbert-Williams, N.J.; Fern, L.A.; Lawal, M.; Gerrand, C.; Windsor, R.; Woodford, J.; Bradley, J.; O'Sullivan, H.; et al. Fear of Cancer Recurrence in Patients with Sarcoma in the United Kingdom. *Cancers* **2023**, *15*, 956. https://doi.org/10.3390/cancers15030956

Academic Editor: Adam C. Berger

Received: 10 January 2023
Revised: 29 January 2023
Accepted: 30 January 2023
Published: 2 February 2023

1. Introduction

Fear of cancer recurrence (FCR) is a highly prevalent and distressing psychological challenge for those living with and beyond cancer [1] and is defined as "fear, worry, or concern that cancer may come back or progress" [2]. FCR is considered one of the most distressing consequences of cancer, demonstrating associations with impaired physical and psychosocial functioning and lower overall quality of life [3–5]. It is reported to occur in 39–97% of cancer survivors; the prevalence of FCR is dependent on how it is measured and the definition of clinical levels of FCR [6,7]. Managing FCR has been highlighted as the number one unmet need among cancer survivors [6]. It does not appear to dissipate with time and, if left unaddressed, can become a complex lifelong concern [4–6].

Research and understanding of FCR have grown rapidly [8]; however, uncertainties remain regarding prevalence by cancer type, underlying mechanisms of action, and the best practices for intervention [9]. As identified by the James Lind Alliance's 'Living with and beyond cancer' research priorities, interventions to best support individuals to cope with FCR are needed [10]. People with sarcoma have been significantly underrepresented within FCR research. This is a concern given that recurrence rates are higher within the sarcoma population than most other solid cancers [11]. In addition, sarcomas are a rare and diverse group of cancers characterized by considerable clinical heterogeneity and significant physical burden on survivors [12]. Thus, sarcoma-specific research is warranted in order to inform intervention development and delivery.

To date, only two reports have included people with sarcoma in their sample [13,14]. First, an observational study in patients with gastrointestinal stromal tumors (GIST) examined fear of progression (not recurrence) and found that approximately 50% of patients had high levels of fear, which were associated with greater psychological distress [13]. Second, a network analysis of FCR among young adult cancer patients included a small number of people with sarcoma in their sample (n = 19; 7.7%) and reported that the majority of their sample scored above the established cut-off for high FCR on their measure (fear of progression questionnaire short form; FoP-Q-SF) [14]. Given the lack of evidence and the need for tailored interventions, additional work is required that is specific to the sarcoma population.

In the broader FCR literature, higher FCR scores have been associated with younger age, female gender, physical symptoms (e.g., pain, fatigue) and greater anxiety, and depression [8]. Potential underlying mechanisms have been explored and include optimism and social support [4,6]. These findings emphasize the potential interplay between the physical and psychological challenges faced by patients that may be impacting FCR and the need for evidence-based supportive interventions. Interventions that have been developed to date to manage FCR have included cognitive behavioral techniques, as well as relaxation, meditation, and other positive psychology-based approaches [15]. Acceptance and commitment therapy (ACT)-based interventions have shown promise in reducing cancer-related distress and FCR [16,17]. However, a greater understanding of specific active mechanisms or intervention components is needed in order to impact FCR for a range of subgroups.

Psychological flexibility is a core component of ACT [13] and may be crucial in understanding how individuals are affected by, and cope with, the significant challenges brought on by cancer and its treatments. Psychological flexibility is defined as the ability to identify and adapt to situational demands in an attempt to improve longer-term outcomes in a way that is personally meaningful [18,19]. It has been associated with improved psychological health, quality of life, and well-being in both clinical and non-clinical populations [20–24], including both distress-related and positive outcomes (e.g., benefit finding) in cancer survivors [25]. Furthermore, psychological flexibility has been shown to be amenable to change over time, presenting a potential target for interventions.

This study aimed to determine the prevalence of FCR among people with sarcoma in the United Kingdom (UK) and explore associated factors specific to the physical and psychological impacts of cancer and psychological flexibility.

2. Materials and Methods

2.1. Participants & Recruitment

Following approval from an institutional research ethics committee (Birmingham City University: Storey/#9678/sub2/R(A)/2021/Jul/BLSS FAEC), patients with sarcoma living in the UK were invited to participate in an online cross-sectional survey. The survey was administered by Quality Health using their in-house online survey software, which was open for 13 weeks (July to October 2021). Patients self-identified to participate after receiving information from sarcoma-specific and cancer charities via newsletters or social media posts. Patients were eligible to participate if they met the following criteria: diagnosis of sarcoma (any type); receiving all or some of their care in the UK; aged 16 or over; literate in English; and provided consent to participate (i.e., submitted survey was implicit of the consent). Confirmation of eligibility was required to proceed with the survey.

2.2. Measures

A bespoke survey was developed in collaboration with an established sarcoma patient advisory group and informed by previous work [26]. The survey included investigator-designed questions and validated measures of FCR, the perceived physical and psychological impact of cancer and psychological flexibility.

2.2.1. Fear of Cancer Recurrence

The fear of cancer recurrence inventory (FCRI) is a 42-item scale, widely established, an in-depth measure of FCR [27]. A total score was obtained (ranging from 0 to 168), with higher scores indicating greater FCR. The FCRI has been utilized in clinical and research practice and has been shown to be valid and reliable [8].

2.2.2. Activities of Daily Living

Two items from the Toronto extremity salvage score (TESS) [28] were used to measure perceived disability status and overall impact on activities of daily living (ADL). Both questions were answered on a 1–5 scale, with the ability to perform ADLs during the past week ranging from 'not at all difficult' to 'impossible to do' and self-reported disability status as 'not at all disabled' to 'completely disabled'. The TESS is widely used as a patient-reported functional assessment following the diagnosis and treatment of upper and lower extremity sarcoma [29–32] and has been tested for validity and reliability [28]. A generic version combining both the upper and lower extremity scale was developed so it could be administered generically to patients with any type of cancer and contained 48 items.

2.2.3. Psychology Impact

The psychological impact of cancer (PIC) scale is a valid and reliable tool for assessing the perceived psychological impact of cancer [33]. This scale contains 12 items answered on a scale of 1–4 ranging from 'definitely does not apply to me' to 'definitely applies to me', which then make up four individual subscales (cognitive distress, cognitive avoidance, fighting spirit, and emotional distress). Higher scores (ranging from 3 to 12) on each subscale represent the greater endorsement of the said factor (e.g., cognitive distress). The PIC has been validated in patients living with and beyond cancer in the UK and Australia [33].

2.2.4. Psychological Flexibility

Psychological flexibility was assessed using the comprehensive assessment of acceptance and commitment therapy processes (CompACT) [34]. This measure is comprised 23 items assessing the key dyadic process of psychological flexibility scored on a 7-point Likert scale, ranging from 0 to 6 ('strongly disagree' to 'strongly agree'). The total sum score ranges from 0 to 138, with higher scores indicating greater psychological flexibility. The CompACT has been shown to be valid and reliable within nonclinical populations [18,35,36], as well as within oncology settings [37].

Personal characteristics were collected to describe the sample and included gender, age, ethnicity, sexual orientation, marital status, employment status, and caregiver status. Cancer-specific characteristics were collected on the type of sarcoma, the year diagnosed, treatments received, amputation status, as well as a history of recurrence and/or metastatic disease. Interest in engaging with support specific to FCR was also queried.

2.3. Analysis

Data were analysed in R (Version 3.6.1). Data were inspected for missing values and then described: normally distributed data by the mean and standard deviation (SD), and binary and categorical variables were presented using frequency and percentages. The prevalence and magnitude of FCR in this sample were described and referenced in relation to available meta-analytic data and were inclusive of multiple cancer types reported in the literature. To explore differences in FCR by sarcoma type (soft tissue, bone, GIST), a one-way ANOVA was conducted, adjusting for multiple comparisons using the Tukey method. Bivariate correlations examined the size and direction of correlation between theoretically hypothesized associated factors. Multiple regression analysis was performed to establish how much variance in FCR scores was explained by physical and psychological impacts of cancer and psychological flexibility in this sample when accounting for relevant personal and cancer-specific characteristics [i.e., age, gender, marital status (single/coupled), time since diagnosis and recurrence status (no/yes)].

3. Results

In total, 229 people with sarcoma aged 18–85 completed the survey. Personal and cancer-specific characteristics are summarized in Table 1. The majority of respondents identified as female (n = 168, 73%), were married or in a long-term relationship (n = 165, 73%), employed (n = 133, 60%), and white (n = 216, 96%). Most participants had been diagnosed with soft tissue sarcoma (n = 165, 74%), received surgery (n = 177, 77%), and were not on active treatment (n = 189, 86%).

Table 1. Patient participant personal and cancer specific characteristics.

Characteristic	Mean (*SD*)/*n* (%)
Age	52.45 (14.7)
Gender	
Male	61 (27%)
Female	168 (73%)
Ethnicity	
White	216 (96%)
Other	9 (4%)
Marital status	
Married/in long-term relationship	165 (73%)
In a relationship but not cohabitating	12 (5%)
Single	32 (14%)
Widowed or divorced	19 (8%)
Employment status	
Employed full time or part time	133 (60%)
Permanently sick/disabled	26 (12%)
Retired	61 (28%)
Caregiver status	
Yes	64 (29%)
No	153 (71%)

Table 1. *Cont.*

Characteristic	Mean (*SD*)/*n* (%)
Type of sarcoma	
Soft tissue sarcoma	167 (74%)
Bone sarcoma	25 (11%)
GIST	33 (15%)
Time since diagnosis	
<1 year	14 (7%)
2 to 5 years	107 (51%)
6 to 10 years	57 (27%)
>10 years	31 (15%)
Treatments received *	
Surgery	177 (77%)
Radiotherapy	16 (7%)
Chemotherapy	28 (12%)
Other	21 (9%)
History of recurrence	
Yes	53 (24%)
No	149 (68%)
Unknown	17 (8%)
History of metastatic disease	
Yes	58 (26%)
No	146 (66%)
Unknown	17 (8%)
Amputation status	
Yes	18 (9%)
No	168 (79%)
Not applicable	26 (12%)
Disability status	
Not at all disabled	87 (45%)
Mildly to moderately disabled	69 (35%)
Severely or completely disabled	39 (20%)

* = multiple responses given; GIST: gastrointestinal stromal tumours.

Bivariate correlations between study variables are displayed in Table 2. Results from the multiple linear regression analysis are summarized in Table 3. Of the covariates included less time since diagnosis and reporting not having had a recurrence were significantly associated with greater FCR. Specific to the psychological impact of cancer, cognitive distress, and emotional distress were positively and significantly associated with higher levels of FCR, whereas cognitive avoidance and fighting spirit were not significantly associated. Psychological flexibility was negatively and significantly associated with lower levels of FCR. No significant associations were found between perceptions of the physical impact of cancer (i.e., disability status and impact on ADL). The model accounted for 56% (95% CI = 0.42, 0.61) of the variance in FCR among people with sarcoma in our sample.

Table 2. Bivariate correlations of study variables.

Variable	1	2	3	4	5	6	7	8	9	10	11	12
1. Age	-											
2. Gender	−0.07	-										
3. Marital status	−0.15	0.13 *	-									
4. Time since diagnosis	0.21 **	−0.04	0.00	-								
5. Recurrence status	0.17 *	0.10	−0.07	0.29 **	-							
6. Disability status	−0.07	−0.06	0.16 *	0.13	−0.02	-						
7. ADL impact	−0.09	−0.04	0.20 **	0.12	−0.02	0.72 **	-					
8. Cognitive distress	−0.09	0.03	0.09 *	−0.10	0.06	0.26 **	0.20 **	-				
9. Cognitive avoidance	0.04	0.09	0.09	0.01	0.07	−0.10	0.02	0.24	-			
10. Emotional distress	−0.19 **	0.22 **	−0.03	−0.16 *	0.12	0.24 **	0.27 **	0.65	0.19 **	-		
11. Fighting spirit	−0.12	0.01	−0.01	−0.06	0.05	0.20 **	0.23 **	0.10	0.24 **	0.27 **	-	
12. Psychological flexibility	0.27 **	−0.06	−0.17 *	0.12	0.10	−0.14 *	−0.24 **	−0.54	−0.22 **	−0.42 **	−0.03	-
13. FCR	−0.18 *	0.19 **	0.06	−0.23 **	0.17 *	0.20 **	0.19 **	0.57 **	0.15 *	0.69 **	0.25 **	−0.47 **

* Indicates $p < 0.05$ ** indicates $p < 0.01$; ADL: activities of daily living; FCR: fear of cancer recurrence.

Table 3. Regression results for FCR.

Factor	*b*	95% CI	*beta*	95% CI	sr^2	95% CI
(Intercept)	63.85 **	[27.28, 100.43]				
Age	0.01	[−0.19, 0.22]	0.01	[−0.11, 0.13]	0.00	[−0.00, 0.00]
Gender	2.84	[−3.71, 9.40]	0.05	[−0.07, 0.17]	0.00	[−0.01, 0.01]
Marital status +	−1.45	[−8.86, 5.96]	−0.02	[−0.14, 0.09]	0.00	[−0.00, 0.00]
Time since diagnosis	−0.63 *	[−1.21, −0.04]	−0.13	[−0.26, −0.01]	0.01	[−0.01, 0.04]
Recurrence status	−9.73 *	[−17.54, −1.93]	−0.15	[−0.27, −0.03]	0.02	[−0.01, 0.05]
Disability status	2.00	[−2.91, 6.91]	0.07	[−0.10, 0.23]	0.00	[−0.01, 0.01]
ADL impact	−2.03	[−6.39, 2.34]	−0.08	[−0.24, 0.09]	0.00	[−0.01, 0.01]
Cognitive distress	2.15 *	[0.26, 4.03]	0.19	[0.02, 0.35]	0.02	[−0.01, 0.04]
Cognitive avoidance	−0.51	[−1.95, 0.93]	−0.04	[−0.16, 0.08]	0.00	[−0.01, 0.01]
Emotional distress	5.39 **	[3.44, 7.34]	0.44	[0.28, 0.61]	0.09	[0.03, 0.15]
Fighting spirit	0.61	[−1.02, 2.23]	0.05	[−0.08, 0.18]	0.00	[−0.01, 0.01]
Psychological flexibility	−0.22 **	[−0.38, −0.05]	−0.19	[−0.33, −0.05]	0.02	[−0.01, 0.05]

Note. A significant *b*-weight indicates that the beta-weight and semi-partial correlation are also significant. *b* represents unstandardized regression weights. *beta* indicates the standardized regression weights. sr^2 represents the semi-partial correlation squared. *r* represents the zero-order correlation. Figures in brackets indicate the lower and upper limits of a confidence interval, respectively. + marital status was dichotomized to 1 = coupled; 2 = uncoupled (single, widowed, or divorced). * Indicates $p < 0.05$. ** indicates $p < 0.01$.

4. Discussion

In this study, we observed the prevalence of FCR among people with sarcoma in the UK, explored associated factors specific to the physical and psychological impact of cancer, and examined the role of psychological flexibility. Compared to the meta-analytic data of common cancer types [8], our study reported high levels of FCR. Participants expressed an interest in engaging in supportive interventions, which highlights the need for support among this population. Our findings demonstrate that the psychological impact of cancer, specifically cognitive and emotional distress, are significantly associated with greater levels of FCR, whereas the physical impact of cancer is insignificantly associated. Lastly, psychological flexibility was found to be negatively associated with FCR, representing a potential target for intervention development.

People with sarcoma have been historically underrepresented in FCR research. When examining the prevalence of FCR, mean scores reported across 10 different cancer types ranging from 39.8 among prostate cancer survivors to 113.5 among gynaecological cancer survivors, with standard deviations ranged from 18.6 to 28.2 using the FCRI [8]. The mean score among our sample of people with sarcoma was 91.4 (*SD* = 26.5). This is higher than the overall combined, weighted mean FCRI-Total score inclusive of all cancer types, which was reported as 65.2 (95% CI: 58.0–72.3) [8]. Only the estimate reported among the sample of gynaecological cancer patients [38] was higher than that observed in our sample. A higher FCR may be attributed to higher rates of recurrence in this population and the impact of sarcoma and its treatment on physical and psychological well-being and quality of life. Interestingly, our sample was predominantly female, in common with the gynaecological sample. Previous research has emphasised gender as a relevant demographic factor [4,6,7]; however, our own findings may be confounded by the unequal gender representation within our sample. Furthermore, a higher prevalence of anxiety has been noted in females [39]; thus, FCR, a form of state anxiety, may also be associated with differences in gender. Nonetheless, the reasons for gender-based differences within FCR severity have yet to be definitively identified and should be explored.

In addition to being identified as an unmet need among cancer survivors [6] and a research priority within the UK [10], FCR has been associated with increased costs to healthcare systems [40,41]. Thus, it is imperative that this field of research focuses on intervention development, testing, and optimization. Findings from this study support the need for interventions aimed at patients with sarcoma, given their high FCR. Furthermore, the high level of interest in supportive care interventions throughout the cancer care continuum highlights the demand for these interventions to be offered continuously starting from the time of diagnosis.

Recent efforts to develop interventions aimed at managing FCR have focused on a mind–body approach, which addresses the physical, cognitive, emotional, and behavioural aspects of the cancer experience [15]. Distress is a complex experience that results from the individual interplay of these aspects and is highly variable within and between individuals. The PIC scale used in this study to assess components of the psychological impact of cancer was developed using items previously forming the Mini-Mental Adjustment to Cancer Scale [42] to provide a brief and conceptually accessible tool with good psychometric properties. However, it is important to note that the psychometric properties of the fighting spirit sub-scale remain poor. In our sample, cognitive distress and emotional distress explained some of the unique variances of FCR. Thus, interventions should consider focusing on strategies that are aimed at reducing cognitive and emotional distress.

Psychological flexibility emerged as a factor associated with reduced FCR in our sample, highlighting the critical role this construct can play in facilitating psychological health and adjustment to cancer. Given the unique profile of people with sarcoma, it is essential that programs respond effectively to the challenges of this diagnosis in pursuit of mitigating long-term goals around health and wellbeing. This study is cross-sectional, so it does not provide any insight into the causality of associations; however, observations are in line with broader theory and evidence within this clinical population and provide justification for the continued exploration of this key construct.

Of the covariates included in our model, the time since diagnosis and recurrence status emerged as associated factors, with lower levels of FCR observed among those further from diagnosis, as well as those who had already experienced recurrence. The type of sarcoma also emerged as an associated factor, with higher levels of FCR observed in those with soft tissue sarcoma. Based on the prior literature [13,14] and clinical experience, it was hypothesised that marital status (as a form of support), age, and gender would account for some of the unique variances in FCR; however, this was not the case in these data. For example, recent work has demonstrated that younger patients may have greater severity of FCR compared to older patients [14]. It could be surmised that this is due to heightened levels of psychological distress in younger age groups, underdeveloped coping strategies, and concerns regarding developmental tasks related to the life stage at the time of diagnosis [14]. Given sample characteristics, findings specific to demographic and clinical characteristics may be due to a lack of representation and should be explored further.

Limitations

There are limitations to consider when interpreting the findings of this study. This study was cross-sectional; thus, causal and temporal inferences are not possible. The sampling approach was open to self-selection bias in that access was limited to those who used participating charities or social media. We recognise that our sample was predominately white and female. However, there is a similar proportion to the types of sarcoma that are represented in the UK. Future work should aim for a more balanced distribution of demographic characteristics through a more targeted sampling strategy. A longitudinal design would allow for the exploration of temporal changes and opportunities to explore causality between the variables assessed in this study. Whilst the selection of variables included in our analytic models was theoretically driven and accounted for the majority of the variance in FCR, there are additional constructs yet to be identified that may provide additional insight into mechanisms of action in FCR. Despite the aforementioned limitations, this is the largest study reporting FCR in patients presenting with sarcoma, and findings from this study provide valuable insight into this understudied population.

5. Conclusions

To our knowledge, this is the first study to explore the prevalence of FCR among those living with and beyond sarcoma in the UK. In comparison to other cancer types, a high prevalence and severity of FCR were observed. The psychological impact of sarcoma and the potential benefit of fostering psychological flexibility when aiming to address FCR demonstrates the importance of addressing cognitive and emotional distress after a sarcoma diagnosis. Interventions targeting these constructs, for example, acceptance and commitment therapy-based approaches, warrant further investigation and hold promise for managing FCR in both the short- and long-term.

Author Contributions: Conceptualization, R.M.T., L.S., L.A.F., M.L., C.G., R.W., J.W. and M.W.; methodology, A.P., L.S., N.J.H.-W. and R.M.T.; validation, A.P., L.S., N.J.H.-W., M.L. and R.M.T.; formal analysis, A.P., L.S., N.J.H.-W. and R.M.T.; investigation, A.P., R.M.T., J.B. and H.O.; resources, R.M.T. and L.S.; data curation, A.P., R.M.T., L.S. and N.J.H.-W.; writing—original draft preparation, A.P.; writing—review and editing, All authors; visualization, A.P., L.S., N.J.H.-W. and R.M.T.; supervision, R.M.T.; project administration, R.M.T., L.S., J.B. and H.O.; funding acquisition, R.M.T., L.S., L.A.F., M.L., C.G., R.W. and J.W. All authors have read and agreed to the published version of the manuscript.

Funding: This work was funded by Sarcoma UK (grant number SUK201.2019); R.M.T. is funded by UCLH Charity, L.A.F. is funded by Teenage Cancer Trust. M.W. acknowledges support from the Imperial Biomedical Research Centre (BRC). The views are the authors and do not necessarily reflect those of Sarcoma UK, UCLH Charity or Teenage Cancer Trust.

Institutional Review Board Statement: This study was approved by Birmingham City University Research Ethics Committee: Storey/#9678/sub2/R(A)/2021/Jul/BLSS FAEC.

Informed Consent Statement: Informed consent was obtained from all subjects involved in the study.

Data Availability Statement: Data are available on request from the authors.

Acknowledgments: We would like to thank all the participants for completing the survey, and Sarcoma UK, Teenage Cancer Trust and the Bone Cancer Research Trust for supporting the circulation of the survey.

Conflicts of Interest: The authors declare no conflict of interest.

References

1. Butow, P.; Sharpe, L.; Thewes, B.; Turner, J.; Gilchrist, J.; Beith, J. Fear of cancer recurrence: A practical guide for clinicians. *Oncology* **2018**, *32*, 32–38. [PubMed]
2. Lebel, S.; Ozakinci, G.; Humphris, G.; Mutsaers, B.; Thewes, B.; Prins, J.; Dinkel, A.; Butow, P. From normal response to clinical problem: Definition and clinical features of fear of cancer recurrence. *Support. Care Cancer* **2016**, *24*, 3265–3268. [CrossRef] [PubMed]
3. Simard, S.; Savard, J. Screening and comorbidity of clinical levels of fear of cancer recurrence. *J. Cancer Surviv.* **2015**, *9*, 481–491. [CrossRef] [PubMed]
4. Koch, L.; Jansen, L.; Brenner, H.; Arndt, V. Fear of recurrence and disease progression in long-term (>=5 years) cancer survivors–a systematic review of quantitative studies. *Psychooncology* **2013**, *22*, 1–11. [CrossRef]
5. Crist, J.V.; Grunfeld, E.A. Factors reported to influence fear of recurrence in cancer patients: A systematic review. *Psychooncology* **2013**, *22*, 978–986. [CrossRef]

6. Simard, S.; Thewes, B.; Humphris, G.; Dixon, M.; Hayden, C.; Mireskandari, S.; Ozakinci, G. Fear of cancer recurrence in adult cancer survivors: A systematic review of quantitative studies. *J. Cancer Surviv.* **2013**, *7*, 300–322. [CrossRef]
7. Yang, Y.; Li, W.; Wen, Y.; Wang, H.; Sun, H.; Liang, W.; Zhang, B.; Humphris, G. Fear of cancer recurrence in adolescent and young adult cancer survivors: A systematic review of the literature. *Psychooncology* **2019**, *28*, 675–686. [CrossRef]
8. Smith, A.B.; Costa, D.; Galica, J.; Lebel, S.; Tauber, N.; van Helmondt, S.J.; Zachariae, R. Spotlight on the fear of cancer recurrence inventory (FCRI). *Psychol. Res. Behav. Manag.* **2020**, *13*, 1257. [CrossRef]
9. Shaw, J.; Kamphuis, H.; Sharpe, L.; Lebel, S.; Smith, A.B.; Hulbert-Williams, N.; Dhillon, H.M.; Butow, P. Setting an international research agenda for fear of cancer recurrence: An online delphi consensus study. *Front. Psychol.* **2021**, *12*, 596682. [CrossRef]
10. Li, F.; Morgan, A.; McCullagh, A.; Johnson, A.; Giles, C.; Greenfield, D.; Crawford, G.; Gath, J.; Lyons, J.; Andreyev, J.; et al. Top 10 living with and beyond cancer research priorities. *Cancer Res.* **2019**, *79* (Suppl. 13), 3348. [CrossRef]
11. Trovik, C.S. Local recurrence of soft tissue sarcoma: A Scandinavian sarcoma group project. *Acta Orthop. Scand.* **2001**, *72*, 1–27.
12. Burns, J.; Wilding, C.P.; Jones, R.L.; Huang, P.H. Proteomic research in sarcomas–current status and future opportunities. In *Seminars in Cancer Biology*; Academic Press: Cambridge, MA, USA, 2020; Volume 61, pp. 56–70.
13. Custers, J.A.E.; Tielen, R.; Prins, J.B.; de Wilt, J.H.W.; Gielissen, M.F.M.; van der Graaf Winette, T.A. Fear of progression in patients with gastrointestinal stromal tumors (GIST): Is extended lifetime related to the sword of damocles? *Acta Oncol.* **2015**, *54*, 1202–1208. [CrossRef] [PubMed]
14. Richter, D.; Clever, K.; Mehnert-Theuerkauf, A.; Schönfelder, A. Fear of Recurrence in Young Adult Cancer Patients—A Network Analysis. *Cancers* **2022**, *14*, 2092. [CrossRef] [PubMed]
15. Hall, D.L.; Luberto, C.M.; Philpotts, L.L.; Song, R.; Park, E.R.; Yeh, G.Y. Mind-body interventions for fear of cancer recurrence: A systematic review and meta-analysis. *Psychooncology* **2018**, *27*, 2546–2558. [CrossRef] [PubMed]
16. Butow, P.N.; Turner, J.; Gilchrist, J.; Sharpe, L.; Smith, A.B.; Fardell, J.E.; Tesson, S.; O'Connell, R.; Girgis, A.; Gebski, V.J.; et al. Randomized trial of ConquerFear: A novel, theoretically based psychosocial intervention for fear of cancer recurrence. *J. Clin. Oncol.* **2017**, *35*, 4066–4077. [CrossRef]
17. Arch, J.J.; Mitchell, J.L.; Genung, S.R.; Fisher, R.; Andorsky, D.J.; Stanton, A.L. A randomized controlled trial of a group acceptance-based intervention for cancer survivors experiencing anxiety at re-entry ('valued living'): Study protocol. *BMC Cancer* **2019**, *19*, 89. [CrossRef]
18. Dawson, D.L.; Golijani-Moghaddam, N. COVID-19: Psychological flexibility, coping, mental health, and wellbeing in the UK during the pandemic. *J. Context. Behav. Sci.* **2020**, *17*, 126–134. [CrossRef]
19. Hayes, S.C.; Luoma, J.B.; Bond, F.W.; Masuda, A.; Lillis, J. Acceptance and commitment therapy: Model, processes and outcomes. *Behav. Res. Ther.* **2006**, *44*, 1–25. [CrossRef]
20. Gloster, A.T.; Klotsche, J.; Chaker, S.; Hummel, K.V.; Hoyer, J. Assessing psychological flexibility: What does it add above and beyond existing constructs? *Psychol. Assess.* **2011**, *23*, 970. [CrossRef]
21. Hulbert-Williams, N.J.; Storey, L.; Wilson, K.G. Psychological interventions for patients with cancer: Psychological flexibility and the potential utility of Acceptance and Commitment Therapy. *Eur. J. Cancer Care* **2015**, *24*, 15–27. [CrossRef]
22. Kashdan, T.B.; Rottenberg, J. Psychological flexibility as a fundamental aspect of health. *Clin. Psychol. Rev.* **2010**, *30*, 865–878. [CrossRef] [PubMed]
23. Masuda, A.; Tully, E.C. The role of mindfulness and psychological flexibility in somatization, depression, anxiety, and general psychological distress in a nonclinical college sample. *J. Evid.-Based Complement. Altern. Med.* **2012**, *17*, 66–71. [CrossRef]
24. McCracken, L.M.; Morley, S. The psychological flexibility model: A basis for integration and progress in psychological approaches to chronic pain management. *J. Pain* **2014**, *15*, 221–234. [CrossRef] [PubMed]
25. Hulbert-Williams, N.J.; Storey, L. Psychological flexibility correlates with patient-reported outcomes independent of clinical or sociodemographic characteristics. *Support. Care Cancer* **2016**, *24*, 2513–2521. [CrossRef]
26. Martins, A.; Bennister, L.; Fern, L.A.; Gerrand, C.; Onasanya, M.; Storey, L.; Wells, M.; Whelan, J.S.; Windsor, R.; Woodford, J.; et al. Development of a patient-reported experience questionnaire for patients with sarcoma: The Sarcoma Assessment Measure (SAM). *Qual. Life Res.* **2020**, *29*, 2287–2297. [CrossRef]
27. Simard, S.; Savard, J. Fear of Cancer Recurrence Inventory: Development and initial validation of a multidimensional measure of fear of cancer recurrence. *Support Care Cancer* **2009**, *17*, 241–251. [CrossRef]
28. Davis, A.M.; Wright, J.G.; Williams, J.I.; Bombardier, C.; Griffin, A.; Bell, R.S. Development of a measure of physical function for patients with bone and soft tissue sarcoma. *Qual. Life Res.* **1996**, *5*, 508–516. [CrossRef]
29. Suresh Nathan, S.; Tan Lay Hua, G.; Mei Yoke, C.; Mann Hong, T.; Peter Pereira, B. Outcome satisfaction in long-term survivors of oncologic limb salvage procedures. *Eur. J. Cancer Care* **2021**, *30*, e13377. [CrossRef]
30. Tang, M.H.; Castle, D.J.; Choong, P.F. Identifying the prevalence, trajectory, and determinants of psychological distress in extremity sarcoma. *Sarcoma* **2015**, *2015*, 745163. [CrossRef]
31. Davis, A.M.; Sennik, S.; Griffin, A.M.; Wunder, J.S.; O'Sullivan, B.; Catton, C.N.; Bell, R.S. Predictors of functional outcomes following limb salvage surgery for lower-extremity soft tissue sarcoma. *J. Surg. Oncol.* **2000**, *73*, 206–211. [CrossRef]
32. Schreiber, D.; Bell, R.S.; Wunder, J.S.; O'Sullivan, B.; Turcotte, R.; Masri, B.A.; Davis, A.M. Evaluating function and health related quality of life in patients treated for extremity soft tissue sarcoma. *Qual. Life Res.* **2006**, *15*, 1439–1446. [CrossRef]

33. Hulbert-Williams, N.J.; Hulbert-Williams, L.; Whelen, L.; Mulcare, H. The Psychological Impact of Cancer (PIC) Scale: Development and comparative psychometric testing against the Mini-MAC Scale in UK and Australian cancer survivors. *J. Psychosoc. Oncol. Res. Pract.* **2019**, *1*, e8. [CrossRef]
34. Francis, A.W.; Dawson, D.L.; Golijani-Moghaddam, N. The development and validation of the Comprehensive assessment of Acceptance and Commitment Therapy processes (CompACT). *J. Context. Behav. Sci.* **2016**, *5*, 134–145. [CrossRef]
35. Kroska, E.B.; Roche, A.I.; Adamowicz, J.L.; Stegall, M.S. Psychological flexibility in the context of COVID-19 adversity: Associations with distress. *J. Contextual. Behav. Sci.* **2020**, *18*, 28–33. [CrossRef]
36. Mallett, R.; Coyle, C.; Kuang, Y.; Gillanders, D.T. Behind the masks: A cross-sectional study on intolerance of uncertainty, perceived vulnerability to disease and psychological flexibility in relation to state anxiety and wellbeing during the COVID-19 pandemic. *J. Context. Behav. Sci.* **2021**, *22*, 52–62. [CrossRef]
37. Sevier-Guy, L.J.; Ferreira, N.; Somerville, C.; Gillanders, D. Psychological flexibility and fear of recurrence in prostate cancer. *Eur. J. Cancer Care* **2021**, *30*, e13483. [CrossRef]
38. Wijayanti, T.; Afiyanti, Y.; Rahmah, H.; Milanti, A. Fear of cancer recurrence and social support among Indonesian gynecological cancer survivors. *Arch. Oncol.* **2018**, *24*, 12–19. [CrossRef]
39. Farhane-Medina, N.Z.; Luque, B.; Tabernero, C.; Castillo-Mayén, R. Factors associated with gender and sex differences in anxiety prevalence and comorbidity: A systematic review. *Sci. Prog.* **2022**, *105*, 00368504221135469. [CrossRef]
40. Lebel, S.; Tomei, C.; Feldstain, A.; Beattie, S.; McCallum, M. Does fear of cancer recurrence predict cancer survivors' health care use? *Support. Care Cancer* **2013**, *21*, 901–906. [CrossRef]
41. Champagne, A.; Ivers, H.; Savard, J. Utilization of health care services in cancer patients with elevated fear of cancer recurrence. *Psycho-Oncol.* **2018**, *27*, 1958–1964. [CrossRef]
42. Kang, J.I.; Chung, H.C.; Kim, S.J.; Choi, H.J.; Ahn, J.B.; Jeung, H.C.; Namkoong, K. Standardization of the Korean version of Mini-Mental Adjustment to Cancer (K-Mini-MAC) scale: Factor structure, reliability and validity. *Psycho-Oncol. J. Psychol. Soc. Behav. Dimens. Cancer* **2008**, *17*, 592–597. [CrossRef] [PubMed]

Review

Opioid-Induced Sexual Dysfunction in Cancer Patients

Bartłomiej Salata, Agnieszka Kluczna and Tomasz Dzierżanowski *

Laboratory of Palliative Medicine, Department of Social Medicine and Public Health, Medical University of Warsaw, ul. Oczki 3, 02-007 Warsaw, Poland
* Correspondence: tomasz.dzierzanowski@wum.edu.pl

Simple Summary: Sexual disorders affect up to 80% of cancer patients, depending on the type of cancer, yet they are commonly overlooked and untreated. Opioid-induced sexual dysfunction (OISD) is reported in half of opioid users. The pathophysiology of OISD—still a subject for research—may include disorders of both the endocrine and nervous systems, expressed in, among other things, erectile dysfunction and declined sexual desire, sexual arousal, orgasm, and general satisfaction with one's sex life. The etiology of sexual dysfunction in cancer patients is usually multifactorial, so the management should be multifaceted and individualized by targeting pathophysiological factors. The treatment options for OISD are few and include testosterone replacement therapy, bupropion, opioid antagonists, phosphodiesterase type 5 inhibitors, plant-derived substances, and non-pharmacological treatments, although the evidence is insufficient. One of the treatment options may also be a choice of an opioid that is less likely to cause sexual dysfunction, yet further research is necessary.

Abstract: Sexual dysfunction is common in patients with advanced cancer, although it is frequently belittled, and thus consistently underdiagnosed and untreated. Opioid analgesics remain fundamental and are widely used in cancer pain treatment. However, they affect sexual functions primarily due to their action on the hypothalamus–pituitary–gonadal axis. Other mechanisms such as the impact on the central and peripheral nervous systems are also possible. The opioid-induced sexual dysfunction includes erectile dysfunction, lack of desire and arousal, orgasmic disorder, and lowered overall sexual satisfaction. Around half of the individuals taking opioids chronically may be affected by sexual dysfunction. The relative risk of sexual dysfunction in patients on chronic opioid therapy and opioid addicts increased two-fold in a large meta-analysis. Opioids differ in their potential to induce sexual dysfunctions. Partial agonists and short-acting opioids may likely cause sexual dysfunction to a lesser extent. Few pharmaceutical therapies proved effective: testosterone replacement therapy, PDE5 inhibitors, bupropion, trazodone, opioid antagonists, and plant-derived medicines such as Rosa damascena and ginseng. Non-pharmacological options, such as psychosexual or physical therapies, should also be considered. However, the evidence is scarce and projected primarily from non-cancer populations, including opioid addicts. Further research is necessary to explore the problem of sexuality in cancer patients and the role of opioids in inducing sexual dysfunction.

Keywords: cancer; pain management; opioid; sexual dysfunction; sexual disorder; erectiledysfunction

Citation: Salata, B.; Kluczna, A.; Dzierżanowski, T. Opioid-Induced Sexual Dysfunction in Cancer Patients. *Cancers* **2022**, *14*, 4046. https://doi.org/10.3390/cancers14164046

Academic Editor: António Araújo

Received: 30 May 2022
Accepted: 17 August 2022
Published: 22 August 2022

1. Introduction

Sexuality is an essential aspect of life, also for cancer patients [1]. Despite this, many of them believe that they do not receive proper care in this sphere of life [2], and only one in ten cancer patients are asked by a doctor about the quality of their sex life [3]. The quality of the sexual life of these patients has often deteriorated, and there are various reasons for this, such as pain and other physical symptoms, deformities of the body due to cancer or after medical interventions, the feeling of being unattractive, medications, or the lack of privacy conditions in long-term care facilities [4–6]. As opioids are often used in this group of patients, their influence on libido and the quality of sexual function is essential. This publication aims to synthesize the current knowledge on the relationship between the use of opioids and the occurrence of sexual dysfunctions.

2. Etiology and Pathophysiology

The etiology of sexual dysfunctions is often multifactorial, making it challenging to distinguish these dysfunctions and then clearly set a proper diagnosis. The fundamental problem is whether a given disorder is due to organic dysfunction or is psychological in origin. The International Classification of Diseases 11th Revision (ICD-11) divides sexual disorders into sexual dysfunctions and sexual pain disorders [7]. Additional coding (HA40.2) is recommended for the conditions associated with the use of opioids (Table 1).

Table 1. ICD-11 classification of sexual disorders possibly associated with opioids [7].

17 Conditions Related to Sexual Health
Sexual dysfunctions HA00 Hypoactive sexual desire dysfunction HA01 Sexual arousal dysfunctions HA02 Orgasmic dysfunctions HA03 Ejaculatory dysfunctions HA0Y Other specified sexual dysfunctions HA0Z Sexual dysfunctions, unspecified Sexual pain disorders HA20 Sexual pain-penetration disorder HA2Y Other specified sexual pain disorders HA2Z Sexual pain disorders, unspecified
HA40 Aetiological considerations in sexual dysfunctions and sexual pain disorders HA40.2 Associated with use of psychoactive substance or medication

In cancer patients, occurrence of sexual dysfunctions depends on the primary tumor location and treatment used. It is associated with, among other things, damage to the vascularization or innervation of the genital organs and their damage or post-surgical scarring, radiotherapy, high-dose chemotherapy, hormonal disorders, chronic fatigue and pain [6,8]. It results in (1) loss of desire; (2) genitourinary atrophy, dryness and pain, (3) difficulty experiencing pleasure and reaching orgasm, and (4) erectile dysfunction. Additionally, they may also be a sequela of psychological problems, such as unacceptance of one's body image. On top of that, stomas in patients with gastrointestinal or urinary tract cancers may impede sexual activity as well.

2.1. The Role of Hormones

Hormones are an essential factor in modulating sexual functions. The role of testosterone, dehydroepiandrosterone (DHEA), and prolactin are best known, while the functions of estrogen, oxytocin, and progesterone are less clear. Testosterone in men increases libido, the degree of excitement, sexual satisfaction, the degree of penile stiffness, and the time of erectile response [9]. In women, it increases desire, excitement, vaginal congestion, and orgasm. These effects in women may, to some extent, be the effect of testosterone conversion to dihydrotestosterone and estradiol [10,11].

DHEA, produced by the adrenal glands, works mainly as a prohormone, and testosterone, dihydrotestosterone, estrone, and estriol are formed as a result of its subsequent transformations. It has a positive effect on desire, arousal, frequency of sexual thoughts, and satisfaction with the physical and emotional aspects of sexuality, among other things [12].

Prolactin most likely inhibits sexual functions. It delays ejaculation and reduces craving. Its concentration increases after the onset of orgasm, and it is responsible for the subsequent refractory period (feeling of sexual satiety and inhibited sexual behavior) [13]. Estrogens affect proper vaginal lubrication in women, increase excitement and, through muscle relaxation, prevent dyspareunia [14]. Oxytocin at high doses probably reduces sexual arousal and causes a refractory period after orgasm in males, whereas, at lower doses, it probably stimulates sexual behavior. Its secretion increases during sexual arousal and probably also stimulates the occurrence of erection—its concentration in the central nervous system is reduced in men with erectile dysfunction [15]. Progesterone is poorly understood in this respect. It probably inhibits sexual functions [14].

The effects of opioids on the endocrine system are probably mainly related to their effects on the hypothalamic–pituitary–gonadal axis (Figure 1). The mu-opioid receptors (MOR) are found in the hypothalamus [16], pituitary [17], testes [18], and ovaries [19]. Therefore, inhibition by opioid agonists of both the hypothalamus's pulsatile secretion of the gonadotropin-releasing hormone (gonadoliberin; GnRH) causing hypogonadotropic hypogonadism, and the testosterone secretion directly in the testes occur. Another mechanism is related to the increase in prolactin secretion by the pituitary gland, which reduces the secretion of testosterone [20]. In addition, the production of DHEA in the adrenal cortex may be reduced [20], and a study on rats in which morphine was used shows that there may be an increase in the mRNA expression of enzymes that break down testosterone [21].

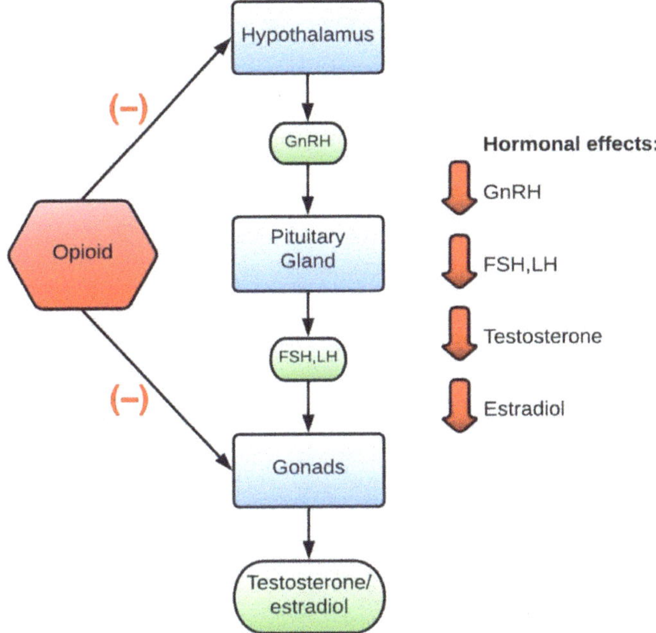

Figure 1. The impact of opioids on the hypothalamus−pituitary−gonadal axis.

2.2. Sexual Desire

Sexual desire can be defined as a subjective psychological state related to the initiation and maintenance of sexual behavior caused by internal or external factors [22] or as the sum of factors that motivate or demotivate a person to engage in sexual activity [23]. It depends on biological, psychological, and social aspects. The biological aspects comprise, inter alia, the actions of the endocrine system and neurotransmitters. The stimulating neurotransmitters include dopamine, norepinephrine, oxytocin and melanocortins (beta-endorphin, adrenocorticotropic hormone, and alpha-melanotropin). The inhibitory ones include serotonin, endogenous cannabinoids, and opioids [24]. In the DSM-5 (Diagnostic and Statistical Manual of Mental Disorders) classification, there is a concept of Female Sexual Interest–Arousal Disorder (FSIAD), defined as the absence or a significant reduction in sexual interest or arousal. It consists of six domains: (1) sexual activity, (2) sexual or erotic thoughts or fantasies, (3) initiation of sexual activity, (4) excitement or pleasure during sexual activity, (5) sexual interest or arousal, and (6) genital or non-genital sensations during sexual activity. The absence, or a significant reduction, of at least three of them for at least six months and 75–100% of the time allows the diagnosis of the disorder [25].

In men, a similar disorder, in the DSM-5 classification, is the reduction of sexual desire in men, defined as Male Hypoactive Sexual Desire Disorder (MHSDD) [25].

2.3. Erection

Male erection is a complex neurovascular process dependent on balancing inhibitory and stimulating factors (sympathetic and parasympathetic systems, respectively). The reflex is caused by the stimulation from the sacral (S2–S4) section of the spinal cord, triggered by stimulation of penile afferents or by higher centers of the central nervous system upon visual, olfactory, and tactile stimulation, or imagination [26–28]. It causes the extension of the penile arteries, and simultaneous pressure on the venous vessels causes blood stagnation in the corpus cavernosum of the penis and, as a result, an erection.

Erectile dysfunction (ED) in a man can be diagnosed when during almost all occasions of sexual activity (75–100% on average), at least one of the following three symptoms occurs: (1) marked difficulty in obtaining an erection during sexual activity, (2) marked difficulty in maintaining an erection until completion of sexual activity, (3) marked decrease in erectile rigidity [25]. The incidence of ED in the general population ranges from 10–15% in men aged 40–49 to 50–70% in men aged 60–79 years [29,30] Risk factors include age, cardiovascular diseases, diabetes, obesity, smoking, and depression [31,32]. The incidence of ED in the cancer patient population may be around 29% at the time of diagnosis and 43% after treatment, but it is strongly dependent on the type of cancer and may be up to 80–90% for prostate, anus or colorectal cancers [33]. The effect of opioids on male erection is not well known. Several mechanisms that may coincide are suggested. First of all, by affecting MOR located in the area of the paraventricular nucleus of the hypothalamus, these drugs can inhibit the synthesis of nitric oxide (NO), which is part of the neurological pathway responsible for inducing an erection [27,34]. Another mechanism is related to the aforementioned reduction of testosterone concentration by inhibiting GnRH secretion in the hypothalamus and directly inhibiting testosterone secretion in the testes [20]. Based on studies in animal models, it can be assumed that there are other mechanisms affecting the peripheral nervous system [35].

2.4. Orgasm

A woman's orgasm can be defined as "a variable, transient peak sensation of intense pleasure creating an altered state of consciousness usually with an initiation accompanied by involuntary, rhythmic contractions of the pelvic striated circumvaginal musculature often with concomitant uterine and anal contractions and myotonia that resolves the sexually-induced vasocongestion (sometimes only partially) and myotonia usually with an induction of well-being and contentment" [36]. In men, orgasm is related to ejaculation and results from stimulation of the vulvar nerve by increasing pressure in the posterior part of the urethra during ejaculation, stimulation of verumontanum, and contraction of the urethra and accessory sexual organs [37]. Physiological changes during a male orgasm are similar to those in the female body: contractions of the pelvic muscles and anal sphincter, hyperventilation, tachycardia, and increased blood pressure [38]. An orgasmic disorder listed in the ICD-10 classification is inhibition of orgasm (anorgasmia). It occurs when orgasm is not achieved despite high levels of excitement, or the intensity of the feeling of orgasm is reduced or delayed. It is more common in women than men [25].

The data on a prevalence of orgasmic disorders in cancer patients and mechanisms of influence of opioids on orgasm is scarce. In women it may be connected with low testosterone level [20].

3. Diagnostic Tools

Apart from the regular history, the most frequently used tools for diagnosing and assessing the effectiveness of sexual disorders treatment are self-report techniques consisting of the patient's self-assessment by answering the questions posed in a given questionnaire. They facilitate a conversation about the sexual domain of the patient's life and consecutive diagnostics. They can be especially beneficial for practitioners inexperienced in collecting sexological history. None of the available tools refer to opioid use, nor have they been validated in the opioid users' population.

3.1. Female Sexual Function Index (FSFI)

For women, the Female Sexual Function Index (FSFI) can be used (Table 2). It consists of 19 items combined in six areas: (1) desire, (2) arousal, (3) lubrication, (4) orgasm, (5) sexual satisfaction, and (6) pain, assessed for the last four weeks [39]. This questionnaire has been validated in cancer survivors [40] and is suggested by the Cancer Patient-Reported Outcomes Measurement Information System (PROMIS) Sexual Function Committee [41]. Additionally, the Brief Sexual Symptom Checklist for Women (SSFF-A) can be used as a primary screening tool of women with cancer [42].

Table 2. Areas assessed by the diagnostic tools in women (Female Sexual Function Index) and men (International Index of Erectile Function) [39,43].

Female Sexual Function Index (FSFI)	International Index of Erectile Function (IIEF)
Desire	Erectile function
Arousal	Orgasmic function
Lubrication	Sexual desire
Orgasm	Intercourse satisfaction
Sexual satisfaction	Overall satisfaction
Pain	

3.2. International Index of Erectile Function (IIEF)

In men, the International Index of Erectile Function (IIEF), which is also suggested by the PROMIS Sexual Function Committee [41], can be used. It is a 15-item questionnaire assessing five areas (Table 2) in the four weeks prior to testing [43]. The IIEF-5 questionnaire, i.e., a shortened version of the IIEF consisting of five items, may be more convenient in clinical practice [44].

Additionally, there are questionnaires developed for specific cancer populations: University of California-Los Angeles Prostate Cancer Index (UCLA PCI) [45], Sexual Function-Vaginal Changes Qeustionnaire [46] for the assessment after gynecological cancer and FSFI adaption for breast cancer patients (FSFI-BC) [47].

4. Epidemiology

4.1. Sexual Disorders in Cancer Patients

Sexual disorders are a common problem in cancer patients, and their occurrence depends on the primary tumor location and treatment used [6,8]. On average, they affect more than half of patients, but they vary widely depending on specific cancers. In female breast cancer, it may be around 66% [48], 65–90% in colorectal cancers [33,48], 78% in gynecological cancers [48] and up to 80% in prostate cancer [49].

4.2. Sexual Dysfunction in Patients Taking Opioids

There has been little research on sexual dysfunction with opioids in people with cancer as yet, so most of the evidence comes from studies in other patient populations.

In a case–control study by Rajagopal et al. [50], the incidence of hypogonadism and sexual dysfunction in men on chronic opioid therapy for cancer pain was assessed. The study and control groups consisted of 20 men each, who had taken at least 200 mg/day of an oral morphine equivalent (OME) dose for at least one year, or placebo, respectively. The total testosterone, FSH, and LH concentrations were measured, and the quality of sexual function was assessed using the Sexual Desire Inventory (SDI) questionnaire. The mean concentration of all three hormones was two to three times lower in the study than in a control group. The mean SDI score was 18.5 vs. 40 in the control group, and it was statistically significant ($p = 0.01$).

In a study by Venkatesh et al. [51], the sexual function of 100 men with a history of at least one year of opioid dependence vs. 50 men in the control group was assessed. Forty-eight percent, vs. eight percent in the control group, had sexual dysfunction according to the Arizona Sexual Experiences Scale (ASEX). Of them, 45% had ED, defined as less than 25 points on the IIEF-5 scale, significantly more often than in the control group (16%). Ninety-two percent of the study group had impairment of at least one of the five functions tested on the IIEF-5 scale, vs. sixteen percent in the control group. Other sexual functions were also impaired vs. the control group: desire (41% vs. 8%), sexual arousal (29% vs. 2%), the ability to achieve orgasm (21% vs. 0%), and satisfaction with orgasm (25% vs. 6%).

In a meta-analysis by Zhao et al. [52] of nine cross-sectional studies and one cohort study involving 8829 patients on chronic opioid therapy, or heroin- or opium-addicted, the relative risk (RR) of ED approached 2. In addition, a strong association between long-term opioid use (>3 years) and ED was reported (RR 2.25), also in men under 50 years (RR 2.21).

Deyo et al. [53] investigated the frequency of prescribing medications or testosterone replacement therapy (TRT) for ED in 11,327 men prescribed opioids for lower back pain. It appeared to be associated with the doses and duration of opioid use, and in the case of long-term opioid use (>120 days or at least ten prescriptions over >90 days), it equaled 13.1% and was higher (19%) in the presence of a high daily opioid dose (OME > 120 mg).

In a prospective observational study (Ajo et al. [54]), opioid-induced sexual disorders were reported in 33% of patients. ED was present in 27.6%, and in 64% of cases it was assessed as severe.

In a study on men by Rubinstein et al. [55], use of long-acting opioids was connected with higher frequency of hypogonadism than in men using short-acting opioids—74% (34/46) vs. 34% (12/35). After controlling for daily dosage and body mass index, men on long-acting opioids had 4.78 times greater odds of becoming hypogonadal than men on short-acting opioids. The studies on long-acting (sustained release) opioids seem to be of great importance, as opioids of that type are a base of cancer pain treatment.

4.2.1. Tramadol

In a case–control study, Hashim et al. [56] compared sexual function in a group of opioid addicts in tramadol, heroin, and control groups of 30 patients each. The mean scores on the IIEF-5 scale regarding erection were 8.6, 15, and 29.9, respectively, and the differences between the groups were statistically significant ($p < 0.001$). Compared to placebo, tramadol worsened orgasm ($p = 0.003$), desire ($p = 0.002$), and overall satisfaction ($p < 0.001$). The concentrations of free testosterone ($p = 0.041$) and LH ($p = 0.004$) were also significantly reduced versus the control group. However, all the assessed indicators were significantly better among tramadol addicts than heroin users ($p < 0.001$).

In a small study by Kabbash et al. [57], ED occurred in 44% of addicts taking tramadol and 10% in the placebo control group ($p = 0.001$). The occurrence of ED was dose-related and equaled 14.3% in the individuals taking $\leq$ 400 mg/day, 48.4% in the case of 400–1000 mg/day, and 50% when the dose exceeded 1000 mg/day, although the differences between the groups were not statistically significant ($p = 0.23$). Notably, a higher incidence of ED was reported when the daily dose exceeded the maximum recommended for regular medical use. Furthermore, the incidence of ED depended on the duration of tramadol use and was 20% for 1–2 years, 30.4% for 2–5 years, and 63.6% for more than 5 years ($p = 0.04$). Serum testosterone concentration was significantly lower in tramadol addicts than in the control group ($p = 0.001$), whereas serum prolactin concentration was significantly higher ($p = 0.001$). Consistently, a higher incidence of decreased libido was noticed in the group taking tramadol than in the control group (48% vs. 16.7%; $p = 0.005$). Noteworthily, 20% of the individuals in this study reported off-label tramadol use for the prevention of premature ejaculation [57]. Based on animal models, low tramadol doses may stimulate ejaculation, while high doses have an inhibitory effect [58]. According to a few studies, it may be effective in treating premature ejaculation [59]; however, the quality of evidence is low. Interestingly, one of reported adverse events was erectile dysfunction.

4.2.2. Morphine

So far, there is no clinical evidence of morphine concerning sexual dysfunction. However, based on animal models, the intraperitoneal administration of morphine to male rats reduces the likelihood of erections proportionally to its dose, and the effect was reversed by naloxone [60]. Interestingly, the administration of naloxone alone at the lowest dose tested (0.1 mg/kg) also inhibited erection, and this effect was not observed at higher doses (1 and 10 mg/kg).

Additionally, the impact of morphine on sexual functions can be extrapolated from diamorphine (diacetylmorphine, heroine), a pro-drug deacetylated to morphine as an active molecule.

4.2.3. Methadone and Buprenorphine

Methadone and buprenorphine, especially as an opioid substitution therapy, are collectively the subject of the largest number of research papers on this topic. In a meta-analysis of 16 studies on the prevalence of sexual dysfunction among male patients on methadone and buprenorphine therapy by Yee et al. [61], 52% of the 1570 methadone-treated patients reported sexual dysfunction. ED was assessed in 12 studies and occurred in 46% of patients. Decreased desire or libido was observed in 51% of patients in four studies collectively.

In the meta-analysis by Zhao et al. [52], the use of methadone was associated with a lower risk of ED (RR = 1.82) compared to other opioids (heroin and opium; RR = 2.04) in this study, which may explain the improvement in sexual function after starting methadone replacement therapy.

In the above-cited meta-analysis by Yee et al. [61], the incidence of sexual dysfunction in patients taking buprenorphine was 24%. A meta-analysis of four studies comparing the incidence of these disorders in patients taking buprenorphine and methadone revealed that methadone was associated with a five-fold higher risk of sexual dysfunction (OR = 4.95). Another study by this author [62] showed that patients taking buprenorphine (average dose 2.4 mg/day) reported a higher degree of sexual desire than patients using methadone (average dose 74.5 mg/day)—7.6 and 6.1 on the IIEF scale, respectively—and higher testosterone concentrations (18.5 vs. 12.5 nmol/L).

In turn, in a small study by Hallinan et al. [63], the mean IIEF score did not differ significantly between buprenorphine and control groups and was better than in the methadone-treated patients (61 vs. 50).

In a study of 258 women, mean age of 38 years, taking methadone (mean 61 mg/day) or buprenorphine (mean 11 mg/day) as opioid maintenance therapy, 56% of patients reported sexual dysfunction [64]. Notably, the patients with sexual dysfunction were characterized, inter alia, by older age, lower levels of education, higher doses of methadone, and worse mental health than patients without sexual dysfunction.

The lower incidence of sexual dysfunction in buprenorphine use compared to other opioids may be explained by its partial agonist/antagonist mode of action [65].

4.2.4. Tapentadol

In a 12-week randomized clinical study, Baron et al. [66] compared the safety of prolonged-release tapentadol with prolonged-release oxycodone/naloxone tablets in patients with severe chronic low back pain with a neuropathic component. The participants were <64 years old and had an initial testosterone concentration within the normal range.

Mean doses of opioids were 362 mg for tapentadol and 83 mg for oxycodone/naloxone. In 45.5% of the oxycodone/naloxone patients, testosterone concentrations decreased below the norm, while this only occurred in 10.5% of the tapentadol group.

An important limiting factor for the presented studies is a lack of control for confounding factors (e.g., pain, mental health, physical health, quality of life) between opioid and non-opioid groups. Additionally, most of the evidence is derived from patients addicted to opioids, and this population may not be representative for patients taking opioids for cancer pain as addiction differs from appropriate prescription opioid use and both populations are also different.

5. Treatment of Opioid-Induced Sexual Dysfunction

There are few treatment options for sexual dysfunction (Table 3), and most of the studies were based on male groups only. The choice of an opioid with a less-negative impact on the endocrine system, such as buprenorphine or tapentadol, seems to be a vital element of management, but clinical evidence for the effectiveness of such a strategy is scarce.

Table 3. Pharmacological treatment options for opioid-induced sexual dysfunction.

Pharmacological Treatment Options
Testosterone Replacement Therapy
Bupropion
Trazodone
Opioid antagonist (naltrexone, nalmefene)
Phosphodiesterase type 5 inhibitors
Plant-derived medicines (damask rose oil, ginseng)

5.1. Testosterone Replacement Ttherapy

In a medium-sized, 3-year prospective observational study, Ajo et al. [54] verified the effectiveness of testosterone replacement therapy (TRT) and the phosphodiesterase type 5 inhibitor (PDE5i) for the treatment of ED in patients using opioids (the mean duration of opioid therapy was 5 years and 6 months, the mean opioid dose was 107.1 mg/day OME). After six months of therapy, 42% of patients experienced a significant improvement as measured by the IIEF questionnaire. A positive correlation was also observed between the improvement in the IIEF score and the quality of sexual life and a reduction of anxiety. However, one systematic review [67] suggests that TRT may be effective only in improving pain and emotional functioning, but not sexual function. The quality of evidence is low, and further research is needed.

5.2. Bupropion and Trazodone

A small-sized, randomized, double-blind, placebo-controlled trial investigated the efficacy of 50 mg bupropion twice a day in the treatment of sexual dysfunction in men using methadone (mean 70 mg for 46 months) [68]. The mean erection quality score measured on the IIEF-15 scale improved from 18.1 to 22.6, and sexual satisfaction from 7 to 8.8, with statistically significant differences, compared to the control group ($p = 0.03$ and $p = 0.02$, respectively).

The efficacy of trazodone on erectile dysfunction in men on methadone maintenance therapy was evaluated in a small study [69]. Patients received 50 mg/day of trazodone for four days, then the dosage was increased to 100 mg/day and maintained for six weeks. The mean score on the Erectile Dysfunction Intensity Scale (EDIS) improved from 12.21 to 16.78 ($p < 0.05$; 5–10—severe ED, 11–15—moderate ED, 16–20—mild ED, 21–25—no ED)

5.3. Opioid Receptor Antagonists

Another therapeutic option for ED is the use of an opioid receptor antagonist. In a study by van Ahlen et al. [70], the efficacy of naltrexone in the treatment of idiopathic ED has been investigated. Patients in the study group took 25 mg of naltrexone for four weeks followed by 50 mg of naltrexone for another four weeks. An improvement in the number of morning spontaneous erections was reported for both the 25 mg and 50 mg naltrexone treatment arms. Such an improvement was not seen in the placebo group. Neither libido, nor FSH, LH, or testosterone, changed in either the research or the placebo groups. The authors attribute the erectile-stimulating effect of naltrexone in the group of patients not taking opioids to the antagonization of endogenous opioids, which may inhibit sexual function and lower LH release. An alternative to naltrexone may be nalmefene, which may also increase FSH, LH, and testosterone levels [71].

In one study the frequency of sexual dysfunctions in patients on buprenorphine and naltrexone maintenance therapy was compared [72]. Erection difficulty and reduction of sexual desire was reported more often by patients in the naltrexone group than in the buprenorphine group (66.7% vs. 43.3% and 46.7% vs. 33.3%) when asked about experiencing them "ever". However, when asked about the last month, both groups reported similar frequency of erection difficulty, and the reduction of sexual desire was higher in the buprenorphine group (10% vs. 26.7%). Additionally, sexual functions were similar between the two groups when measured with Brief Male Sexual Functioning Inventory (BMSFI) and asked about the last month. The quality of evidence is rather poor.

There are no studies on peripheral restricted opioid receptor antagonists as a treatment for opioid-induced sexual dysfunction.

5.4. *Plant-Derived Medicines*

There is limited evidence for the effectiveness of herbal products, including damask rose oil. It has been tested in a small, randomized, double-blind, placebo-controlled trial of women undergoing methadone replacement therapy [73]. Improvement was demonstrated in all domains of the FSFI scale after eight weeks of treatment.

Another randomized controlled trial shows that ginseng may also have a positive effect on methadone-induced sexual function, both in men and women [74].

5.5. *Non-Pharmacological Methods*

There are no studies on the effectiveness of non-pharmacological methods for the treatment of opioid-induced sexual dysfunction. However, as these drugs may be only one of the etiological factors, the use of standard non-pharmacological methods in these disorders should be considered. These include, among others, psychotherapy or psychosexual therapy [75,76], physiotherapy [77], physical therapy with the use of instruments [78] and the use of lubricants in women [75], and the use of erection aids in men [79].

6. Conclusions

Sexual disorders are common among cancer patients undergoing opioid therapy, yet they are frequently overlooked and untreated. The etiology of these disorders is multifactorial and, therefore, difficult to interpret, as the disease itself and its treatment could be confounding variables not controlled in the presented studies. However, despite the scarce evidence, opioids may be one of the etiological factors not widely known so far. The pathophysiology of this phenomenon is not yet clear and should be a subject for future research. It may include disorders of both the endocrine and nervous systems, resulting in, among other things, ED and deteriorating sexual desire, sexual arousal, orgasm, and general satisfaction with sex life. The treatment options for opioid-induced sexual dysfunction include testosterone replacement therapy, phosphodiesterase type 5 inhibitors, bupropion, opioid antagonists, and plant-derived means, such as damask rose oil and ginseng. However, most of them were studied in males, and further research on treatment for women is needed. Although opioids are believed to be an important cause, the etiology is usually multifactorial, so management should be multifaceted and individualized by targeting pathophysiological factors. One of the treatment options may also be a choice of an opioid that is less likely to cause sexual dysfunction, yet further research is necessary.

Author Contributions: Conceptualization, B.S. and T.D.; evidence query, B.S.; writing—original draft preparation, B.S.; writing—review and editing, A.K. and T.D.; visualization, B.S. and TD.; supervision, T.D.; project administration, T.D.; funding acquisition, T.D. All authors have read and agreed to the published version of the manuscript.

Funding: This research received no external funding.

Conflicts of Interest: The authors declare no conflict of interest.

References

1. Rosales, A.R.; Byrne, D.; Burnham, C.; Watts, L.; Clifford, K.; Zuckerman, D.S.; Beck, T. Comprehensive Survivorship Care with Cost and Revenue Analysis. *J. Oncol. Pract.* **2013**, *10*, e81–e85. [CrossRef] [PubMed]
2. Schover, L.R.; van der Kaaij, M.; van Dorst, E.; Creutzberg, C.; Huyghe, E.; Kiserud, C.E. Sexual Dysfunction and Infertility as Late Effects of Cancer Treatment. *Eur. J. Cancer Suppl.* **2014**, *12*, 41–53. [CrossRef] [PubMed]
3. Lemieux, L.; Kaiser, S.; Pereira, J.; Meadows, L.M. Sexuality in Palliative Care: Patient Perspectives. *Palliat. Med.* **2004**, *18*, 630–637. [CrossRef]
4. Ussher, J.M.; Perz, J.; Gilbert, E. Perceived Causes and Consequences of Sexual Changes after Cancer for Women and Men: A Mixed Method Study. *BMC Cancer* **2015**, *15*, 268. [CrossRef] [PubMed]
5. Shell, J.A. Sexual Issues in the Palliative Care Population. *Semin. Oncol. Nurs.* **2008**, *24*, 131–134. [CrossRef]
6. Bond, C.B.; Jensen, P.T.; Groenvold, M.; Johnsen, A.T. Prevalence and Possible Predictors of Sexual Dysfunction and Self-Reported Needs Related to the Sexual Life of Advanced Cancer Patients. *Acta Oncol.* **2019**, *58*, 769–775. [CrossRef] [PubMed]
7. ICD-11 for Mortality and Morbidity Statistics. Available online: https://icd.who.int/browse11/l-m/en (accessed on 22 May 2022).
8. Schover, L.R. Sexual Quality of Life in Men and Women after Cancer. *Climacteric* **2019**, *22*, 553–557. [CrossRef]
9. Gannon, J.R.; Walsh, T.J. Testosterone and Sexual Function. *Urol. Clin. N. Am.* **2016**, *43*, 217–222. [CrossRef]
10. Basson, R.; Brotto, L.A.; Petkau, A.J.; Labrie, F. Role of Androgens in Women's Sexual Dysfunction. *Menopause* **2010**, *17*, 962–971. [CrossRef]
11. Davis, S.R.; Wahlin-Jacobsen, S. Testosterone in Women-the Clinical Significance. *Lancet Diabetes Endocrinol.* **2015**, *3*, 980–992. [CrossRef]
12. Traish, A.M.; Paco Kang, H.; Saad, F.; Guay, A.T. Dehydroepiandrosterone (DHEA)—A Precursor Steroid or an Active Hormone in Human Physiology. *J. Sex. Med.* **2011**, *8*, 2960–2982. [CrossRef] [PubMed]
13. Krüger, T.H.C.; Haake, P.; Hartmann, U.; Schedlowski, M.; Exton, M.S. Orgasm-Induced Prolactin Secretion: Feedback Control of Sexual Drive? *Neurosci. Biobehav. Rev.* **2002**, *26*, 31–44. [CrossRef]
14. Stuckey, B.G.A. Female Sexual Function and Dysfunction in the Reproductive Years: The Influence of Endogenous and Exogenous Sex Hormones. *J. Sex. Med.* **2008**, *5*, 2282–2290. [CrossRef] [PubMed]
15. Lee, H.J.; Macbeth, A.H.; Pagani, J.H.; Scott Young, W. Oxytocin: The Great Facilitator of Life. *Prog. Neurobiol.* **2009**, *88*, 127–151. [CrossRef] [PubMed]
16. Bedos, M.; Antaramian, A.; Gonzalez-Gallardo, A.; Paredes, R.G. Paced Mating Increases the Expression of μ Opioid Receptors in the Ventromedial Hypothalamus of Male Rats. *Behav. Brain Res.* **2019**, *359*, 401–407. [CrossRef]
17. Carretero, J.; Bodego, P.; Rodríguez, R.E.; Rubio, M.; Blanco, E.; Burks, D.J. Expression of the μ-Opioid Receptor in the Anterior Pituitary Gland Is Influenced by Age and Sex. *Neuropeptides* **2004**, *38*, 63–68. [CrossRef]
18. Estomba, H.; Muñoa-Hoyos, I.; Gianzo, M.; Urizar-Arenaza, I.; Casis, L.; Irazusta, J.; Subirán, N. Expression and Localization of Opioid Receptors in Male Germ Cells and the Implication for Mouse Spermatogenesis. *PLoS ONE* **2016**, *11*, e0152162. [CrossRef]
19. Kaminski, T. The Involvement of Protein Kinases in Signalling of Opioid Agonist FK 33-824 in Porcine Granulosa Cells. *Anim. Reprod. Sci.* **2006**, *91*, 107–122. [CrossRef]
20. Katz, N.; Mazer, N.A. The Impact of Opioids on the Endocrine System. *Clin. J. Pain* **2009**, *25*, 170–175. [CrossRef]
21. Aloisi, A.M.; Ceccarelli, I.; Fiorenzani, P.; Maddalena, M.; Rossi, A.; Tomei, V.; Sorda, G.; Danielli, B.; Rovini, M.; Cappelli, A.; et al. Aromatase and 5-Alpha Reductase Gene Expression: Modulation by Pain and Morphine Treatment in Male Rats. *Mol. Pain* **2010**, *6*, 69. [CrossRef]
22. Nimbi, F.M.; Tripodi, F.; Rossi, R.; Navarro-Cremades, F.; Simonelli, C. Male Sexual Desire: An Overview of Biological, Psychological, Sexual, Relational, and Cultural Factors Influencing Desire. *Sex. Med. Rev.* **2020**, *8*, 59–91. [CrossRef] [PubMed]
23. Dosch, A.; Rochat, L.; Ghisletta, P.; Favez, N.; van der Linden, M. Psychological Factors Involved in Sexual Desire, Sexual Activity, and Sexual Satisfaction: A Multi-Factorial Perspective. *Arch. Sex. Behav.* **2016**, *45*, 2029–2045. [CrossRef] [PubMed]
24. Kingsberg, S.A.; Clayton, A.H.; Pfaus, J.G. The Female Sexual Response: Current Models, Neurobiological Underpinnings and Agents Currently Approved or under Investigation for the Treatment of Hypoactive Sexual Desire Disorder. *CNS Drugs* **2015**, *29*, 915–933. [CrossRef] [PubMed]
25. McCabe, M.P.; Sharlip, I.D.; Atalla, E.; Balon, R.; Fisher, A.D.; Laumann, E.; Lee, S.W.; Lewis, R.; Segraves, R.T. Definitions of Sexual Dysfunctions in Women and Men: A Consensus Statement from the Fourth International Consultation on Sexual Medicine. *J. Sex. Med.* **2016**, *13*, 135–143. [CrossRef]
26. Jung, J.; Jo, H.W.; Kwon, H.; Jeong, N.Y. Clinical Neuroanatomy and Neurotransmitter-Mediated Regulation of Penile Erection. *Int. Neurourol. J.* **2014**, *18*, 58. [CrossRef] [PubMed]

27. Andersson, K.-E. Mechanisms of Penile Erection and Basis for Pharmacological Treatment of Erectile Dysfunction. *Pharmacol. Rev.* **2011**, *63*, 811–859. [CrossRef]
28. Andersson, K.-E. Erectile Physiological and Pathophysiological Pathways Involved in Erectile Dysfunction. *J. Urol.* **2003**, *170*, S6–S14. [CrossRef]
29. Mckinlay, J. The Worldwide Prevalence and Epidemiology of Erectile Dysfunction. *Int. J. Impot. Res.* **2000**, *12*, S6–S11. [CrossRef]
30. Chew, K.K. Prevalence of Erectile Dysfunction in Community-Based Studies. *Int. J. Impot. Res.* **2004**, *16*, 201–202. [CrossRef]
31. Grover, S.A.; Lowensteyn, I.; Kaouache, M.; Marchand, S.; Coupal, L.; DeCarolis, E.; Zoccoli, J.; Defoy, I. The Prevalence of Erectile Dysfunction in the Primary Care Setting: Importance of Risk Factors for Diabetes and Vascular Disease. *Arch. Intern. Med.* **2006**, *166*, 213–219. [CrossRef]
32. Nguyen, H.M.T.; Gabrielson, A.T.; Hellstrom, W.J.G. Erectile Dysfunction in Young Men—A Review of the Prevalence and Risk Factors. *Sex. Med. Rev.* **2017**, *5*, 508–520. [CrossRef] [PubMed]
33. Pizzol, D.; Xiao, T.; Smith, L.; Sanchez, G.F.L.; Garolla, A.; Parris, C.; Barnett, Y.; Ilie, P.C.; Soysal, P.; Shin, J.I.; et al. Prevalence of Erectile Dysfunction in Male Survivors of Cancer: A Systematic Review and Meta-Analysis of Cross-Sectional Studies. *Br. J. Gen. Pract.* **2021**, *71*, e372. [CrossRef] [PubMed]
34. Toda, N.; Kishioka, S.; Hatano, Y.; Toda, H. Interactions between Morphine and Nitric Oxide in Various Organs. *J. Anesth.* **2009**, *23*, 554–568. [CrossRef]
35. Agmo, A.; Rojas, J.; Vazquez, P. Inhibitory Effect of Opiates on Male Rat Sexual Behavior May Be Mediated by Opiate Receptors Outside the Central Nervous System. *Psychopharmacol. Berl.* **1992**, *107*, 89–96. [CrossRef] [PubMed]
36. Meston, C.M.; Hull, E.; Levin, R.J.; Sipski, M. Disorders of Orgasm in Women. *J. Sex. Med.* **2004**, *1*, 66–68. [CrossRef]
37. Rowland, D.; McMahon, C.G.; Abdo, C.; Chen, J.; Jannini, E.; Waldinger, M.D.; Ahn, T.Y. Disorders of Orgasm and Ejaculation in Men. *J. Sex. Med.* **2010**, *7*, 1668–1686. [CrossRef]
38. Alwaal, A.; Breyer, B.N.; Lue, T.F. Normal Male Sexual Function: Emphasis on Orgasm and Ejaculation. *Fertil Steril* **2015**, *104*, 1051. [CrossRef]
39. Meston, C.M.; Freihart, B.K.; Handy, A.B.; Kilimnik, C.D.; Rosen, R.C. Scoring and Interpretation of the FSFI: What Can Be Learned From 20 Years of Use? *J. Sex. Med.* **2020**, *17*, 17–25. [CrossRef]
40. Baser, R.E.; Li, Y.; Carter, J. Psychometric Validation of the Female Sexual Function Index (FSFI) in Cancer Survivors. *Cancer* **2012**, *118*, 4606–4618. [CrossRef]
41. Jeffery, D.D.; Tzeng, J.P.; Keefe, F.J.; Porter, L.S.; Hahn, E.A.; Flynn, K.E.; Reeve, B.B.; Weinfurt, K.P. Initial Report of the Cancer Patient-Reported Outcomes Measurement Information System (PROMIS) Sexual Function Committee: Review of Sexual Function Measures and Domains Used in Oncology. *Cancer* **2009**, *115*, 1142–1153. [CrossRef]
42. Denlinger, C.S.; Carlson, R.W.; Are, M.; Baker, K.S.; Davis, E.; Edge, S.B.; Friedman, D.L.; Goldman, M.; Jones, L.; King, A.; et al. Survivorship: Sexual Dysfunction (Female), Version 1.2013: Clinical Practice Guidelines in Oncology. *J. Natl. Compr. Cancer Netw.* **2014**, *12*, 184. [CrossRef]
43. Rosen, R.C.; Riley, A.; Wagner, G.; Osterloh, I.H.; Kirkpatrick, J.; Mishra, A. The International Index of Erectile Function (IIEF): A Multidimensional Scale For Assesment of Erectile Dysfunction. *J. Sex. Med.* **1997**, *49*, 822–830. [CrossRef]
44. Rosen, R.; Cappelleri, J.; Lipsky, J.; Pen, Ä.B. Development and Evaluation of an Abridged, 5-Item Version of the International Index of Erectile Function (IIEF-5) as a Diagnostic Tool for Erectile Dysfunction. *Nature* **1999**, *11*, 319–326. [CrossRef] [PubMed]
45. Litwin, M.S.M.M.; Hays, R.D.P.; Fink, A.P.; Ganz, P.A.M.; Leake, B.P.; Brook, R.H.M.S. The UCLA Prostate Cancer Index: Development, Reliability, and Validity of a Health-Related Quality of Life Measure. *Med. Care* **1998**, *36*, 1002–1012. [CrossRef] [PubMed]
46. Jensen, P.T.; Klee, M.C.; Thranov, I.; Groenvold, M. Validation of a Questionnaire for Self-Assessment of Sexual Function and Vaginal Changes after Gynaecological Cancer. *Psycho-Oncology* **2004**, *13*, 577–592. [CrossRef] [PubMed]
47. Bartula, I.; Sherman, K.A. Development and Validation of the Female Sexual Function Index Adaptation for Breast Cancer Patients (FSFI-BC). *Breast Cancer Res. Treat.* **2015**, *152*, 477–488. [CrossRef] [PubMed]
48. Maiorino, M.I.; Chiodini, P.; Bellastella, G.; Giugliano, D.; Esposito, K. Sexual Dysfunction in Women with Cancer: A Systematic Review with Meta-Analysis of Studies Using the Female Sexual Function Index. *Endocrine* **2016**, *54*, 329–341. [CrossRef]
49. Bruner, D.W.; Calvano, T. The Sexual Impact of Cancer and Cancer Treatments in Men. *Nurs. Clin. N. Am.* **2007**, *42*, 555–580. [CrossRef]
50. Rajagopal, A.; Vassilopoulou-Sellin, R.; Palmer, J.L.; Kaur, G.; Bruera, E. Symptomatic Hypogonadism in Male Survivors of Cancer with Chronic Exposure to Opioids. *Cancer* **2004**, *100*, 851–858. [CrossRef]
51. Venkatesh, K.; Mattoo, S.K.; Grover, S. Sexual Dysfunction in Men Seeking Treatment for Opioid Dependence: A Study from India. *J. Sex. Med.* **2014**, *11*, 2055–2064. [CrossRef]
52. Zhao, S.; Deng, T.; Luo, L.; Wang, J.; Li, E.; Liu, L.; Li, F.; Luo, J.; Zhao, Z. Association Between Opioid Use and Risk of Erectile Dysfunction: A Systematic Review and Meta-Analysis. *J. Sex. Med.* **2017**, *14*, 1209–1219. [CrossRef] [PubMed]
53. Deyo, R.A.; Smith, D.H.M.; Johnson, E.S.; Tillotson, C.J.; Donovan, M.; Yang, X.; Petrik, A.; Morasco, B.J.; Dobscha, S.K. Prescription Opioids for Back Pain and Use of Medications for Erectile Dysfunction. *Spine Phila Pa 1976* **2013**, *38*, 909–915. [CrossRef] [PubMed]
54. Ajo, R.; Segura, A.; Inda, M.-D.-M.; Margarit, C.; Ballester, P.; Martínez, E.; Ferrández, G.; Sánchez-Barbie, Á.; Peiró, A.M. Erectile Dysfunction in Patients with Chronic Pain Treated with Opioids. *Med. Clín.* **2017**, *149*, 49–54. [CrossRef]

55. Rubinstein, A.L.; Carpenter, D.M.; Minkoff, J.R. Hypogonadism in Men With Chronic Pain Linked to the Use of Long-Acting Rather Than Short-Acting Opioids. *Clin. J. Pain* **2013**, *29*, 840–845. [CrossRef] [PubMed]
56. Hashim, M.A.; el Rasheed, A.H.; Ismail, G.A.W.; Awaad, M.I.; el Habiby, M.M.; Mohsen Ibrahim, N.M.; Abdeen, M.S. Sexual Dysfunction in Tramadol Hydrochloride Use Disorder Male Patients: A Case-Control Study. *Int. Clin. Psychopharmacol.* **2020**, *35*, 42–48. [CrossRef] [PubMed]
57. Kabbash, A.; el Kelany, R.; Oreby, M.; el Gameel, D. Effect of Tramadol Dependence on Male Sexual Dysfunction. *Interdiscip. Toxicol.* **2019**, *12*, 157. [CrossRef]
58. Kozyrev, N.; Coolen, L.M. Activation of Mu or Delta Opioid Receptors in the Lumbosacral Spinal Cord Is Essential for Ejaculatory Reflexes in Male Rats. *PLoS ONE* **2015**, *10*, e0121130. [CrossRef]
59. Martyn-St James, M.; Cooper, K.; Kaltenthaler, E.; Dickinson, K.; Cantrell, A.; Wylie, K.; Frodsham, L.; Hood, C. Tramadol for Premature Ejaculation: A Systematic Review and Meta-Analysis. *BMC Urol.* **2015**, *15*, 6. [CrossRef]
60. Gomez-Marrero, J.; Feria, M.; Mas, M. Stimulation of Opioid Receptors Suppresses Penile Erectile Reflexes and Seminal Emission in Rats. *Pharmacol. Biochem. Behav.* **1988**, *31*, 393–396. [CrossRef]
61. Yee, A.; Loh, H.S.; Maria, H.; Hashim, H.; Ng, C.G. The Prevalence of Sexual Dysfunction among Male Patients on Methadone and Buprenorphine Treatments: A Meta-Analysis Study. *J. Sex. Med.* **2013**, *11*, 22–32. [CrossRef]
62. Yee, A.; Loh, H.S.; Loh, H.H.; Riahi, S.; Ng, C.G.; Sulaiman, A.H. bin A Comparison of Sexual Desire in Opiate-Dependent Men Receiving Methadone and Buprenorphine Maintenance Treatment. *Ann. Gen. Psychiatry* **2019**, *18*, 25. [CrossRef] [PubMed]
63. Hallinan, R.; Byrne, A.; Agho, K.; McMahon, C.; Tynan, P.; Attia, J. Erectile Dysfunction in Men Receiving Methadone and Buprenorphine Maintenance Treatment. *J. Sex. Med.* **2007**, *5*, 684–692. [CrossRef] [PubMed]
64. Zamboni, L.; Franceschini, A.; Portoghese, I.; Morbioli, L.; Lugoboni, F. Sexual Functioning and Opioid Maintenance Treatment in Women. Results from a Large Multicentre Study. *Front. Behav. Neurosci.* **2019**, *13*, 97. [CrossRef]
65. Varma, A.; Sapra, M.; Iranmanesh, A. Impact of Opioid Therapy on Gonadal Hormones: Focus on Buprenorphine. *Horm. Mol. Biol. Clin. Investig.* **2018**, *36*, 20170080. [CrossRef] [PubMed]
66. Baron, R.; Jansen, J.P.; Binder, A.; Pombo-Suarez, M.; Kennes, L.; Müller, M.; Falke, D.; Steigerwald, I. Tolerability, Safety, and Quality of Life with Tapentadol Prolonged Release (PR) Compared with Oxycodone/Naloxone PR in Patients with Severe Chronic Low Back Pain with a Neuropathic Component: A Randomized, Controlled, Open-Label, Phase 3b/4 Trial. *Pain Pract.* **2016**, *16*, 600–619. [CrossRef] [PubMed]
67. AminiLari, M.; Manjoo, P.; Craigie, S.; Couban, R.; Wang, L.; Busse, J.W. Hormone Replacement Therapy and Opioid Tapering for Opioid-Induced Hypogonadism among Patients with Chronic Noncancer Pain: A Systematic Review. *Pain Med.* **2019**, *20*, 301–313. [CrossRef]
68. Yee, A.; Loh, H.S.; Ong, T.A.; Guan Ng, C.; Sulaiman, A.H. Randomized, Double-Blind, Parallel-Group, Placebo-Controlled Trial of Bupropion as Treatment for Methadone-Emergent Sexual Dysfunction in Men. *Am. J. Men Health* **2018**, *12*, 1705–1718. [CrossRef]
69. Tatari, F.; Farnia, V.; Nasiri, R.F.; Najafi, F. Trazodone in Methandone Induced Erectile Dysfunction. *Iran. J. Psychiatry* **2010**, *5*, 164.
70. van Ahlen, H.; Piechota, H.J.; Kiasa, H.J.; Brennemann', W.; Klingmiillerb, D. Opiate Antagonists in Erectile Dysfunction: A Possible New Treatment Option? Results of a Pilot Study with Naltrexone. *Eur. Urol.* **1995**, *28*, 246–250. [CrossRef]
71. Billington, C.J.; Shafer, R.B.; Morley, J.E. Effects of Opioid Blockade with Nalmefene in Older Impotent Men. *Life Sci.* **1990**, *47*, 799–805. [CrossRef]
72. Ramdurg, S.; Ambekar, A.; Lal, R. Sexual Dysfunction among Male Patients Receiving Buprenorphine and Naltrexone Maintenance Therapy for Opioid Dependence. *J. Sex. Med.* **2012**, *9*, 3198–3204. [CrossRef] [PubMed]
73. Farnia, V.; Tatari, F.; Alikhani, M.; Yazdchi, K.; Taghizadeh, M.; Sadeghi Bahmani, D.; Karbasizadeh, H.; Holsboer-Trachsler, E.; Brand, S. Rosa Damascena Oil Improved Methadone-Related Sexual Dysfunction in Females with Opioid Use Disorder under Methadone Maintenance Therapy—Results from a Double-Blind, Randomized, and Placebo-Controlled Trial. *J. Psychiatr. Res.* **2017**, *95*, 260–268. [CrossRef] [PubMed]
74. Farnia, V.; Alikhani, M.; Ebrahimi, A.; Golshani, S.; Sadeghi Bahmani, D.; Brand, S. Ginseng Treatment Improves the Sexual Side Effects of Methadone Maintenance Treatment. *Psychiatry Res.* **2019**, *276*, 142–150. [CrossRef] [PubMed]
75. Basson, R.; Wierman, M.E.; van Lankveld, J.; Brotto, L. Summary of the Recommendations on Sexual Dysfunctions in Women. *J. Sex. Med.* **2010**, *7*, 314–326. [CrossRef]
76. McCarthy, B.; McDonald, D. Assessment, Treatment, and Relapse Prevention: Male Hypoactive Sexual Desire Disorder. *J. Sex Marital. Ther.* **2009**, *35*, 58–67. [CrossRef]
77. Stein, A.; Sauder, S.K.; Reale, J. The Role of Physical Therapy in Sexual Health in Men and Women: Evaluation and Treatment. *Sex. Med. Rev.* **2019**, *7*, 46–56. [CrossRef]
78. Schroder, M.A.; Mell, L.K.; Hurteau, J.A.; Collins, Y.C.; Rotmensch, J.; Waggoner, S.E.; Yamada, S.D.; Small, W.; Mundt, A.J. Clitoral Therapy Device for Treatment of Sexual Dysfunction in Irradiated Cervical Cancer Patients. *Int. J. Radiat. Oncol. Biol. Phys.* **2005**, *61*, 1078–1086. [CrossRef]
79. Wassersug, R.; Wibowo, E. Non-Pharmacological and Non-Surgical Strategies to Promote Sexual Recovery for Men with Erectile Dysfunction. *Transl. Androl. Urol.* **2017**, *6*, S776–S794. [CrossRef]

Perspective

Tackling Insomnia Symptoms through Vestibular Stimulation in Patients with Breast Cancer: A Perspective Paper

Joy Perrier [1,*], Melvin Galin [1,2], Pierre Denise [2], Bénédicte Giffard [1] and Gaëlle Quarck [2]

1 Neuropsychologie et Imagerie de la Mémoire Humaine U1077, EPHE, INSERM, CHU de Caen, GIP Cyceron, PSL Université, Normandie Univ, Université de Caen Normandie, 14000 Caen, France
2 COMETE U1075, INSERM, CYCERON, CHU de Caen, Normandie Univ, Université de Caen Normandie, 14000 Caen, France
* Correspondence: joy.perrier@unicaen.fr; Tel.: +33-(0)-2-31-56-83-83

Simple Summary: Patients with breast cancer frequently complaint from insomnia difficulties that can affect quality of life and cancer progression. Such difficulties may result from rest-activity (i.e., 24 h alternation of sleep and wake) rhythm alterations consistently reported in this pathology. Currently proposed approaches to counter insomnia difficulties in patients with breast cancer have positive effects only on sleep complaints and well-being. Moreover, such approaches may be difficult to implement shortly after chemotherapy. Innovatively, vestibular stimulation would be particularly suited to tackling insomnia symptoms in patients with breast cancer. Indeed, recent reports have shown that vestibular stimulation could improve rest-activity rhythm and sleep in healthy volunteers. This perspective paper aims to support the evidence of using vestibular stimulation to improve rest-activity rhythms and reduce insomnia symptoms in patients with BC, with beneficial effects on quality of life and, potentially, survival.

Abstract: Insomnia symptoms are common among patients with breast cancer (BC; 20–70%) and are predictors of cancer progression and quality of life. Studies have highlighted sleep structure modifications, including increased awakenings and reduced sleep efficiency and total sleep time. Such modifications may result from circadian rhythm alterations consistently reported in this pathology and known as carcinogenic factors, including lower melatonin levels, a flattened diurnal cortisol pattern, and lower rest-activity rhythm amplitude and robustness. Cognitive behavioral therapy and physical activity are the most commonly used non-pharmacological interventions to counter insomnia difficulties in patients with BC. However, their effects on sleep structure remain unclear. Moreover, such approaches may be difficult to implement shortly after chemotherapy. Innovatively, vestibular stimulation would be particularly suited to tackling insomnia symptoms. Indeed, recent reports have shown that vestibular stimulation could resynchronize circadian rhythms and improve deep sleep in healthy volunteers. Moreover, vestibular dysfunction has been reported following chemotherapy. This perspective paper aims to support the evidence of using galvanic vestibular stimulation to resynchronize circadian rhythms and reduce insomnia symptoms in patients with BC, with beneficial effects on quality of life and, potentially, survival.

Keywords: sleep; insomnia; breast cancer; circadian rhythms; vestibular stimulation; chemotherapy

Citation: Perrier, J.; Galin, M.; Denise, P.; Giffard, B.; Quarck, G. Tackling Insomnia Symptoms through Vestibular Stimulation in Patients with Breast Cancer: A Perspective Paper. *Cancers* **2023**, *15*, 2904. https://doi.org/10.3390/cancers15112904

Academic Editor: António Araújo

Received: 28 March 2023
Revised: 5 May 2023
Accepted: 9 May 2023
Published: 25 May 2023

1. Sleep Is Often Overlooked as a Target to Improve Quality of Life and Survival in Patients with Cancer

We spend around one-third of our lives sleeping, which is critical for learning and memory, the immune system, brain energy, and plasticity [1,2]. Inadequate sleep is associated with numerous illnesses detrimental to metabolic and cardiovascular health, including a predisposition to obesity, diabetes, heart disease, and depression [3,4]. Sleep complaints, in particular, insomnia, are now well recognized in patients with non-central nervous system (CNS) cancers, not only following chemotherapy [5,6] but also before the occurrence of treatments [7–9]. Despite the high prevalence of complaints related to sleep disturbances in patients with cancer, such disturbances remain under-evaluated using gold-standard measures and are, thus, misunderstood in this population.

Sleep disturbances have been associated with increased risks of cancer [10–12] and tumor progression [13,14] in mouse models. Cancer could also indirectly influence sleep through various pathophysiological processes, including inflammation [15–17]. Therefore, a bi-directional relationship between sleep disturbances and inflammatory-associated cancer processes may exist. Sleep is part of one circadian rhythm (i.e., physiological processes regulated over 24 h) called the rest-activity rhythm. Previous reports have consistently shown rest-activity rhythm, and more generally, circadian rhythm, alterations in patients with cancer [18–21] that have been brought to the fore as a potential carcinogenic factor and are now considered a potential therapeutic target to improve survival in patients with cancer. Targeting not only sleep difficulties but also, more broadly, circadian rhythm alterations in patients with cancer would therefore offer the opportunity to improve their quality of life and also their survival.

2. Insomnia Symptoms and Sleep Structure Modifications in Patients with Breast Cancer

Insomnia is among the most common sleep complaints in patients with cancer, particularly in those with breast cancer (BC; 20–70%) compared to those with other non-CNS cancers and the general population (30%) [22,23]. Moreover, BC has become the most commonly diagnosed cancer in women worldwide [24]. Therefore, this perspective paper will focus on BC. The following section is intended to provide an overview of insomnia difficulties in patients with BC. An in-depth review of such difficulties is beyond the scope of this paper and has been specifically conducted previously [25–27].

Recent studies have shown that insomnia complaints are present in patients with BC even before and at diagnosis [28], before treatment initiation [25], and following treatment, mainly radiotherapy and chemotherapy [25,28,29]. Based on these results highlighting sleep complaints in patients with BC, and to tackle such difficulties, there is a need to describe the actual sleep structure modifications associated with BC and its treatments. Sleep structure disturbances in patients with BC have been mainly described using actigraphy, which indirectly measures sleep quality and quantity. However, while polysomnography (PSG) remains the gold standard for evaluating sleep quantity and quality, it has been less used in patients with cancer [30,31].

Two studies using actigraphy reported worse total sleep time and sleep efficiency before initiating chemotherapy treatments for BC [32,33] compared to normative values established in the US using actigraphy data [34]. Kreutz et al. [35] evaluated sleep parameters using actigraphy in patients with BC starting chemotherapy compared to normative data and stratified patients as good and poor sleepers. Patients did not differ from normative data, and no difference was found between good and poor sleepers. Unfortunately, no information was available about the time since the start of chemotherapy at the time of evaluation. However, the significance of these results is limited by the lack of a control group.

Two recent studies [36,37] evaluated sleep parameters using actigraphy in patients with BC at least 12 months after treatment completion (e.g., surgery, radiotherapy, or chemotherapy) or within 12 months after chemotherapy. Trivedi et al. did not find differences in sleep parameters between healthy controls and patients with BC 12 months after chemotherapy completion. Given the variability in treatments in their sample, this study does not permit us to disentangle the effects of each treatment modality. Moreover, since more than half of their sample took endocrine therapy, this could have driven their results. Ratcliff et al. found that patients with BC experienced relatively long waking after sleep onset, poor sleep efficiency, and short total sleep time compared to non-clinical norms [38]. However, no control group was included in this study.

Other studies have performed longitudinal protocols before, during, and/or after chemotherapy [39–46], of which only two included a control group without a history of cancer [42,43]. Their results showed longer total sleep time and nap time during chemotherapy compared to before treatment [39,40,44] and controls [42,43]. Conversely, Li et al. showed that chemotherapy initiation was associated with less sleep time, more arousal, and lower sleep quality compared with pretreatment and at the end of chemotherapy [46]. Beck et al. also reported a shorter sleep time on the first night after initiating chemotherapy than before and after [45]. Finally, Kuo et al. found no difference between assessments before and during chemotherapy [41]. Therefore, reports about the effects of chemotherapy on sleep parameters measured in actimetry are highly variable and do not allow us to draw a definitive conclusion. Nevertheless, a literature review published in 2015 [47] concluded that chemotherapy might accentuate sleep difficulties already present before the initiation of treatment for BC.

To our knowledge, only three studies have investigated the effects of endocrine therapy (anti-aromatase) [48,49] and radiotherapy [50] on sleep parameters measured using actimetry in patients with BC. One study showed no changes in sleep efficiency, total sleep time, and nocturnal arousals at endocrine therapy initiation compared to before [48]. This study's lack of a control group could explain the lack of significant changes in sleep patterns with endocrine therapy. Conversely, Martin et al. [49] found that patients treated with endocrine therapy had lower sleep efficiency, more time awake, and higher activity levels at night than patients treated only with surgery and radiotherapy. Another study showed that almost half of the patients treated with radiotherapy had wakefulness and total sleep times outside the normal values [50].

Due to its ease of use, actigraphy is the primary approach used to evaluate sleep in patients with cancer. However, it does not facilitate a deep characterization of sleep and its pathophysiological modifications. In contrast, PSG is considered the gold standard for sleep evaluation. It comprises a multimodal recording (electroencephalography, electrooculography, electromyography, cardiac activity, and oximetry) and enables a deeper understanding of sleep modifications.

Previous experimental studies using PSG to evaluate sleep in patients with BC are scarce and have shown either a lack of sleep alterations following chemotherapy [51,52] or deleterious effects of chemotherapy and/or radiotherapy [53,54]. One study reported a longer deep sleep duration after radiotherapy in patients with BC than in healthy controls [53]. However, in this latest study, the delay between the PSG recording and the end of radiotherapy was unclear. In their pioneering study, Silberfab et al. [51] showed no differences between patients with BC and control subjects without a history of cancer for the primary sleep parameters of efficiency, number of awakenings, duration of stages, and sleep latency. Roscoe et al. [52] showed that patients slept more after than before chemotherapy. This result could be explained by poor sleep quality at baseline due to the stress associated with the cancer diagnosis and apprehension about chemotherapy or by an accumulation of fatigue and sleep deprivation during treatment leading to compensation after treatment.

Parker et al. compared the sleep structure of patients treated for advanced non-CNS cancers (stage 3 or 4 cancers), including 32 patients treated for BC (28% of the cohort) [54]. Their results showed that these patients had reduced sleep efficiency and amounts of the different sleep stages outside the norms established by Williams et al. in healthy participants [55]. Tag Eldin et al. [53] analyzed the sleep architecture of treated and untreated patients with BC and control subjects with no history of cancer. Patients treated with chemotherapy or radiotherapy had lower sleep time and efficiency and higher sleep latency and deep sleep duration than untreated patients and controls. Finally, the time spent in rapid eye movement (REM) sleep was lower in these patients.

These studies have shown changes in sleep structure in patients treated for cancer. Such sleep modifications may partly result from the circadian process alterations found in this population.

3. Circadian Rhythms Alterations in Patients with Cancer

Circadian rhythm alterations in patients with BC have been identified through physiological markers, including melatonin production, temperature, and cortisol rhythms over 24 h, and also using actigraphy. The latter is particularly suited to quantifying rest-activity rhythm (see [56] for rest-activity parameter definitions) over 24 h and is easier to use than physiological measures, although both approaches are complementary. Previous reports have indicated lower melatonin levels before and during chemotherapy administration [46]. Circadian disruptions such as hot flushes (i.e., a sudden wave of mild or intense body heat caused by rapid hormonal changes) [57] and flattened diurnal cortisol patterns [58,59] have been reported, supporting the view of circadian rhythm alterations in patients with BC. Using actigraphy, a lower amplitude, mean activity, and robustness of the rest-activity rhythm were identified in patients with BC during chemotherapy [42,46,60]. For example, Li et al. [46] showed that the first administration of chemotherapy was associated with an altered rest-activity rhythm, with a decrease in its amplitude, mean activity, and robustness compared with pretreatment. These changes appeared to attenuate as the treatment cycles were administered. Conversely, Sultan et al. [61] showed a worsening of alterations in the rest-activity rhythm, with a decrease in mean activity and amplitude and a shift in peak activity over chemotherapy cycles. However, the lack of measurements before chemotherapy initiation leaves open the extent to which these alterations existed before chemotherapy.

In a longitudinal study, Liu et al. found alterations in rest-activity rhythm in patients with BC compared with control subjects before and during chemotherapy [40]. Before and after chemotherapy, the patients' rhythm showed reduced amplitude, mean activity, and robustness compared to controls. In patients, chemotherapy treatment was also associated with decreased amplitude and mean activity of the rest-activity rhythm compared to before chemotherapy initiation. In addition, two recent studies based on the same cohort showed that changes in rest-activity rhythm associated with chemotherapy, reflected by its decreased amplitude, mean activity, and intradaily variability, appeared to persist for up to five years after BC diagnosis [62,63]. In these two studies, the mean diurnal activity was lower, and its intradaily variability was larger in patients with BC than in healthy controls, suggesting an alteration in the rest-activity rhythm over 24 h in patients, even at a distance from their treatment. Overall, these results suggest that chemotherapy negatively impacts circadian rhythms, notably on the rest-activity rhythm parameters, which appear to persist years after chemotherapy. More recently, our team has shown that the amplitude of the rest-activity rhythm was reduced in patients with BC not treated with chemotherapy compared to healthy controls beyond the effects of endocrine therapy [49]. These results suggest that rest-activity modifications may be altered even before the initiation of adjuvant treatments and could be further exacerbated by chemotherapy. These results are summarized in Table 1.

Table 1. The main results related to circadian rhythm modifications in patients with breast cancer (BC), including melatonin, hot flushes, cortisol, and actigraphy-derived (rest-activity rhythm) parameters.

Authors, Year	Measures	N	Groups	Cancer Stages	Treatments	Time of Assessments	Main Results
Li et al., 2019 [46]	Urinary melatonin	180	BC	I–III	Surgery and CT	Before and during first and last cycles of CT	Mean levels of melatonin *Before CT < after CT*
Carpenter et al., 2004 [57]	Hot flashes using skin conductance	21	BC	0–III	Surgery and RT and/or CT	After treatments, >5 years post-diagnosis	High variability in the rhythm of hot flashes
Abercrombie et al., 2004 [58,59]	Salivary cortisol	17 31	BC HC	Metastatic	NA	NA	Rhythm over 24 h *BC < HC*
Hsiao et al., 2017 [58,59]	Salivary cortisol	85	BC	0–III	Surgery and RT and/or CT	Before (T0) and after treatments (T0 + 2 months; T0 + 5 months; T0 + 8 months; T0 + 14 months)	Flatter cortisol levels tend to recover *over the course of treatments*
Ancoli-Israël et al., 2014 [42,46,60]	Actigraphy (3 h)	68	BC HC	I–III	Surgery and RT and/or CT	Before CT and at Cycle 4 of CT and 1 year post-CT	Naps time before CT *BC > HC* Robustness * *BC during Cycle 4 < before CT and HC* Naps time and robustness 1 year post-CT *BC = HC*
Li et al., 2019 [42,46,60]	Actigraphy (7 days)	180	BC	I–III	Surgery and CT	Before and during first and last cycles of CT	Amplitude, mean activity, and robustness * *Before CT > after CT*
Savard et al., 2009 [42,46,60]	Actigraphy (3 days)	95	BC	I–III	Surgery and CT + ET (27% of the sample)	Before CT and at Cycle 1 and 4 of CT	Amplitude, mean activity, and robustness * *First week of each CT cycle < before CT*
Roveda et al., 2019 [62]	Actigraphy (7 days)	15 13	BC HC	NA	NA	5 years post-diagnosis	Mean activity and amplitude *BC < HC*
Sultan et al., 2016 [62]	Actigraphy (3–4 days)	25	BC	II	CT	Cycles 1, 3, and 6 of CT	Mean activity and amplitude decreased *over the course of CT* Shift in peak activity *over the course of CT*

Table 1. *Cont.*

Authors, Year	Measures	N	Groups	Cancer Stages	Treatments	Time of Assessments	Main Results
Liu et al., 2013 [40]	Actigraphy (3 days)	79 61	BC HC	I–III	Surgery or/and CT	Before CT and at the end of Cycle 4 of CT	Amplitude and mean activity *Before CT > Cycle 4 of CT* Amplitude, robustness *, mean activity *Both before and after Cycle 4 of CT, BC < HC*
Galasso et al., 2019 [63]	Actigraphy (7 days)	15 13	BC HC	NA	NA	5 years post-diagnosis	Intradaily variability * *BC > HC*
Martin et al., 2021 [49]	Actigraphy (15 days)	18 18 16	BC with ET BC without ET HC	0–II	Surgery and RT	6 months post-RT	Amplitude *BC < HC* Interdaily stability * *ET− < HC*

Note: N, number of participants in each group; BC, patients with breast cancer; HC, healthy controls; NA, unavailable; CT, chemotherapy; RT, radiotherapy; ET−, with/without endocrine therapy; *, the robustness of the rest-activity rhythm refers to the relative amplitude that is the relative difference between the 10 h with maximal activity and the 5 h with minimal activity. The interdaily stability is the ratio between the variance of the average 24 h pattern around the mean and the overall variance, with higher values indicative of rhythm stability. The intradaily variability is calculated as the ratio of the mean squares of the difference between successive hours and the mean squares around the grand mean. Comparisons are given in italic for ease of reading. See [56] for rest-activity measure definitions.

Rest-activity rhythm modifications might be partly responsible for sleep disturbances in patients with BC. Indeed, associations have been shown between greater fatigue and altered circadian rhythms [64] but with greater time spent in bed [65] in nap periods [39]. These results suggest that fatigue would lead to an adaptation of sleep time over 24 h. Increased fatigue following cancer and its treatment could therefore induce alterations in the rest-activity rhythm and poor sleep habits (i.e., less activity time and more napping time), leading to the development or persistence of sleep disturbances. Moreover, circadian rhythm alterations have been associated with lower survival, calling for further studies to evaluate and address these modifications to improve sleep, quality of life, and also survival.

4. Associations with Survival

The hypothesis that activity-rest rhythm disorders could be carcinogenic (i.e., contribute to the occurrence of cancer) has emerged due to epidemiological studies showing an increased risk of developing BC in individuals performing shift work. It has been proposed that nocturnal exposure to light would suppress melatonin (a key regulator of central and peripheral oscillators) secretion by the pineal gland [12,66,67]. The model developed by Sephton et al. [68,69] has proposed a more precise interconnection about potential pathways through which circadian dysregulation could mediate psychosocial effects on cancer progression. They proposed that altered circadian rhythms and anxiety-depressive factors might contribute to tumor growth by deregulating glucocorticoid secretion by the hypothalamic–pituitary–adrenal axis [70]. In support of this model, a cortisol spike has been observed in patients with advanced BC compared to healthy controls. Such a higher nighttime peak was associated with a poorer prognosis, including more rapid development of metastases [71].

It has also been proposed that altered sleep may be one modifiable factor contributing to breast oncogenesis [10–12]. For example, a recent study on patients with advanced BC found that better sleep quality and fewer arousals, measured by actimetry, were associated with significantly reduced mortality risk over six years of follow-up [72]. Sleep disorders appear to be associated with an increased risk and/or higher aggressiveness of cancer [73,74]. However, our understanding of the underlying mechanisms needs further development and requires a multidisciplinary approach.

Screening for sleep disorders could be performed more systematically in patients with BC in the context of research protocols or clinical practice when they complain of difficulties in their daytime functioning. This screening would allow us to better understand to what extent sleep disorders influence the occurrence, aggressiveness, and recovery of BC and non-cerebral cancers in general and to treat sleep pathologies that potentially impact patients' quality of life.

5. Perspective of Using Vestibular Stimulation to Improve Sleep in Patients with BC

Commonly prescribed drugs such as benzodiazepines have well-known adverse side effects [75–77]. Therefore, non-pharmacological interventions have gained increasing attention as an alternative first-line approach in recent years. Current non-drug approaches have been tested and partly validated to improve the quality of life and sleep complaints of patients with BC during and after treatments [78]. Such therapies include adapted physical activity or cognitive behavioral therapy that appear to positively affect quality of life, self-esteem [79–82], and subjective reports of insomnia symptoms even in the long term [83–88]. However, previous studies have provided a low quality of evidence [89].

However, while positively affecting well-being and self-reported sleep difficulties, these approaches are not explicitly targeting potential carcinogenic modifiable factors such as circadian rhythms and sleep. A challenge also remains when making non-drug therapies available and easily accessible to patients to improve adherence [90,91], highlighting the need for other approaches, such as vestibular stimulation. Indeed, previous reports suggest that chemotherapy for BC may affect vestibular function [92–95]. Moreover, considering the implication of the vestibular system in circadian regulation, its stimulation may offer a unique opportunity to regulate circadian rhythms and improve sleep in patients with BC.

The vestibular system is located within the inner ear next to the cochlear organ. It comprises three semi-circular canals detecting three-dimensional angular head velocity and two otolithic sensors detecting linear acceleration.

Besides its known role in detecting head movements and orientation, promising findings from lesion studies have also suggested that the vestibular system could constitute an input to the circadian clock or be involved in circadian regulation and synchronization [96–98]. Fuller et al. reported an association between the vestibular system and the circadian pacemaker based on animal studies [99–101]. In support of this hypothesis, bilateral vestibular lesions in rats led to a sharp fall in their core temperature and a disruption in their daily rhythmicity [96]. Similarly, a study in healthy volunteers reported a significant phase advance in the rest-activity rhythm using vestibular stimulation through a rotary chair at 18:00 compared to sham stimulation (i.e., lack of chair inclination and movement) [102]. These results support the hypothesis of the involvement of vestibular inputs in rest-activity rhythm regulation.

Since sleep is intimately associated with the rest-activity rhythm, these results suggested a potential association between sleep and the vestibular system. Patients with bilateral vestibular loss had abnormal sleep patterns and shorter sleep duration than healthy controls [97]. A recent study used actigraphy to quantify sleep in patients with unilateral vestibular hypofunction, finding that they slept less and took longer to fall asleep than healthy controls [103]. Recent epidemiological evidence has also shown abnormal sleep duration in patients with vestibular vertigo [104]. In addition, several studies have highlighted a potential association between sleep apnea (i.e., one of the most frequent sleep disorders) and vestibular impairments. However, this association remains to be clearly established [105,106].

The putative causal link between the vestibular system and sleep is strengthened by recent reports showing that the rehabilitation or stimulation of the vestibular system could promote sleep. In-hospital vestibular rehabilitation for chronic dizziness improved sleep complaints compared to before therapy [107]. Participants were taught to perform the rehabilitation program for 30 min four times a day over five days when they were in the hospital, and this program was conducted in groups. Similarly, compared to a stationary condition, continuous rocking (at 0.25 Hz) during an afternoon nap or the night promoted sleep by reducing latency into and maintenance of deep sleep (non-REM) in healthy volunteers [108,109]. Finally, a study evaluating the effects of a recliner chair with a rocking motion on sleep in healthy volunteers reported a decrease in light sleep and an increase in deep sleep when the chair moved compared to being stationary [110].

Positive effects of vestibular stimulation appear to be mediated by a decrease in the arousal level resulting from cholinergic tonus modulation and rhythmic entrainment of thalamocortical activity [111]. Older adults are more prone to experience sleep difficulties and are therefore of particular interest in using vestibular stimulation as a sleep-promoting intervention. However, three recent studies did not report sleep improvements using night or afternoon nap vestibular stimulation through rocking [112–114]. While this lack of significant effects of rocking on sleep may be due to the already high sleep efficiency of the participants, it could also be argued that the stimulation protocol was inadequate for efficiently stimulating the vestibular system.

Altogether, these results partly support (1) the involvement of the vestibular system as an input to circadian regulation and (2) the positive effects of vestibular rehabilitation and stimulation on sleep. The mechanisms involved remain to be understood, which will be required to form a consensus about such beneficial effects. Indeed, it might be argued that sleep and circadian disorders are not specific to vestibular lesions. Therefore, there is a need to highlight functional associations between the vestibular system and cortical/sub-cortical regions involved in sleep regulation. The association between the vestibular system and circadian rhythms is currently supported by neuro-anatomical pathways between the median vestibular nuclei and the suprachiasmatic nucleus (i.e., circadian pacemaker) [98,115].

Considering that patients with BC could have both circadian rhythm and sleep difficulties that could negatively influence their recovery after chemotherapy, innovative approaches targeting both phenomena are needed. Galvanic vestibular stimulation (GVS) could be particularly well suited for this purpose [116]. Indeed, GVS is based on stimulation of the peripheral vestibular organ via direct activation of vestibular nerve afferent fibers at the spike trigger zone, bypassing hair cell synapses [117,118] using electrical stimulation at the back of each ear. The question of which parts of the vestibule GVS activates has been debated over the past decade. A recent review proposed that GVS's effects would result from a central semi-circular canal-otolith signal convergence and integration, further converging in central vestibular neurons [119]. Based on previous literature, working hypotheses can be proposed for GVS's specific effects on circadian rhythms and sleep. Notably, the orexinergic system shares a bi-directional relationship with the vestibular system [119] and could mediate these effects. A functional hypothesis is that the vestibular system monitors the daily motion quantity and informs the orexinergic neurons, influencing the sleep–wake state switch [98].

Altogether, these previous reports support using GVS to limit circadian rhythms and sleep disturbances in patients with BC. Compared to other non-drug approaches currently available, GVS has the additional benefit of offsetting the decrease in vestibular stimulation resulting from hair cell impairments caused by chemotherapy agents.

6. Conclusions and Research Agenda

This perspective paper has highlighted sleep and circadian rhythm disturbances in patients with BC before and after chemotherapy. Several studies also suggest sleep difficulties and circadian rhythm alterations due to chemotherapy effects. Moreover, results from previous studies have repeatedly shown that sleep difficulties are associated with a lower quality of life and that circadian rhythm alterations and, to a lesser extent, sleep difficulties are associated with lower survival. These results call for approaches targeting circadian rhythm alterations and sleep difficulties that could be implemented shortly after chemotherapy in patients with BC.

Given the beneficial effects of vestibular stimulation on circadian rhythms and sleep and knowing the potential vestibulotoxic effects of chemotherapy, we have proposed using vestibular stimulation as a new non-pharmacological intervention. Vestibular stimulation is expected to specifically reduce circadian rhythm alterations and sleep difficulties in patients with BC after chemotherapy. GVS would be of particular interest since it has three potential benefits. Firstly, it allows remote vestibular stimulation to be performed at home. Patients with BC may have difficulties attending oncological support due to increased fatigue or distance between home and the hospital. Therefore, having a remote approach is of interest to increase adherence. Secondly, it is a safe and easy-to-use approach. Several systems can currently be used to pre-program stimulation and avoid patients doing so, avoiding accidental mishandling. Thirdly, videoconferencing during the stimulation could be considered to ensure social support and increase patient adherence.

Overall, this perspective paper supports using vestibular stimulation, particularly GVS, in patients with BC to resynchronize their circadian rhythms and improve their sleep, benefiting their quality of life and, potentially, survival.

7. Research Agenda

In laboratory settings:

1. Quantify the beneficial effects of GVS on circadian rhythms and sleep in patients with BC after chemotherapy as a proof of concept.
2. Compare the quality-of-life outcomes between patients using GVS and those not using stimulation.

3. Compare the survival outcomes between patients using GVS and those not using stimulation.
4. Determine the best time to use GVS in patients (e.g., before, during, or after [and how long after] chemotherapy).
5. Determine and quantify such beneficial effects in other cancer populations.
6. Explore the physiological and neurofunctional correlates of GVS to clarify its underlying mechanisms.

The above steps will further facilitate the implementation of GVS in oncological support in close partnership with clinical oncologists.

This research agenda requires close collaboration between several scientific disciplines (i.e., neuroscientists, neuropsychologists, and neurophysiologists) and clinicians that frequently work in different departments and speak different scientific languages. Moreover, this topic would benefit from translational studies that are currently rarely conducted, mainly for the same reasons.

Author Contributions: J.P. wrote the manuscript. G.Q. and J.P. initiated the purpose of the paper. P.D., M.G., B.G. and G.Q. provided feedback on the manuscript. All authors have read and agreed to the published version of the manuscript.

Funding: This work was supported by the Région Normandie (Chair of Excellence, 2021–2024, 20E04944; Réseaux d'Intérêts Normands—RIN, 2022–2025) and the Fondation ARC for Cancer Research—Fondation Arc pour al recherche sur le cancer (ARCPJA2021060003783).

Institutional Review Board Statement: Not applicable.

Informed Consent Statement: Not applicable.

Data Availability Statement: Not applicable.

Conflicts of Interest: The authors declare no conflict of interest.

References

1. Frank, M.G.; Heller, H.C. The Function(s) of Sleep. *Handb. Exp. Pharmacol.* **2019**, *253*, 3–34. [CrossRef] [PubMed]
2. Krueger, J.M.; Frank, M.G.; Wisor, J.P.; Roy, S. Sleep Function: Toward Elucidating an Enigma. *Sleep Med. Rev.* **2016**, *28*, 46–54. [CrossRef] [PubMed]
3. Maquet, P. Sleep Function(s) and Cerebral Metabolism. *Behav. Brain Res.* **1995**, *69*, 75–83. [CrossRef] [PubMed]
4. van Mill, J.G.; Vogelzangs, N.; van Someren, E.J.W.; Hoogendijk, W.J.G.; Penninx, B.W.J.H. Sleep Duration, but Not Insomnia, Predicts the 2-Year Course of Depressive and Anxiety Disorders. *J. Clin. Psychiatry* **2014**, *75*, 119–126. [CrossRef]
5. Lowery-Allison, A.E.; Passik, S.D.; Cribbet, M.R.; Reinsel, R.A.; O'Sullivan, B.; Norton, L.; Kirsh, K.L.; Kavey, N.B. Sleep Problems in Breast Cancer Survivors 1–10 Years Posttreatment. *Palliat. Support. Care* **2018**, *16*, 325–334. [CrossRef]
6. Howell, D.; Oliver, T.K.; Keller-Olaman, S.; Davidson, J.R.; Garland, S.; Samuels, C.; Savard, J.; Harris, C.; Aubin, M.; Olson, K.; et al. Sleep Disturbance in Adults with Cancer: A Systematic Review of Evidence for Best Practices in Assessment and Management for Clinical Practice. *Ann. Oncol.* **2014**, *25*, 791–800. [CrossRef]
7. Fox, R.S.; Ancoli-Israel, S.; Roesch, S.C.; Merz, E.L.; Mills, S.D.; Wells, K.J.; Sadler, G.R.; Malcarne, V.L. Sleep Disturbance and Cancer-Related Fatigue Symptom Cluster in Breast Cancer Patients Undergoing Chemotherapy. *Support. Care Cancer* **2020**, *28*, 845–855. [CrossRef]
8. Van Onselen, C.; Paul, S.M.; Lee, K.; Dunn, L.; Aouizerat, B.E.; West, C.; Dodd, M.; Cooper, B.; Miaskowski, C. Trajectories of Sleep Disturbance and Daytime Sleepiness in Women Before and After Surgery for Breast Cancer. *J. Pain Symptom Manag.* **2013**, *45*, 244–260. [CrossRef]
9. Fortner, B.V.; Stepanski, E.J.; Wang, S.C.; Kasprowicz, S.; Durrence, H.H. Sleep and Quality of Life in Breast Cancer Patients. *J. Pain Symptom Manag.* **2002**, *24*, 471–480. [CrossRef]
10. Erren, T.C.; Morfeld, P.; Foster, R.G.; Reiter, R.J.; Groß, J.V.; Westermann, I.K. Sleep and Cancer: Synthesis of Experimental Data and Meta-Analyses of Cancer Incidence among Some 1,500,000 Study Individuals in 13 Countries. *Chronobiol. Int.* **2016**, *33*, 325–350. [CrossRef]
11. Mogavero, M.P.; DelRosso, L.M.; Fanfulla, F.; Bruni, O.; Ferri, R. Sleep Disorders and Cancer: State of the Art and Future Perspectives. *Sleep Med. Rev.* **2021**, *56*, 101409. [CrossRef]
12. Samuelsson, L.B.; Bovbjerg, D.H.; Roecklein, K.A.; Hall, M.H. Sleep and Circadian Disruption and Incident Breast Cancer Risk: An Evidence-Based and Theoretical Review. *Neurosci. Biobehav. Rev.* **2018**, *84*, 35–48. [CrossRef] [PubMed]
13. Papagiannakopoulos, T.; Bauer, M.R.; Davidson, S.M.; Heimann, M.; Subbaraj, L.; Bhutkar, A.; Bartlebaugh, J.; Vander Heiden, M.G.; Jacks, T. Circadian Rhythm Disruption Promotes Lung Tumorigenesis. *Cell Metab.* **2016**, *24*, 324–331. [CrossRef] [PubMed]

14. De Lorenzo, B.H.P.; E Brito, R.R.N.; Paslar Leal, T.; Piqueira Garcia, N.; Martins Dos Santos, R.M.; Alvares-Saraiva, A.M.; Perez Hurtado, E.C.; Braga Dos Reis, T.C.; Duarte Palma, B. Chronic Sleep Restriction Impairs the Antitumor Immune Response in Mice. *Neuroimmunomodulation* **2018**, *25*, 59–67. [CrossRef] [PubMed]
15. Irwin, M.R.; Olmstead, R.; Carroll, J.E. Sleep Disturbance, Sleep Duration, and Inflammation: A Systematic Review and Meta-Analysis of Cohort Studies and Experimental Sleep Deprivation. *Biol. Psychiatry* **2016**, *80*, 40–52. [CrossRef]
16. Raison, C.L.; Rye, D.B.; Woolwine, B.J.; Vogt, G.J.; Bautista, B.M.; Spivey, J.R.; Miller, A.H. Chronic Interferon-Alpha Administration Disrupts Sleep Continuity and Depth in Patients with Hepatitis C: Association with Fatigue, Motor Slowing, and Increased Evening Cortisol. *Biol. Psychiatry* **2010**, *68*, 942–949. [CrossRef]
17. Besedovsky, L.; Lange, T.; Haack, M. The Sleep-Immune Crosstalk in Health and Disease. *Physiol. Rev.* **2019**, *99*, 1325–1380. [CrossRef]
18. Perrier, J.; Duivon, M.; Rauchs, G.; Giffard, B. Le sommeil dans les cancers non-cérébraux: Revue de la littérature, mécanismes potentiels et perspectives pour mieux comprendre les troubles cognitifs associés. *Médecine Du Sommeil* **2021**, *18*, 90–103. [CrossRef]
19. Ancoli-Israel, S.; Liu, L.; Natarajan, L.; Rissling, M.; Neikrug, A.B.; Youngstedt, S.D.; Mills, P.J.; Sadler, G.R.; Dimsdale, J.E.; Parker, B.A.; et al. Reductions in Sleep Quality and Circadian Activity Rhythmicity Predict Longitudinal Changes in Objective and Subjective Cognitive Functioning in Women Treated for Breast Cancer. *Support. Care Cancer* **2022**, *30*, 3187–3200. [CrossRef]
20. Lin, H.-H.; Farkas, M.E. Altered Circadian Rhythms and Breast Cancer: From the Human to the Molecular Level. *Front. Endocrinol.* **2018**, *9*, 219. [CrossRef]
21. Cash, E.; Sephton, S.E.; Chagpar, A.B.; Spiegel, D.; Rebholz, W.N.; Zimmaro, L.A.; Tillie, J.M.; Dhabhar, F.S. Circadian Disruption and Biomarkers of Tumor Progression in Breast Cancer Patients Awaiting Surgery. *Brain Behav. Immun.* **2015**, *48*, 102–114. [CrossRef] [PubMed]
22. Roth, T. Insomnia: Definition, Prevalence, Etiology, and Consequences. *J. Clin. Sleep Med.* **2007**, *3*, s7–s10. [CrossRef] [PubMed]
23. Fiorentino, L.; Ancoli-Israel, S. Insomnia and Its Treatment in Women with Breast Cancer. *Sleep Med. Rev.* **2006**, *10*, 419–429. [CrossRef] [PubMed]
24. Sung, H.; Ferlay, J.; Siegel, R.L.; Laversanne, M.; Soerjomataram, I.; Jemal, A.; Bray, F. Global Cancer Statistics 2020: GLOBOCAN Estimates of Incidence and Mortality Worldwide for 36 Cancers in 185 Countries. *CA Cancer J. Clin.* **2021**, *71*, 209–249. [CrossRef]
25. Leysen, L.; Lahousse, A.; Nijs, J.; Adriaenssens, N.; Mairesse, O.; Ivakhnov, S.; Bilterys, T.; Van Looveren, E.; Pas, R.; Beckwée, D. Prevalence and Risk Factors of Sleep Disturbances in Breast Cancersurvivors: Systematic Review and Meta-Analyses. *Support. Care Cancer* **2019**, *27*, 4401–4433. [CrossRef]
26. Ancoli-Israel, S. Sleep Disturbances in Cancer: A Review. *Sleep Med. Res.* **2015**, *6*, 45–49. [CrossRef]
27. Savard, J.; Savard, M.-H. Insomnia and Cancer: Prevalence, Nature, and Nonpharmacologic Treatment. *Sleep Med. Clin.* **2013**, *8*, 373–387. [CrossRef]
28. Fleming, L.; Randell, K.; Stewart, E.; Espie, C.A.; Morrison, D.S.; Lawless, C.; Paul, J. Insomnia in Breast Cancer: A Prospective Observational Study. *Sleep* **2019**, *42*, zsy245. [CrossRef]
29. Costa, A.R.; Fontes, F.; Pereira, S.; Gonçalves, M.; Azevedo, A.; Lunet, N. Impact of Breast Cancer Treatments on Sleep Disturbances—A Systematic Review. *Breast* **2014**, *23*, 697–709. [CrossRef]
30. Otte, J.L.; Carpenter, J.S.; Manchanda, S.; Rand, K.L.; Skaar, T.C.; Weaver, M.; Chernyak, Y.; Zhong, X.; Igega, C.; Landis, C. Systematic Review of Sleep Disorders in Cancer Patients: Can the Prevalence of Sleep Disorders Be Ascertained? *Cancer Med.* **2015**, *4*, 183–200. [CrossRef]
31. Jakobsen, G.; Gjeilo, K.H.; Hjermstad, M.J.; Klepstad, P. An Update on Prevalence, Assessment, and Risk Factors for Sleep Disturbances in Patients with Advanced Cancer-Implications for Health Care Providers and Clinical Research. *Cancers* **2022**, *14*, 3933. [CrossRef] [PubMed]
32. Ancoli-Israel, S.; Liu, L.; Marler, M.R.; Parker, B.A.; Jones, V.; Sadler, G.R.; Dimsdale, J.; Cohen-Zion, M.; Fiorentino, L. Fatigue, Sleep, and Circadian Rhythms Prior to Chemotherapy for Breast Cancer. *Support. Care Cancer* **2006**, *14*, 201–209. [CrossRef] [PubMed]
33. Berger, A.M.; Farr, L.A.; Kuhn, B.R.; Fischer, P.; Agrawal, S. Values of Sleep/Wake, Activity/Rest, Circadian Rhythms, and Fatigue Prior to Adjuvant Breast Cancer Chemotherapy. *J. Pain Symptom Manag.* **2007**, *33*, 398–409. [CrossRef]
34. Barreira, T.V.; Schuna, J.M.; Chaput, J.-P. Normative reference values for actigraphy-measured total nocturnal sleep time in the us population. *Am. J. Epidemiol.* **2022**, *191*, 360–362. [CrossRef] [PubMed]
35. Kreutz, C.; Müller, J.; Schmidt, M.E.; Steindorf, K. Comparison of Subjectively and Objectively Assessed Sleep Problems in Breast Cancer Patients Starting Neoadjuvant Chemotherapy. *Support. Care Cancer* **2021**, *29*, 1015–1023. [CrossRef] [PubMed]
36. Trivedi, R.; Man, H.; Madut, A.; Mather, M.; Elder, E.; Dhillon, H.M.; Brand, A.; Howle, J.; Mann, G.; DeFazio, A.; et al. Irregular Sleep/Wake Patterns Are Associated with Reduced Quality of Life in Post-Treatment Cancer Patients: A Study Across Three Cancer Cohorts. *Front. Neurosci.* **2021**, *15*, 700923. [CrossRef]
37. Ratcliff, C.G.; Zepeda, S.G.; Hall, M.H.; Tullos, E.A.; Fowler, S.; Chaoul, A.; Spelman, A.; Arun, B.; Cohen, L. Patient Characteristics Associated with Sleep Disturbance in Breast Cancer Survivors. *Support. Care Cancer* **2021**, *29*, 2601–2611. [CrossRef]
38. Natale, V.; Plazzi, G.; Martoni, M. Actigraphy in the Assessment of Insomnia: A Quantitative Approach. *Sleep* **2009**, *32*, 767–771. [CrossRef]

39. Liu, L.; Rissling, M.; Natarajan, L.; Fiorentino, L.; Mills, P.J.; Dimsdale, J.E.; Sadler, G.R.; Parker, B.A.; Ancoli-Israel, S. The Longitudinal Relationship between Fatigue and Sleep in Breast Cancer Patients Undergoing Chemotherapy. *Sleep* **2012**, *35*, 237–245. [CrossRef]
40. Liu, L.; Rissling, M.; Neikrug, A.; Fiorentino, L.; Natarajan, L.; Faierman, M.; Sadler, G.R.; Dimsdale, J.E.; Mills, P.J.; Parker, B.A.; et al. Fatigue and Circadian Activity Rhythms in Breast Cancer Patients Before and After Chemotherapy: A Controlled Study. *Fatigue* **2013**, *1*, 12–26. [CrossRef]
41. Kuo, H.-H.; Chiu, M.-J.; Liao, W.-C.; Hwang, S.-L. Quality of Sleep and Related Factors during Chemotherapy in Patients with Stage I/II Breast Cancer. *J. Formos. Med. Assoc.* **2006**, *105*, 64–69. [CrossRef] [PubMed]
42. Ancoli-Israel, S.; Liu, L.; Rissling, M.; Natarajan, L.; Neikrug, A.B.; Palmer, B.W.; Mills, P.J.; Parker, B.A.; Sadler, G.R.; Maglione, J. Sleep, Fatigue, Depression, and Circadian Activity Rhythms in Women with Breast Cancer before and after Treatment: A 1-Year Longitudinal Study. *Support. Care Cancer* **2014**, *22*, 2535–2545. [CrossRef] [PubMed]
43. Payne, J.; Piper, B.; Rabinowitz, I.; Zimmerman, B. Biomarkers, Fatigue, Sleep, and Depressive Symptoms in Women with Breast Cancer: A Pilot Study. *Oncol. Nurs. Forum* **2006**, *33*, 775–783. [CrossRef]
44. Liu, L.; Mills, P.J.; Rissling, M.; Fiorentino, L.; Natarajan, L.; Dimsdale, J.E.; Sadler, G.R.; Parker, B.A.; Ancoli-Israel, S. Fatigue and Sleep Quality Are Associated with Changes in Inflammatory Markers in Breast Cancer Patients Undergoing Chemotherapy. *Brain Behav. Immun.* **2012**, *26*, 706–713. [CrossRef]
45. Beck, S.L.; Berger, A.M.; Barsevick, A.M.; Wong, B.; Stewart, K.A.; Dudley, W.N. Sleep Quality after Initial Chemotherapy for Breast Cancer. *Support. Care Cancer* **2010**, *18*, 679–689. [CrossRef]
46. Li, W.; Kwok, C.C.-H.; Chan, D.C.-W.; Ho, A.W.-Y.; Ho, C.-S.; Zhang, J.; Wing, Y.K.; Wang, F.; Tse, L.A. Disruption of Sleep, Sleep-Wake Activity Rhythm, and Nocturnal Melatonin Production in Breast Cancer Patients Undergoing Adjuvant Chemotherapy: Prospective Cohort Study. *Sleep Med.* **2019**, *55*, 14–21. [CrossRef] [PubMed]
47. Madsen, M.T.; Huang, C.; Gögenur, I. Actigraphy for Measurements of Sleep in Relation to Oncological Treatment of Patients with Cancer: A Systematic Review. *Sleep Med. Rev.* **2015**, *20*, 73–83. [CrossRef]
48. Bhave, M.A.; Speth, K.A.; Kidwell, K.M.; Lyden, A.; Alsamarraie, C.; Murphy, S.L.; Henry, N.L. Effect of Aromatase Inhibitor Therapy on Sleep and Activity Patterns in Early-Stage Breast Cancer. *Clin. Breast Cancer* **2018**, *18*, 168–174.e2. [CrossRef]
49. Martin, T.; Duivon, M.; Bessot, N.; Grellard, J.M.; Emile, G.; Polvent, S.; Raoul, L.; Viader, F.; Eustache, F.; Joly, F.; et al. Rest Activity Rhythms Characteristics of Breast Cancer Women Following Endocrine Therapy. *Sleep* **2021**, *45*, zsab248. [CrossRef]
50. Dhruva, A.; Paul, S.M.; Cooper, B.A.; Lee, K.; West, C.; Aouizerat, B.E.; Dunn, L.B.; Swift, P.S.; Wara, W.; Miaskowski, C. A Longitudinal Study of Measures of Objective and Subjective Sleep Disturbance in Patients with Breast Cancer before, during, and after Radiation Therapy. *J. Pain Symptom Manag.* **2012**, *44*, 215–228. [CrossRef]
51. Silberfarb, P.M.; Hauri, P.J.; Oxman, T.E.; Schnurr, P. Assessment of Sleep in Patients with Lung Cancer and Breast Cancer. *J. Clin. Oncol.* **1993**, *11*, 997–1004. [CrossRef] [PubMed]
52. Roscoe, J.A.; Perlis, M.L.; Pigeon, W.R.; O'Neill, K.H.; Heckler, C.E.; Matteson-Rusby, S.E.; Palesh, O.G.; Shayne, M.; Huston, A. Few Changes Observed in Polysomnographic-Assessed Sleep before and after Completion of Chemotherapy. *J. Psychosom. Res.* **2011**, *71*, 423–428. [CrossRef]
53. Tag Eldin, E.-S.; Younis, S.G.; Aziz, L.M.A.E.; Eldin, A.T.; Erfan, S.T. Evaluation of Sleep Pattern Disorders in Breast Cancer Patients Receiving Adjuvant Treatment (Chemotherapy and/or Radiotherapy) Using Polysomnography. *J. Buon* **2019**, *24*, 529–534. [PubMed]
54. Parker, K.P.; Bliwise, D.L.; Ribeiro, M.; Jain, S.R.; Vena, C.I.; Kohles-Baker, M.K.; Rogatko, A.; Xu, Z.; Harris, W.B. Sleep/Wake Patterns of Individuals with Advanced Cancer Measured by Ambulatory Polysomnography. *J. Clin. Oncol. Off. J. Am. Soc. Clin. Oncol.* **2008**, *26*, 2464–2472. [CrossRef] [PubMed]
55. Williams, R.L.; Karacan, I.; Hursch, C.J. *Electroencephalography (EEG) of Human Sleep: Clinical Applications*; John Wiley & Sons: Hoboken, NJ, USA, 1974.
56. Gonçalves, B.; Adamowicz, T.; Louzada, F.M.; Moreno, C.R.; Araujo, J.F. A Fresh Look at the Use of Nonparametric Analysis in Actimetry. *Sleep Med. Rev.* **2015**, *20*, 84–91. [CrossRef] [PubMed]
57. Carpenter, J.S.; Gautam, S.; Freedman, R.R.; Andrykowski, M. Circadian Rhythm of Objectively Recorded Hot Flashes in Postmenopausal Breast Cancer Survivors. *Menopause* **2001**, *8*, 181–188. [CrossRef]
58. Abercrombie, H.C.; Giese-Davis, J.; Sephton, S.; Epel, E.S.; Turner-Cobb, J.M.; Spiegel, D. Flattened Cortisol Rhythms in Metastatic Breast Cancer Patients. *Psychoneuroendocrinology* **2004**, *29*, 1082–1092. [CrossRef]
59. Hsiao, F.-H.; Jow, G.-M.; Kuo, W.-H.; Wang, M.-Y.; Chang, K.-J.; Lai, Y.-M.; Chen, Y.-T.; Huang, C.-S. A Longitudinal Study of Diurnal Cortisol Patterns and Associated Factors in Breast Cancer Patients from the Transition Stage of the End of Active Cancer Treatment to Post-Treatment Survivorship. *Breast* **2017**, *36*, 96–101. [CrossRef]
60. Savard, J.; Liu, L.; Natarajan, L.; Rissling, M.B.; Neikrug, A.B.; He, F.; Dimsdale, J.E.; Mills, P.J.; Parker, B.A.; Sadler, G.R.; et al. Breast Cancer Patients Have Progressively Impaired Sleep-Wake Activity Rhythms during Chemotherapy. *Sleep* **2009**, *32*, 1155–1160. [CrossRef]
61. Sultan, A.; Choudhary, V.; Parganiha, A. Worsening of Rest-Activity Circadian Rhythm and Quality of Life in Female Breast Cancer Patients along Progression of Chemotherapy Cycles. *Chronobiol. Int.* **2017**, *34*, 609–623. [CrossRef]

62. Roveda, E.; Bruno, E.; Galasso, L.; Mulè, A.; Castelli, L.; Villarini, A.; Caumo, A.; Esposito, F.; Montaruli, A.; Pasanisi, P. Rest-Activity Circadian Rhythm in Breast Cancer Survivors at 5 Years after the Primary Diagnosis. *Chronobiol. Int.* **2019**, *36*, 1156–1165. [CrossRef] [PubMed]
63. Galasso, L.; Montaruli, A.; Mulè, A.; Castelli, L.; Bruno, E.; Pasanisi, P.; Caumo, A.; Esposito, F.; Roveda, E. Rest-Activity Rhythm in Breast Cancer Survivors: An Update Based on Non-Parametric Indices. *Chronobiol. Int.* **2020**, *37*, 946–951. [CrossRef] [PubMed]
64. De Rooij, B.H.; Ramsey, I.; Clouth, F.J.; Corsini, N.; Heyworth, J.S.; Lynch, B.M.; Vallance, J.K.; Boyle, T. The Association of Circadian Parameters and the Clustering of Fatigue, Depression, and Sleep Problems in Breast Cancer Survivors: A Latent Class Analysis. *J. Cancer Surviv.* **2022**, 1–11. [CrossRef]
65. Komarzynski, S.; Huang, Q.; Lévi, F.A.; Palesh, O.G.; Ulusakarya, A.; Bouchahda, M.; Haydar, M.; Wreglesworth, N.I.; Morère, J.-F.; Adam, R.; et al. The Day after: Correlates of Patient-Reported Outcomes with Actigraphy-Assessed Sleep in Cancer Patients at Home (InCASA Project). *Sleep* **2019**, *42*, zsz146. [CrossRef] [PubMed]
66. Stehle, J.H.; von Gall, C.; Korf, H.-W. Melatonin: A Clock-Output, a Clock-Input. *J. Neuroendocr.* **2003**, *15*, 383–389. [CrossRef]
67. Cherrie, J.W. Shedding Light on the Association between Night Work and Breast Cancer. *Ann. Work Expo. Health* **2019**, *63*, 608–611. [CrossRef]
68. Sephton, S.; Spiegel, D. Circadian Disruption in Cancer: A Neuroendocrine-Immune Pathway from Stress to Disease? *Brain Behav. Immun.* **2003**, *17*, 321–328. [CrossRef]
69. Sephton, S.E.; Sapolsky, R.M.; Kraemer, H.C.; Spiegel, D. Diurnal Cortisol Rhythm as a Predictor of Breast Cancer Survival. *J. Natl. Cancer Inst.* **2000**, *92*, 994–1000. [CrossRef]
70. Eismann, E.A.; Lush, E.; Sephton, S.E. Circadian Effects in Cancer-Relevant Psychoneuroendocrine and Immune Pathways. *Psychoneuroendocrinology* **2010**, *35*, 963–976. [CrossRef]
71. Zeitzer, J.M.; Nouriani, B.; Rissling, M.B.; Sledge, G.W.; Kaplan, K.A.; Aasly, L.; Palesh, O.; Jo, B.; Neri, E.; Dhabhar, F.S.; et al. Aberrant Nocturnal Cortisol and Disease Progression in Women with Breast Cancer. *Breast Cancer Res. Treat.* **2016**, *158*, 43–50. [CrossRef]
72. Palesh, O.; Aldridge-Gerry, A.; Zeitzer, J.M.; Koopman, C.; Neri, E.; Giese-Davis, J.; Jo, B.; Kraemer, H.; Nouriani, B.; Spiegel, D. Actigraphy-Measured Sleep Disruption as a Predictor of Survival among Women with Advanced Breast Cancer. *Sleep* **2014**, *37*, 837–842. [CrossRef] [PubMed]
73. Lu, C.; Sun, H.; Huang, J.; Yin, S.; Hou, W.; Zhang, J.; Wang, Y.; Xu, Y.; Xu, H. Long-Term Sleep Duration as a Risk Factor for Breast Cancer: Evidence from a Systematic Review and Dose-Response Meta-Analysis. *Biomed. Res. Int.* **2017**, *2017*, 4845059. [CrossRef] [PubMed]
74. Song, C.; Zhang, R.; Wang, C.; Fu, R.; Song, W.; Dou, K.; Wang, S. Sleep Quality and Risk of Cancer: Findings from the English Longitudinal Study of Aging. *Sleep* **2021**, *44*, zsaa192. [CrossRef]
75. Rosenberg, R.P.; Benca, R.; Doghramji, P.; Roth, T. A 2023 Update on Managing Insomnia in Primary Care: Insights From an Expert Consensus Group. *Prim. Care Companion CNS Disord.* **2023**, *25*, 22nr03385. [CrossRef] [PubMed]
76. Yue, J.-L.; Chang, X.-W.; Zheng, J.-W.; Shi, L.; Xiang, Y.-J.; Que, J.-Y.; Yuan, K.; Deng, J.-H.; Teng, T.; Li, Y.-Y.; et al. Efficacy and Tolerability of Pharmacological Treatments for Insomnia in Adults: A Systematic Review and Network Meta-Analysis. *Sleep Med. Rev.* **2023**, *68*, 101746. [CrossRef] [PubMed]
77. Amato, J.-N.; Marie, S.; Lelong-Boulouard, V.; Paillet-Loilier, M.; Berthelon, C.; Coquerel, A.; Denise, P.; Bocca, M.-L. Effects of Three Therapeutic Doses of Codeine/Paracetamol on Driving Performance, a Psychomotor Vigilance Test, and Subjective Feelings. *Psychopharmacology* **2013**, *228*, 309–320. [CrossRef] [PubMed]
78. Greenlee, H.; DuPont-Reyes, M.J.; Balneaves, L.G.; Carlson, L.E.; Cohen, M.R.; Deng, G.; Johnson, J.A.; Mumber, M.; Seely, D.; Zick, S.M.; et al. Clinical Practice Guidelines on the Evidence-Based Use of Integrative Therapies during and after Breast Cancer Treatment. *CA A Cancer J. Clin.* **2017**, *67*, 194–232. [CrossRef]
79. Lahart, I.M.; Metsios, G.S.; Nevill, A.M.; Carmichael, A.R. Physical Activity for Women with Breast Cancer after Adjuvant Therapy. *Cochrane Database Syst. Rev.* **2018**, *1*, CD011292. [CrossRef]
80. Landry, S.; Chasles, G.; Pointreau, Y.; Bourgeois, H.; Boyas, S. Influence of an Adapted Physical Activity Program on Self-Esteem and Quality of Life of Breast Cancer Patients after Mastectomy. *Oncology* **2018**, *95*, 188–191. [CrossRef]
81. Savard, J.; Ivers, H.; Savard, M.-H.; Morin, C.M. Long-Term Effects of Two Formats of Cognitive Behavioral Therapy for Insomnia Comorbid with Breast Cancer. *Sleep* **2016**, *39*, 813–823. [CrossRef]
82. Zhang, M.; Huang, L.; Feng, Z.; Shao, L.; Chen, L. Effects of Cognitive Behavioral Therapy on Quality of Life and Stress for Breast Cancer Survivors: A Meta-Analysis. *Minerva Med.* **2017**, *108*, 84–93. [CrossRef] [PubMed]
83. Bernard, P.; Savard, J.; Steindorf, K.; Sweegers, M.G.; Courneya, K.S.; Newton, R.U.; Aaronson, N.K.; Jacobsen, P.B.; May, A.M.; Galvao, D.A.; et al. Effects and Moderators of Exercise on Sleep in Adults with Cancer: Individual Patient Data and Aggregated Meta-Analyses. *J. Psychosom. Res.* **2019**, *124*, 109746. [CrossRef] [PubMed]
84. Amidi, A.; Buskbjerg, C.R.; Damholdt, M.F.; Dahlgaard, J.; Thorndike, F.P.; Ritterband, L.; Zachariae, R. Changes in Sleep Following Internet-Delivered Cognitive-Behavioral Therapy for Insomnia in Women Treated for Breast Cancer: A 3-Year Follow-up Assessment. *Sleep Med.* **2022**, *96*, 35–41. [CrossRef] [PubMed]
85. Kreutz, C.; Schmidt, M.E.; Steindorf, K. Effects of Physical and Mind-Body Exercise on Sleep Problems during and after Breast Cancer Treatment: A Systematic Review and Meta-Analysis. *Breast Cancer Res. Treat.* **2019**, *176*, 1–15. [CrossRef] [PubMed]

86. Mercier, J.; Savard, J.; Bernard, P. Exercise Interventions to Improve Sleep in Cancer Patients: A Systematic Review and Meta-Analysis. *Sleep Med. Rev.* **2017**, *36*, 43–56. [CrossRef] [PubMed]
87. Ma, Y.; Hall, D.L.; Ngo, L.H.; Liu, Q.; Bain, P.A.; Yeh, G.Y. Efficacy of Cognitive Behavioral Therapy for Insomnia in Breast Cancer: A Meta-Analysis. *Sleep Med. Rev.* **2021**, *55*, 101376. [CrossRef] [PubMed]
88. Qiu, H.; Ren, W.; Yang, Y.; Zhu, X.; Mao, G.; Mao, S.; Lin, Y.; Shen, S.; Li, C.; Shi, H.; et al. Effects of Cognitive Behavioral Therapy for Depression on Improving Insomnia and Quality of Life in Chinese Women with Breast Cancer: Results of a Randomized, Controlled, Multicenter Trial. *Neuropsychiatr. Dis. Treat.* **2018**, *14*, 2665–2673. [CrossRef]
89. Gao, Y.; Liu, M.; Yao, L.; Yang, Z.; Chen, Y.; Niu, M.; Sun, Y.; Chen, J.; Hou, L.; Sun, F.; et al. Cognitive Behavior Therapy for Insomnia in Cancer Patients: A Systematic Review and Network Meta-Analysis. *J. Evid. Based Med.* **2022**, *15*, 216–229. [CrossRef]
90. Matthews, E.E.; Arnedt, J.T.; McCarthy, M.S.; Cuddihy, L.J.; Aloia, M.S. Adherence to Cognitive Behavioral Therapy for Insomnia: A Systematic Review. *Sleep Med. Rev.* **2013**, *17*, 453–464. [CrossRef]
91. Johnson, J.A.; Rash, J.A.; Campbell, T.S.; Savard, J.; Gehrman, P.R.; Perlis, M.; Carlson, L.E.; Garland, S.N. A Systematic Review and Meta-Analysis of Randomized Controlled Trials of Cognitive Behavior Therapy for Insomnia (CBT-I) in Cancer Survivors. *Sleep Med. Rev.* **2016**, *27*, 20–28. [CrossRef]
92. Wampler, M.A.; Topp, K.S.; Miaskowski, C.; Byl, N.N.; Rugo, H.S.; Hamel, K. Quantitative and Clinical Description of Postural Instability in Women with Breast Cancer Treated with Taxane Chemotherapy. *Arch. Phys. Med. Rehabil.* **2007**, *88*, 1002–1008. [CrossRef] [PubMed]
93. Medina, H.N.; Liu, Q.; Cao, C.; Yang, L. Balance and Vestibular Function and Survival in US Cancer Survivors. *Cancer* **2021**, *127*, 4022–4029. [CrossRef] [PubMed]
94. Monfort, S.M.; Pan, X.; Patrick, R.; Singaravelu, J.; Loprinzi, C.L.; Lustberg, M.B.; Chaudhari, A.M.W. Natural History of Postural Instability in Breast Cancer Patients Treated with Taxane-Based Chemotherapy: A Pilot Study. *Gait Posture* **2016**, *48*, 237–242. [CrossRef] [PubMed]
95. Winters-Stone, K.M.; Torgrimson, B.; Horak, F.; Eisner, A.; Nail, L.; Leo, M.C.; Chui, S.; Luoh, S.-W. Identifying Factors Associated With Falls in Postmenopausal Breast Cancer Survivors: A Multi-Disciplinary Approach. *Arch. Phys. Med. Rehabil.* **2011**, *92*, 646–652. [CrossRef] [PubMed]
96. Martin, T.; Mauvieux, B.; Bulla, J.; Quarck, G.; Davenne, D.; Denise, P.; Philoxène, B.; Besnard, S. Vestibular Loss Disrupts Daily Rhythm in Rats. *J. Appl. Physiol.* **2015**, *118*, 310–318. [CrossRef]
97. Martin, T.; Moussay, S.; Bulla, I.; Bulla, J.; Toupet, M.; Etard, O.; Denise, P.; Davenne, D.; Coquerel, A.; Quarck, G. Exploration of Circadian Rhythms in Patients with Bilateral Vestibular Loss. *PLoS ONE* **2016**, *11*, e0155067. [CrossRef]
98. Besnard, S.; Tighilet, B.; Chabbert, C.; Hitier, M.; Toulouse, J.; Le Gall, A.; Machado, M.-L.; Smith, P.F. The Balance of Sleep: Role of the Vestibular Sensory System. *Sleep Med. Rev.* **2018**, *42*, 220–228. [CrossRef]
99. Fuller, P.M.; Jones, T.A.; Jones, S.M.; Fuller, C.A. Neurovestibular Modulation of Circadian and Homeostatic Regulation: Vestibulohypothalamic Connection? *Proc. Natl. Acad. Sci. USA* **2002**, *99*, 15723–15728. [CrossRef]
100. Fuller, P.M.; Jones, T.A.; Jones, S.M.; Fuller, C.A. Evidence for Macular Gravity Receptor Modulation of Hypothalamic, Limbic and Autonomic Nuclei. *Neuroscience* **2004**, *129*, 461–471. [CrossRef]
101. Fuller, P.M.; Fuller, C.A. Genetic Evidence for a Neurovestibular Influence on the Mammalian Circadian Pacemaker. *J. Biol. Rhythm.* **2006**, *21*, 177–184. [CrossRef]
102. Pasquier, F.; Bessot, N.; Martin, T.; Gauthier, A.; Bulla, J.; Denise, P.; Quarck, G. Effect of Vestibular Stimulation Using a Rotatory Chair in Human Rest/Activity Rhythm. *Chronobiol. Int.* **2020**, *37*, 1244–1251. [CrossRef] [PubMed]
103. Micarelli, A.; Viziano, A.; Pistillo, R.; Granito, I.; Micarelli, B.; Alessandrini, M. Sleep Performance and Chronotype Behavior in Unilateral Vestibular Hypofunction. *Laryngoscope* **2021**, *131*, 2341–2347. [CrossRef] [PubMed]
104. Albathi, M.; Agrawal, Y. Vestibular Vertigo Is Associated with Abnormal Sleep Duration. *J. Vestib. Res. Equilib. Orientat.* **2017**, *27*, 127–135. [CrossRef] [PubMed]
105. Pace, A.; Milani, A.; Rossetti, V.; Iannella, G.; Maniaci, A.; Cocuzza, S.; Alunni Fegatelli, D.; Vestri, A.; Magliulo, G. Evaluation of Vestibular Function in Patients Affected by Obstructive Sleep Apnea Performing Functional Head Impulse Test (FHIT). *Nat. Sci. Sleep* **2022**, *14*, 475–482. [CrossRef]
106. Cheung, I.C.W.; Thorne, P.R.; Hussain, S.; Neeff, M.; Sommer, J.U. The Relationship between Obstructive Sleep Apnea with Hearing and Balance: A Scoping Review. *Sleep Med.* **2022**, *95*, 55–75. [CrossRef]
107. Sugaya, N.; Arai, M.; Goto, F. The Effect of Vestibular Rehabilitation on Sleep Disturbance in Patients with Chronic Dizziness. *Acta Oto-Laryngol.* **2017**, *137*, 275–278. [CrossRef]
108. Bayer, L.; Constantinescu, I.; Perrig, S.; Vienne, J.; Vidal, P.-P.; Mühlethaler, M.; Schwartz, S. Rocking Synchronizes Brain Waves during a Short Nap. *Curr. Biol.* **2011**, *21*, R461–R462. [CrossRef]
109. Perrault, A.A.; Khani, A.; Quairiaux, C.; Kompotis, K.; Franken, P.; Muhlethaler, M.; Schwartz, S.; Bayer, L. Whole-Night Continuous Rocking Entrains Spontaneous Neural Oscillations with Benefits for Sleep and Memory. *Curr. Biol.* **2019**, *29*, 402–411.e3. [CrossRef]
110. Baek, S.; Yu, H.; Roh, J.; Lee, J.; Sohn, I.; Kim, S.; Park, C. Effect of a Recliner Chair with Rocking Motions on Sleep Efficiency. *Sensors* **2021**, *21*, 8214. [CrossRef]
111. Kompotis, K.; Hubbard, J.; Emmenegger, Y.; Perrault, A.; Mühlethaler, M.; Schwartz, S.; Bayer, L.; Franken, P. Rocking Promotes Sleep in Mice through Rhythmic Stimulation of the Vestibular System. *Curr. Biol.* **2019**, *29*, 392–401.e4. [CrossRef]

112. Van Sluijs, R.M.; Rondei, Q.J.; Schluep, D.; Jäger, L.; Riener, R.; Achermann, P.; Wilhelm, E. Effect of Rocking Movements on Afternoon Sleep. *Front. Neurosci.* **2019**, *13*, 1446. [CrossRef]
113. Van Sluijs, R.; Wilhelm, E.; Rondei, Q.; Omlin, X.; Crivelli, F.; Straumann, D.; Jäger, L.; Riener, R.; Achermann, P. Gentle Rocking Movements during Sleep in the Elderly. *J. Sleep Res.* **2020**, *29*, e12989. [CrossRef] [PubMed]
114. Omlin, X.; Crivelli, F.; Näf, M.; Heinicke, L.; Skorucak, J.; Malafeev, A.; Fernandez Guerrero, A.; Riener, R.; Achermann, P. The Effect of a Slowly Rocking Bed on Sleep. *Sci. Rep.* **2018**, *8*, 2156. [CrossRef]
115. Horowitz, S.S.; Blanchard, J.; Morin, L.P. Medial Vestibular Connections with the Hypocretin (Orexin) System. *J. Comp. Neurol.* **2005**, *487*, 127–146. [CrossRef]
116. Cellini, N.; Mednick, S.C. Stimulating the Sleeping Brain: Current Approaches to Modulating Memory-Related Sleep Physiology. *J. Neurosci. Methods* **2019**, *316*, 125–136. [CrossRef] [PubMed]
117. Fitzpatrick, R.C.; Day, B.L. Probing the Human Vestibular System with Galvanic Stimulation. *J. Appl. Physiol.* **2004**, *96*, 2301–2316. [CrossRef]
118. Utz, K.S.; Dimova, V.; Oppenländer, K.; Kerkhoff, G. Electrified Minds: Transcranial Direct Current Stimulation (TDCS) and Galvanic Vestibular Stimulation (GVS) as Methods of Non-Invasive Brain Stimulation in Neuropsychology—A Review of Current Data and Future Implications. *Neuropsychologia* **2010**, *48*, 2789–2810. [CrossRef] [PubMed]
119. Dlugaiczyk, J.; Gensberger, K.D.; Straka, H. Galvanic Vestibular Stimulation: From Basic Concepts to Clinical Applications. *J. Neurophysiol.* **2019**, *121*, 2237–2255. [CrossRef]

 cancers

Review

An Update on Prevalence, Assessment, and Risk Factors for Sleep Disturbances in Patients with Advanced Cancer—Implications for Health Care Providers and Clinical Research

Gunnhild Jakobsen [1,2,*], Kari Hanne Gjeilo [1,3], Marianne Jensen Hjermstad [4,5,6] and Pål Klepstad [7,8]

1 Department of Public Health and Nursing, Faculty of Medicine and Health Sciences, Norwegian University of Science and Technology (NTNU), 7491 Trondheim, Norway
2 Cancer Clinic, St. Olavs Hospital, Trondheim University Hospital, 7006 Trondheim, Norway
3 Clinic of Cardiology and Department of Cardiothoracic Surgery, St. Olavs Hospital, Trondheim University Hospital, 7006 Trondheim, Norway
4 Regional Advisory Unit in Palliative Care, Department of Oncology, Oslo University Hospital, 0424 Oslo, Norway
5 European Palliative Care Research Centre, Department of Oncology, Oslo University Hospital, 0424 Oslo, Norway
6 Institute of Clinical Medicine, University of Oslo, 0318 Oslo, Norway
7 Department of Anaesthesiology and Intensive Care Medicine, St. Olavs Hospital, Trondheim University Hospital, 7006 Trondheim, Norway
8 Department of Circulation and Medical Imaging, Faculty of Medicine and Health Sciences, Norwegian University of Science and Technology (NTNU), 7491 Trondheim, Norway
* Correspondence: gunnhild.jakobsen@ntnu.no

Citation: Jakobsen, G.; Gjeilo, K.H.; Hjermstad, M.J.; Klepstad, P. An Update on Prevalence, Assessment, and Risk Factors for Sleep Disturbances in Patients with Advanced Cancer—Implications for Health Care Providers and Clinical Research. *Cancers* **2022**, *14*, 3933. https://doi.org/10.3390/cancers14163933

Academic Editor: António Araújo

Received: 28 June 2022
Accepted: 11 August 2022
Published: 15 August 2022

Simple Summary: This review focuses on sleep in patients with advanced cancer. Cancer patients experience multiple symptoms and they receive concomitant medications. These are all factors that may affect sleep. In this paper, we present recommendations on sleep assessment in patients with advanced cancer and highlight cancer-related factors that may contribute to insomnia. Sleep is an essential aspect of health-related quality of life; therefore, it is important for health care providers to focus on sleep to improve patient care.

Abstract: Patients with advanced cancer experience multiple symptoms, with fluctuating intensity and severity during the disease. They use several medications, including opioids, which may affect sleep. Sleep disturbance is common in cancer patients, decreases the tolerability of other symptoms, and impairs quality of life. Despite its high prevalence and negative impact, poor sleep quality often remains unrecognized and undertreated. Given that sleep is an essential aspect of health-related quality of life, it is important to extend both the knowledge base and awareness among health care providers in this field to improve patient care. In this narrative review, we provide recommendations on sleep assessment in patients with advanced cancer and highlight cancer-related factors that contribute to insomnia. We also present direct implications for health care providers working in palliative care and for future research.

Keywords: sleep; sleep disturbances; insomnia; advanced cancer; palliative; palliative care

1. Introduction

Despite advances in treatment, cancer continues to cause substantial morbidity and mortality. For patients with a life-threatening disease, issues regarding quality of life for their remaining lifetime are critical [1]. Patients with advanced cancer experience multiple symptoms of fluctuating intensity and severity during the disease trajectory [2,3]. They normally use multiple concomitant medications, including opioids, which together with adverse symptoms may affect sleep [4–7]. Sleep is an essential aspect of health-related quality of life, and thus it is important to gain knowledge in this field to improve patient care. In this narrative review, we present current knowledge that relates to sleep disturbances and sleep assessment in patients with advanced cancer, defined as cancer that is unlikely to be cured and that may have spread from the original site to other parts of the body [8]. In addition, we highlight cancer-related and other factors which may contribute to insomnia in these patients. We focus on the implications for health care providers working in palliative as well as recommendations for future clinical research.

2. Sleep Disorders and Sleep Disturbances

According to the International Classification of Sleep Disorders, third edition (ICSD-3) of the American Academy of Sleep Medicine, sleep disorders are grouped into six major categories: insomnia, sleep-related breathing disorders, central disorders of hypersomnolence, circadian rhythm sleep–wake disorders, parasomnias, and sleep-related movement disorders [9]. In this paper, we use the term "insomnia in the context of cancer", as proposed by Savard and Morin: [10]

(1) Difficulty initiating sleep (greater than 30 min to sleep onset) and/or difficulty maintaining sleep (greater than 30 min nocturnal waking time);
(2) Sleep difficulty at least 3 nights per week;
(3) Sleep difficulty that causes significant impairment of daytime functioning.

Although there is a clear statement of insomnia as a sleep disorder, several non-specific terms are used for sleep by researchers, clinicians, and the public. The term «sleep disturbances» is used to designate insufficient or excessive sleep duration or poor self-reported sleep quality and may refer to sleep related symptoms and signs regardless of whether they fulfil criteria for specific diagnoses or not [11,12]. Another term which also lacks definitional consensus is «sleep quality». Consequently, sleep continuity measures such as sleep latency, awakenings, wake after sleep onset, and sleep efficiency are used as indicators of sleep quality [13]. For instance, shorter sleep latencies, fewer awakenings, and reduced wake after sleep onset indicate good sleep quality. The patient's subjective experience of sleep quality, as for instance reported on a numerical rating scale, can also be considered to describe sleep quality. Poor sleep quality is a subjective phenomenon and may be described by individual patients as a disruption of their habitual sleep pattern, difficulty falling asleep, frequent awakening, or nonrestorative sleep [14]. This review embraces both aspects of sleep quality; the patient-reported overall global approach of each night's sleep and quantitative aspects of sleep, such as total sleep time and sleep onset latency (i.e., how many minutes it takes to fall asleep starting from the moment of intention to fall asleep). However, health care providers should be aware of other sleep-related issues as the ones mentioned above, such as excessive daytime sleepiness, circadian rhythm disorders, or sleep-disordered breathing in cancer patients.

3. Sleep Assessment

To obtain detailed information on sleep disturbances, it is recommended to examine sleep by combining subjective methods using patient-reported outcome measures (PROMs) and objective registrations such as polysomnography (PSG) and actigraphy [15–18].

PROMs of sleep include sleep diaries and questionnaires [19]. A structured sleep diary is used by patients to register their bedtime hour, time to fall asleep, number and duration of awakenings during the night, and time of morning awakening and arising from bed [15,20]. In an expert consensus statement Carney et al. concluded that standardized, patient-informed sleep diaries are the standard for subjective sleep assessments [20]. In routine clinical care, a questionnaire such as the revised Edmonton Symptom Assessment System (ESAS-r) is recommended to screen for sleep disturbances in patients with advanced cancer [21]. ESAS-r is a valid and reliable questionnaire for the assessment of the intensity of symptoms in cancer populations, where the severity of each symptom is rated from 0 to 10 on a numerical scale, with 0 meaning that symptom is absent and 10 meaning that it is of the worst possible severity [22]. The ESAS-r consists of nine core symptoms (pain, tiredness, nausea, depression, anxiety, drowsiness, appetite, feeling of well-being, shortness of breath, and an optional 10th symptom to be selected by patients). Today, sleep is not a part of the ESAS-r symptoms, and the optional 10th symptom is often used to assess sleep. For screening purposes, Yennurajalingam et al. suggests that a cut-off of greater than or equal to four should generate further assessment of sleep [21,23].

Another questionnaire, which is validated and widely used to assess sleep quality in patients with advanced cancer is the Pittsburgh Sleep Quality Index (PSQI) [15,24]. It includes seven components of sleep: sleep quality, sleep latency, sleep duration, sleep efficiency, sleep disturbances, use of sleep medications, and daytime dysfunction. The component scores are summed to obtain a global sleep score ranging from 0 to 21, with higher scores indicating worse sleep quality [24]. Using this tool might improve the understanding of sleep difficulties experienced by cancer patients [25]. It covers multiple aspects relevant to sleep quality and might clarify the effect of sleep disturbances on patients' daily life. In addition, it is simple to use in clinical practice with a completion time of 5 to 10 min [26]. Other examples are the Insomnia Severity Index (ISI) and Athens Insomnia Scale (AIS). The ISI measures patients' perception of insomnia [27]. It is composed of seven items that evaluate the severity of sleep-onset, sleep maintenance, early morning awakening, satisfaction with current sleep pattern, interference with daily functioning, noticeability of impairment attributed to sleep problems, and level of distress caused by the sleep problems. The AIS is a self-assessment instrument, designed for quantifying sleep difficulty based on the ICD-10 criteria [28]. Health care providers in palliative care can use either ISI or AIS for quick identification of potential sleep problems in an individual cancer patient [29]. Thus, the use of such questionnaires in routine clinical care may help health care providers to gain insights into the patients' sleep problems.

The specific PROMs for the assessment of sleep vary in relation to which period they are designed to cover. For instance, the ESAS-r is typically used for the assessment of sleep last night [21], while the PSQI is designed to assess sleep last month [24]. The time interval in the ISI is the last two weeks [27], and the AIS is during the last month, or some other period of time, whose length depends on the purpose of a given study [28]. Thus, different studies with different aims and time frames may use different PROMs for sleep assessments.

Today, PSG is the gold standard for measuring sleep [15,30]. However, PSG is a comprehensive assessment method that provides overnight measures of brain waves, eye movement, muscle tension, electrocardiogram, and respiratory parameters. The PSG instrument is a complex monitoring device which requires specially trained personnel to attach the patients to its multiple sensors. As such, this method is usually too demanding for patients with a hight symptom burden, even in a study setting, let alone in routine care [31]. However, several studies have used actigraphy in the monitoring of sleep in patients with advanced cancer [32–34]. An actigraph, also known as an actometer, is worn on the wrist or ankle to record acceleration or deceleration of body movements, which indirectly indicates the state of sleep or wakefulness [35,36]. Advantages of actigraphy over PSG include ease of use, inexpensive recordings over extended periods of days, weeks, or months, and usefulness in cognitive impaired patients where PSG is not possible [37]. For seriously ill patients, such as patients with advanced cancer, actigraphy has become a valuable tool for objective sleep assessment [17].

Actigraphy is also a validated method to evaluate circadian rhythms both in research and clinical settings [16,38]. A recent review analysed the rest-activity circadian rhythm disruption in advanced cancer patients [39]. Circadian disruption was reported to be prevalent in this patient population. The disruption was manifested as lower activity levels during the day, more frequent and longer daytime naps, and fragmented night-time sleep. The circadian process is an internal rhythm or clock that dictates periods of activity (wakefulness) and inactivity (sleep) based in a light–dark cycle, and sleep is one of many bodily functions under control of the circadian clock [40,41]. As altered patterns have been described for several circadian rhythms in cancer [42], it is important to evaluate circadian rhythms in these patients. In fact, a study among patients with advanced cancer reported statistically significant and clinically meaningful associations between circadian rest–activity rhythm alterations and the severity of fatigue and anorexia, as well as impairment of physical and social dimensions of health-related quality of life [43]. This supports the need to develop interventions that target the circadian clock to improve symptom control in these patients.

4. Prevalence of Poor Sleep Quality

Sleep disturbances are prevalent in cancer [44–47]. A recent meta-analysis on the prevalence of sleep disturbances in patients with cancer reported an overall prevalence of 60.7%, suggesting that more than half of the cancer patients experience sleep disturbances [44]. Most importantly, the prevalence was even higher in patients with advanced cancer, with an overall prevalence of 70.8% [44]. Insomnia is considered an underdiagnosed and undertreated health problem in palliative care [48], as about one third of patients with cancer has insomnia symptoms. This is about three times higher than in the general population [45,49,50].

At the same time, the prevalence of patient-reported sleep disturbances in advanced cancer differs largely across studies. Table 1 provides examples of studies that have examined patient-reported sleep prevalence rates in patients with advanced cancer and the different assessment tools being used [21,25,51–59]. Such differences may be due to different study methods, designs and aims, assessment tools used, and population characteristics. In addition, and as mentioned above, the term «sleep disturbances» is non-specific and may contribute to the different prevalence rates across studies.

Table 1. Examples of studies investigating sleep quality in patients with advanced cancer [21,25,51–59].

Author, Country (Year)	N	Prevalence of Poor Sleep [1]	Questionnaire
Mercadante, Italy (2021) [59]	182	50%	Athens Insomnia Scale
Jakobsen, Denmark, Germany, Lithuania, Norway, Switzerland (2018) [51]	604	78%	PSQI
Collins, USA (2017) [52]	292	59%	PSQI
Yennurajalingam USA (2017) [21]	180	62%	PSQI
George, USA (2016) [53]	256	64%	PSQI
Akman, Turkey (2015) [25]	314	40%	PSQI
Nishiura, Tokyo (2015) [55]	50	56%	Athens Insomnia Scale
Mercadante, Italy (2015) [54]	820	61%	Athens Insomnia Scale
Davis, USA (2014) [56]	715	14%	Insomnia Severity Index
Yennurajalingam, USA (2013) [57]	442	75%	Sleep item on a 10-point scale
Delgado-Guay, USA (2011) [58]	101	85%	PSQI

[1] Patient-reported poor sleep prevalence rate in per cent, PSQI = Pittsburgh Sleep Quality Index.

Jakobsen et al. demonstrated that the majority (78%) of 604 adult patients with cancer pain using WHO Step III opioids reported poor sleep quality using the PSQI [51]. All components of sleep quality were affected suggesting that patients with advanced cancer experience a mixture of sleep disturbances, including difficulty initiating sleep, staying asleep, early awakenings, and that external factors such as pain, having to use the bathroom, inability to breath comfortably, or feeling too cold or hot disturbed sleep [51]. In line with other studies in palliative care [52,53,58,60], the mean PSQI global score was 8.8 (±4.2; range 0–20). Overall, studies demonstrate that sleep disturbances in patients with advanced cancer are prevalent and represent a complex clinical situation in palliative care.

5. Predisposing Factors for Insomnia in Advanced Cancer and Consequences for Other Symptoms

The potential causes of sleep disturbances in patients with advanced cancer are many, varied, and complex [10,46,61]. Clearly, the cancer disease and cancer treatment, place patients at increased risk for disruption of normal behaviors, habits, and physiological states that normally lead to restful sleep. For insomnia, several etiologic factors are involved in patients with advanced cancer. These are grouped into three main categories: predisposing factors, precipitating factors, and perpetuating factors [10,46,48,62,63]. Figure 1 illustrates some of these factors.

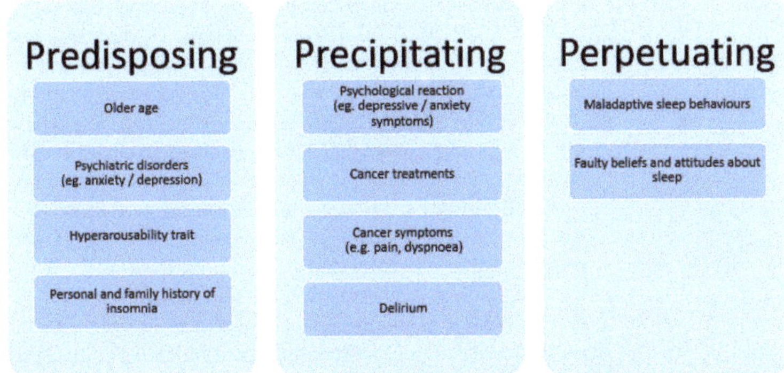

Figure 1. Insomnia in the context of advanced cancer. Examples of predisposing, precipitating, and perpetuating factors involved in the development of insomnia in advanced cancer [10,46,48,62,63].

Predisposing factors increase the individual's general vulnerability to develop insomnia, among these are older age, hyperarousability as trait, and personal or familiar history of insomnia [10,62,63]. Patients who have had insomnia prior to their cancer are at increased risk of experiencing insomnia when they are faced with cancer [62]. On the other hand, contrary to the general population, where female gender is a known predictor of insomnia, gender does not seem to be a predictor if insomnia in patients with advanced cancer [48].

Precipitating factors or situational conditions trigger the onset of insomnia, in which cancer is characterized by a succession of severe stressors that can trigger insomnia at any time during the cancer trajectory [48,64,65]. Precipitating factors include cancer treatments that can alter levels of inflammatory cytokines, disrupt circadian rhythms or sleep–wake cycles or cause menopause. Moreover, hospitalization, in itself, disturbs sleep [66]. Finally, medications used to treat or manage side effects and cancer-related symptoms, such as opioids or corticosteroids, will influence sleep [10,46,64]. However, in patients with advanced cancer, it may be difficult to differentiate, for instance, corticosteroid adverse effects from symptoms related to a progressive malignant disease [57,67]. To illustrate, treatment with methylprednisolone 16 mg twice daily for 7 days in patients with advanced cancer did not result in more patient-reported sleep problems as measured by the European Organisation for Research and Treatment of Cancer Quality of Life Questionnaire Core 30 in a randomized, placebo-controlled, double-blind trial using a standardized dose of corticosteroids [68].

In recent years, there has been an increased interest among researchers to understand the association between sleep problems and cancer-related symptoms. These symptoms are also referred to as precipitating factors for insomnia [10,62,69,70]. Patients with advanced cancer often report high levels of several co-occurring symptoms, in which pain, fatigue, nausea and vomiting, dyspnoea, constipation, loss of appetite, and depression are among the most common symptoms [71,72]. Pain is one of the most frequently reported cancer-related symptoms in association with sleep disturbances in patients with advanced cancer. In line with other studies in palliative care [5,21,56,58], an international multi-centre study reported that more pain was significantly associated with poor sleep as measured by the PSQI, and that pain intensity was a statistically significant predictor of poor sleep in patients with advanced cancer [51].

Another cancer-related symptom, psychological distress, is associated with sleep disturbances in patients with advanced cancer. Sleep quality, as assessed by the PSQI, was associated with emotional functioning in patients with advanced cancer using WHO Step III opioids [51], suggesting that patients with lower scores in emotional function, i.e., feeling tense, being worried, being irritable, and feeling depressed, reported more sleep disturbances. These results tie well with previous studies in palliative care which have demonstrated that sleep disturbances are associated with depression and reduced quality of life [58,73–76]. To illustrate, excessive rumination, potentially involved in increased psychological distress in palliative care, was associated with insomnia in patients with advanced cancer [74]. In addition, sleep disturbance has been suggested as a mediator of the relationship between respiratory symptoms and quality of life in patients with advanced lung cancer [77]. Overall, these findings suggest that emotional function and sleep may be related. However, it is important to recognize that these studies all report associations. A causal effect is therefore not established. For instance, pain may induce disturbed sleep, disturbed sleep may increase the experience of pain, or a shared factor may cause both disturbed sleep and increased pain.

Perpetuating factors include behavioral factors such as excessive daytime sleeping, and maladaptive cognitions, i.e., inaccurate appraisal of sleep [46]. Patients with insomnia might have several faulty beliefs and attitudes about sleep and sleepiness that may contribute to maintaining the sleep problem over time [10]. For cancer patients, this might lead to thoughts such as "If I don't sleep well, my cancer will come back" [10]. Savard and Morin suggested that maladaptive sleep habits, which develop in response to sleep disturbances, are the most salient factors in the maintenance of insomnia [10]. These factors are responsible for increasing physiological, cognitive, and emotional arousal and performance (the pressure to sleep) [62]. To overcome cancer-related fatigue, patients are often advised by health care providers to rest during the day. A single short afternoon nap may not have negative impact on night-time sleep. However, extensive daytime napping and increased time spent in bed might result in irregular sleep–wake patterns. Furthermore, the long-term consequences involve desynchronization of the sleep–wake cycle [10,62]. Thus, excessive daytime sleeping may contribute to the maintenance of insomnia.

6. Implications for Clinical Practice and Future Research

Overall, this review of sleep quality in patients with advanced cancer highlights the importance of sleep in several relevant areas for health care providers and researchers working in oncology and palliative care.

6.1. *Prevalence of Poor Sleep Quality*

Knowledge of the large proportion of patients experiencing poor sleep quality in advanced cancer is important for health care providers. This cohort represents a large number of patients all over the word. Thus, the high prevalence of poor sleep and the mixture of sleep disturbances in these patients call for awareness of sleep quality in daily routine care. Given the high level of physical and psychological symptoms in these patients, health care providers should be aware of the prognostic consequences of sleep disturbances. A recent review indicated that disturbed sleep during oncological treatment might be a relevant behavioral marker of poor cancer prognosis [78]. In detail, disturbances in sleep and sleep–wake activity immediately prior to or during treatment were associated with reduced overall survival, poorer response to treatment, and shorter time to progression [78]. Moreover, sleep disorder prevalence data might be helpful for future development of interventions in the treatment of sleep disturbances. Thus, in palliative care research, sleep prevalence data should be elicited from large samples of patients with advanced cancer to reflect the multitude of cancer-related factors that might affect sleep quality. Clinical studies should incorporate sleep questionnaires to advance the knowledge in this field and to improve care.

6.2. Cancer-Related Factors for Insomnia in Advanced Cancer

This review highlights important cancer-related factors that contribute to poor sleep quality in patients with advanced cancer. Despite previous findings of a relationship between sleep and cancer-related symptoms [53–55,60,69], it is difficult to predict which patients will develop sleep disturbances in palliative care. Individual factors, including both cancer-related and other symptoms as well as psychological factors, interfere with how the patient handles the cancer disease and how they manage their sleep. This also influences to what extent they report sleep disturbances as a problem and how much attention and treatment they want. Taken together, all of this has an impact on sleep quality. However, health care providers may use this knowledge to identify vulnerable patients with an increased risk of sleep problems. Moreover, findings on the associated and predictive factors of sleep quality are of importance when developing appropriate management and/or preventive strategies. Knowledge of how sleep affects daytime functioning is important in patients with life-threatening illness, as restorative sleep is necessary for healing, recovery, and to fight and resist infections [79].

To improve the scientific knowledge of sleep, and to identify risk factors for poor sleep quality in advanced cancer patients, it is useful to examine the relationship between sleep and cancer-related factors in palliative care research. However, interpretations of causality are not possible from cross-sectional designs, which uses estimation of association between variables. In cross-sectional studies it is difficult to establish if sleep problems are simply associated with cancer-related symptoms, or whether a sleep problem in itself elicits symptoms and should be the main target to alleviate symptoms like pain, depression, anxiety, and distress. Hence, the impact of poor sleep on daytime functioning is difficult to establish. To illustrate, the relationship between pain and sleep is reported as bidirectional and reciprocal [80]. Therefore, it might be useful to ask whether this represents a vicious circle in patients with advanced cancer, in which poor sleep quality affects daytime functioning, or even daytime symptom intensity, which in turn affects night-time sleep quality.

Thus, future research in palliative care should investigate daytime consequences of poor sleep quality in advanced cancer. Here, symptom clusters are relevant. Insomnia was recently identified, together with pain and emotional functioning, in terminally ill patients with cancer [81]. Another study identified insomnia as part of a neuropsychological cluster together with depression and anxiety [82]. Future studies should investigate if these symptoms could be treated concomitantly. Consequences of poor sleep will be important to establish, using longitudinal design, to avoid the risk that research on sleep in patients with advanced cancer is limited to correlational science in cross-sectional studies and the low level of evidence in clinical decision making from such studies.

6.3. Sleep Assessment

The knowledge of the complexity of sleep disturbance in advanced cancer, including difficulty initiating sleep, staying asleep, and early awakening is relevant for the understanding of how to categorize poor sleep to address each patient's individual sleep disturbances. Thus, to identify and treat patients with sleep disturbances, health care providers working in oncology and palliative care should routinely assess sleep problems [83].

Oncology nurses can play a leading role in addressing sleep problems, as they often spend more time with patients experiencing cancer-related symptoms than any other health professionals. Thus, it is of utmost importance that oncology nurses have knowledge of sleep assessment to provide good symptom control. However, a study on sleep assessment in patients living with cancer, discovered that few nurses assessed sleep patterns, undertook further assessment and investigations for patient's sleep problems, or reassessed the patients sleep patterns in case the patient complained of non-efficacy of the interventions [84].

Several reasons might explain the lack of sleep assessment among health care providers. Sleep disturbances might be considered a low-priority problem compared to the cancer itself, or because of a lack of sleep assessment protocols or guidelines [84,85]. Interestingly, a literature review on nurses' perceptions of sleep in the intensive care unit revealed that intensive care unit nurses lack a complete understanding of the importance of sleep [86]. Moreover, Ye et al. identified limited understanding of the importance of sleep during hospitalization, the lack of standardized assessment tools for sleep, lack of education in sleep evaluation, inadequate interdisciplinary communication, and lack of supportive hospital infrastructure as barriers to the effective management of sleep [87]. Knowledge about sleep and its physiology is, in many cases, based on personal experience and common sense rather than being evidence based [88]. Thus, one important aspect in sleep assessment is that health care providers have insufficient knowledge about sleep. Fortunately, the problems are now being addressed, and several studies argue that sleep should be a topic included in nursing education and training [89–91].

Although, beyond the scope of this review, we would like to mention that the management of sleep disturbances in patients with advanced cancer lack evidence-based knowledge in palliative care [46,48,92]. Nevertheless, it is important to be aware of this gap in knowledge. The sparse evidence and clear guidelines for treatment of sleep problems in patients with advanced cancer may contribute to health care personnel's reluctance to address sleep problems. In addition, it may also explain the lack of systematic assessment in the first place. One might question whether screening for sleep disturbances is meaningful when there is limited treatment to offer. Some will even find it unethical to systematically screen for sleep problems given the lack of evidence-based knowledge of pharmacological and non-pharmacological treatment in palliative care. On the other hand, we recommend that sleep problems should be assessed as an inherent part of most other prevalent symptoms in this group of patients.

Patients themselves might also contribute to the underassessment of sleep disturbances. Despite many patients expressing concerns about sleep, these problems are not always discussed with health professions during oncology appointments [93]. Patients do not usually report their sleep problems to health professions because in general it is seen as less significant than the cancer [94]. In addition, there is a perception by some patients that health professionals do not want to hear about it [94]. Some patients even believe that healthcare workers are too busy to treat such an insignificant problem [95]. To examine these thoughts might give valuable insights into how patients cope with sleep disturbances. A qualitative study revealed that patients with chronic heart failure used different self-care strategies to promote sleep [96]. However, these strategies were based on common knowledge, and did not follow any common methods. It is important that health care providers are informed about such self-care strategies. This is useful information in order to guide patients about the benefits of using more evidence-based approaches [97]. Thus, patients with advanced cancer should be asked about sleep. Patients should be encouraged to discuss their sleep problems with members of the healthcare team in palliative care.

Future studies on sleep in patients with advanced cancer care should combine PROMs and objective registrations of sleep. As the use of both actigraphy and PROMs is recommended in patients with advanced cancer, a sole use of PROMs can result in a lack of important information on sleep [30–32,98]. In addition, to gain knowledge of patients' perspective in palliative care research, clinical evaluation of insomnia should incorporate qualitative assessments of issues relevant to the patient's subjective experience of insomnia [99,100].

6.4. Treatment

The treatment plan for sleep disturbances in patients with advanced cancer should address the multifactorial and treatable causes. Thus, a combined stepwise pharmacological and non-pharmacological approach is recommended [92]. Symptom control should be the first step to remove the causative condition if possible (e.g., pain, dyspnoea, and anxiety) [101]. The second step should include non-pharmacological sleep interventions with cognitive and behavioral therapy for insomnia (CBT-I) [46]. This treatment incorporates cognitive and behavior-change techniques and targets dysfunctional attitudes, beliefs, and habits involving sleep [102]. Bright-light therapy is also used to improve sleep, but to our knowledge not formally tested in advanced cancer patients [103]. Short-term pharmacological treatment may be necessary until CBI-I takes effect or for those being refractory to CBT-I. A recent systematic review of the treatment of insomnia in palliative care identified hypnotics, antidepressants, and antihistamines as pharmacological treatment options of sleep disturbances [48]. However, evidence-based knowledge about the best pharmacological treatments for insomnia in patients with advanced cancer are scarce. For practical purposes, the palliative care network of Wisconsin has provided an overview of the pharmacological treatment of insomnia [104]. When pharmacological treatment is used, the choice of the specific agent within a class should be directed by factors such as symptom pattern, treatment goals, past treatment responses, and the presence and significance of contraindications [46].

7. Conclusions

The overall aim of this review is to contribute to evidence-based knowledge of sleep in patients with advanced cancer. The high prevalence of poor sleep quality and the mixture of sleep disturbances in these patients calls for awareness in health care providers. To identify and treat patients with sleep disturbances, health care providers should routinely assess sleep problems using PROMs. Patients with advanced cancer should be asked about sleep. More importantly, we should encourage patients to discuss their sleep problems and sleep-related concerns with formal and informal caregivers in hospital as well as at home. Further characterization of sleep disturbances in patients with advanced cancer is needed, with particular emphasis on contributing factors, such as cancer-related symptoms. Thus, more research using robust longitudinal designs with a comprehensive assessment of sleep is necessary. A better understanding of the relationship between cancer-related symptoms and sleep enhances the possibilities of developing more targeted interventions, which will increase the scientific basis for knowledge in the treatment of sleep disturbances in palliative and oncology care.

Author Contributions: Conception of the manuscript, G.J. and P.K.; initial draft, G.J.; review and editing, G.J., K.H.G., M.J.H. and P.K. All authors have read and agreed to the published version of the manuscript.

Funding: This research was funded by The Liaison Committee for education, research, and innovation in Central Norway between the Regional Health Authority and the Norwegian University of Science and Technology [46083200] and the Cancer Foundation, St. Olavs hospital, Trondheim University Hospital, Trondheim, Norway [15/9116-115/NISLIN].

Institutional Review Board Statement: Not applicable.

Informed Consent Statement: Not applicable.

Data Availability Statement: Not applicable.

Conflicts of Interest: The authors declare no conflict of interest.

References

1. Jordan, K.; Aapro, M.; Kaasa, S.; Ripamonti, C.I.; Scotte, F.; Strasser, F.; Young, A.; Bruera, E.; Herrstedt, J.; Keefe, D.; et al. European Society for Medical Oncology (ESMO) position paper on supportive and palliative care. *Ann. Oncol.* **2018**, *29*, 36–43. [CrossRef] [PubMed]
2. Teunissen, S.C.; Wesker, W.; Kruitwagen, C.; de Haes, H.C.; Voest, E.E.; de Graeff, A. Symptom prevalence in patients with incurable cancer: A systematic review. *J. Pain Symptom Manag.* **2007**, *34*, 94–104. [CrossRef] [PubMed]

3. Cleeland, C.S.; Zhao, F.; Chang, V.T.; Sloan, J.A.; O'Mara, A.M.; Gilman, P.B.; Weiss, M.; Mendoza, T.R.; Lee, J.W.; Fisch, M.J. The symptom burden of cancer: Evidence for a core set of cancer-related and treatment-related symptoms from the Eastern Cooperative Oncology Group Symptom Outcomes and Practice Patterns study. *Cancer* **2013**, *119*, 4333–4340. [CrossRef] [PubMed]
4. Klepstad, P.; Fladvad, T.; Skorpen, F.; Bjordal, K.; Caraceni, A.; Dale, O.; Davies, A.; Kloke, M.; Lundstrom, S.; Maltoni, M.; et al. Influence from genetic variability on opioid use for cancer pain: A European genetic association study of 2294 cancer pain patients. *Pain* **2011**, *152*, 1139–1145. [CrossRef] [PubMed]
5. Mystakidou, K.; Parpa, E.; Tsilika, E.; Pathiaki, M.; Gennatas, K.; Smyrniotis, V.; Vassiliou, I. The relationship of subjective sleep quality, pain, and quality of life in advanced cancer patients. *Sleep* **2007**, *30*, 737–742. [CrossRef] [PubMed]
6. Parker, K.P.; Bliwise, D.L.; Ribeiro, M.; Jain, S.R.; Vena, C.I.; Kohles-Baker, M.K.; Rogatko, A.; Xu, Z.; Harris, W.B. Sleep/Wake patterns of individuals with advanced cancer measured by ambulatory polysomnography. *J. Clin. Oncol.* **2008**, *26*, 2464–2472. [CrossRef] [PubMed]
7. Good, P.; Pinkerton, R.; Bowler, S.; Craig, J.; Hardy, J. Impact of Opioid Therapy on Sleep and Respiratory Patterns in Adults With Advanced Cancer Receiving Palliative Care. *J. Pain Symptom Manag.* **2018**, *55*, 962–967. [CrossRef]
8. National Cancer Institute. Dictionary of Cancer Terms. Available online: https://www.cancer.gov/publications/dictionaries/cancer-terms/def/advanced-cancer (accessed on 24 June 2022).
9. American Academy of Sleep Medicine. The International Classification of Sleep Disorders. In *Diagnostic and Coding Manual*, 3rd ed.; American Academy of Sleep Medicine: Darien, IL, USA, 2014.
10. Savard, J.; Morin, C.M. Insomnia in the context of cancer: A review of a neglected problem. *J. Clin. Oncol.* **2001**, *19*, 895–908. [CrossRef]
11. Grandner, M.A. Addressing sleep disturbances: An opportunity to prevent cardiometabolic disease? *Int. Rev. Psychiatry* **2014**, *26*, 155–176. [CrossRef] [PubMed]
12. Anothaisintawee, T.; Reutrakul, S.; Van Cauter, E.; Thakkinstian, A. Sleep disturbances compared to traditional risk factors for diabetes development: Systematic review and meta-analysis. *Sleep Med. Rev.* **2016**, *30*, 11–24. [CrossRef] [PubMed]
13. Ohayon, M.; Wickwire, E.M.; Hirshkowitz, M.; Albert, S.M.; Avidan, A.; Daly, F.J.; Dauvilliers, Y.; Ferri, R.; Fung, C.; Gozal, D.; et al. National Sleep Foundation's sleep quality recommendations: First report. *Sleep Health* **2017**, *3*, 6–19. [CrossRef] [PubMed]
14. Kvale, E.A.; Shuster, J.L. Sleep disturbance in supportive care of cancer: A review. *J. Palliat. Med.* **2006**, *9*, 437–450. [CrossRef] [PubMed]
15. Chen, D.; Yin, Z.; Fang, B. Measurements and status of sleep quality in patients with cancers. *Support. Care Cancer* **2018**, *26*, 405–414. [CrossRef] [PubMed]
16. Madsen, M.T.; Huang, C.; Gogenur, I. Actigraphy for measurements of sleep in relation to oncological treatment of patients with cancer: A systematic review. *Sleep Med. Rev.* **2015**, *20*, 73–83. [CrossRef] [PubMed]
17. Berger, A.M.; Wielgus, K.K.; Young-McCaughan, S.; Fischer, P.; Farr, L.; Lee, K.A. Methodological challenges when using actigraphy in research. *J. Pain Symptom Manag.* **2008**, *36*, 191–199. [CrossRef]
18. Langford, D.J.; Lee, K.; Miaskowski, C. Sleep disturbance interventions in oncology patients and family caregivers: A comprehensive review and meta-analysis. *Sleep Med. Rev.* **2012**, *16*, 397–414. [CrossRef] [PubMed]
19. Beck, S.L.; Schwartz, A.L.; Towsley, G.; Dudley, W.; Barsevick, A. Psychometric evaluation of the Pittsburgh Sleep Quality Index in cancer patients. *J. Pain Symptom Manag.* **2004**, *27*, 140–148. [CrossRef]
20. Carney, C.E.; Buysse, D.J.; Ancoli-Israel, S.; Edinger, J.D.; Krystal, A.D.; Lichstein, K.L.; Morin, C.M. The consensus sleep diary: Standardizing prospective sleep self-monitoring. *Sleep* **2012**, *35*, 287–302. [CrossRef]
21. Yennurajalingam, S.; Balachandran, D.; Pedraza Cardozo, S.L.; Berg, E.A.; Chisholm, G.B.; Reddy, A.; DeLa Cruz, V.; Williams, J.L.; Bruera, E. Patient-reported sleep disturbance in advanced cancer: Frequency, predictors and screening performance of the Edmonton Symptom Assessment System sleep item. *BMJ Support. Palliat. Care* **2017**, *7*, 274–280. [CrossRef]
22. Watanabe, S.M.; Nekolaichuk, C.; Beaumont, C.; Johnson, L.; Myers, J.; Strasser, F. A multicenter study comparing two numerical versions of the Edmonton Symptom Assessment System in palliative care patients. *J. Pain Symptom Manag.* **2011**, *41*, 456–468. [CrossRef]
23. Yennurajalingam, S.; Barla, S.R.; Arthur, J.; Chisholm, G.B.; Bruera, E. Frequency and characteristics of drowsiness, somnolence, or daytime sleepiness in patients with advanced cancer. *Palliat. Support. Care* **2019**, *17*, 459–463. [CrossRef] [PubMed]
24. Buysse, D.J.; Reynolds, C.F., 3rd; Monk, T.H.; Berman, S.R.; Kupfer, D.J. The Pittsburgh Sleep Quality Index: A new instrument for psychiatric practice and research. *Psychiatry Res.* **1989**, *28*, 193–213. [CrossRef]
25. Akman, T.; Yavuzsen, T.; Sevgen, Z.; Ellidokuz, H.; Yilmaz, A.U. Evaluation of sleep disorders in cancer patients based on Pittsburgh Sleep Quality Index. *Eur. J. Cancer Care* **2015**, *24*, 553–559. [CrossRef] [PubMed]
26. Mollayeva, T.; Thurairajah, P.; Burton, K.; Mollayeva, S.; Shapiro, C.M.; Colantonio, A. The Pittsburgh sleep quality index as a screening tool for sleep dysfunction in clinical and non-clinical samples: A systematic review and meta-analysis. *Sleep Med. Rev.* **2016**, *25*, 52–73. [CrossRef] [PubMed]
27. Bastien, C.H.; Vallieres, A.; Morin, C.M. Validation of the Insomnia Severity Index as an outcome measure for insomnia research. *Sleep Med.* **2001**, *2*, 297–307. [CrossRef]
28. Soldatos, C.R.; Dikeos, D.G.; Paparrigopoulos, T.J. Athens Insomnia Scale: Validation of an instrument based on ICD-10 criteria. *J. Psychosom. Res.* **2000**, *48*, 555–560. [CrossRef]

29. Lin, C.Y.; Cheng, A.S.K.; Nejati, B.; Imani, V.; Ulander, M.; Browall, M.; Griffiths, M.D.; Broström, A.; Pakpour, A.H. A thorough psychometric comparison between Athens Insomnia Scale and Insomnia Severity Index among patients with advanced cancer. *J. Sleep Res.* **2020**, *29*, e12891. [CrossRef]
30. Marino, M.; Li, Y.; Rueschman, M.N.; Winkelman, J.W.; Ellenbogen, J.M.; Solet, J.M.; Dulin, H.; Berkman, L.F.; Buxton, O.M. Measuring sleep: Accuracy, sensitivity, and specificity of wrist actigraphy compared to polysomnography. *Sleep* **2013**, *36*, 1747–1755. [CrossRef]
31. Jakobsen, G.; Engstrom, M.; Thronaes, M.; Lohre, E.T.; Kaasa, S.; Fayers, P.; Hjermstad, M.J.; Klepstad, P. Sleep quality in hospitalized patients with advanced cancer: An observational study using self-reports of sleep and actigraphy. *Support. Care Cancer* **2020**, *28*, 2015–2023. [CrossRef]
32. Palesh, O.; Haitz, K.; Levi, F.; Bjarnason, G.A.; Deguzman, C.; Alizeh, I.; Ulusakarya, A.; Packer, M.M.; Innominato, P.F. Relationship between subjective and actigraphy-measured sleep in 237 patients with metastatic colorectal cancer. *Qual. Life Res.* **2017**, *26*, 2783–2791. [CrossRef]
33. Bernatchez, M.S.; Savard, J.; Savard, M.H.; Aubin, M.; Ivers, H. Sleep-wake difficulties in community-dwelling cancer patients receiving palliative care: Subjective and objective assessment. *Palliat. Support. Care* **2018**, *16*, 756–766. [CrossRef] [PubMed]
34. Ma, C.L.; Chang, W.P.; Lin, C.C. Rest/activity rhythm is related to the coexistence of pain and sleep disturbance among advanced cancer patients with pain. *Support. Care Cancer* **2014**, *22*, 87–94. [CrossRef] [PubMed]
35. Boyne, K.; Sherry, D.D.; Gallagher, P.R.; Olsen, M.; Brooks, L.J. Accuracy of computer algorithms and the human eye in scoring actigraphy. *Sleep Breath. Schlaf Atm.* **2013**, *17*, 411–417. [CrossRef] [PubMed]
36. Sadeh, A. The role and validity of actigraphy in sleep medicine: An update. *Sleep Med. Rev.* **2011**, *15*, 259–267. [CrossRef] [PubMed]
37. Ancoli-Israel, S.; Martin, J.L.; Blackwell, T.; Buenaver, L.; Liu, L.; Meltzer, L.J.; Sadeh, A.; Spira, A.P.; Taylor, D.J. The SBSM Guide to Actigraphy Monitoring: Clinical and Research Applications. *Behav. Sleep Med.* **2015**, *13* (Suppl. 1), S4–S38. [CrossRef]
38. Ancoli-Israel, S.; Cole, R.; Alessi, C.; Chambers, M.; Moorcroft, W.; Pollak, C.P. The role of actigraphy in the study of sleep and circadian rhythms. *Sleep* **2003**, *26*, 342–392. [CrossRef]
39. Milanti, A.; Chan, D.N.S.; Li, C.; So, W.K.W. Actigraphy-measured rest-activity circadian rhythm disruption in patients with advanced cancer: A scoping review. *Support. Care Cancer* **2021**, *29*, 7145–7169. [CrossRef] [PubMed]
40. Luyster, F.S.; Strollo, P.J., Jr.; Zee, P.C.; Walsh, J.K. Sleep: A health imperative. *Sleep* **2012**, *35*, 727–734. [CrossRef]
41. Borbely, A.A.; Daan, S.; Wirz-Justice, A.; Deboer, T. The two-process model of sleep regulation: A reappraisal. *J. Sleep Res.* **2016**, *25*, 131–143. [CrossRef]
42. Innominato, P.F.; Roche, V.P.; Palesh, O.G.; Ulusakarya, A.; Spiegel, D.; Levi, F.A. The circadian timing system in clinical oncology. *Ann. Med.* **2014**, *46*, 191–207. [CrossRef]
43. Innominato, P.F.; Komarzynski, S.; Palesh, O.G.; Dallmann, R.; Bjarnason, G.A.; Giacchetti, S.; Ulusakarya, A.; Bouchahda, M.; Haydar, M.; Ballesta, A.; et al. Circadian rest-activity rhythm as an objective biomarker of patient-reported outcomes in patients with advanced cancer. *Cancer Med.* **2018**, *7*, 4396–4405. [CrossRef] [PubMed]
44. Al Maqbali, M.; Al Sinani, M.; Alsayed, A.; Gleason, A.M. Prevalence of Sleep Disturbance in Patients With Cancer: A Systematic Review and Meta-Analysis. *Clin. Nurs. Res.* **2022**, *31*, 1107–1123. [CrossRef] [PubMed]
45. Sateia, M.J.; Lang, B.J. Sleep and cancer: Recent developments. *Curr. Oncol. Rep.* **2008**, *10*, 309–318. [CrossRef] [PubMed]
46. Howell, D.; Oliver, T.K.; Keller-Olaman, S.; Davidson, J.R.; Garland, S.; Samuels, C.; Savard, J.; Harris, C.; Aubin, M.; Olson, K.; et al. Sleep disturbance in adults with cancer: A systematic review of evidence for best practices in assessment and management for clinical practice. *Ann. Oncol.* **2014**, *25*, 791–800. [CrossRef] [PubMed]
47. Divani, A.; Heidari, M.E.; Ghavampour, N.; Parouhan, A.; Ahmadi, S.; Narimani Charan, O.; Shahsavari, H. Effect of cancer treatment on sleep quality in cancer patients: A systematic review and meta-analysis of Pittsburgh Sleep Quality Index. *Support. Care Cancer* **2022**, *30*, 4687–4697. [CrossRef] [PubMed]
48. Nzwalo, I.; Aboim, M.A.; Joaquim, N.; Marreiros, A.; Nzwalo, H. Systematic Review of the Prevalence, Predictors, and Treatment of Insomnia in Palliative Care. *Am. J. Hosp. Palliat. Care* **2020**, *37*, 957–969. [CrossRef]
49. Ohayon, M.M.; Partinen, M. Insomnia and global sleep dissatisfaction in Finland. *J. Sleep Res.* **2002**, *11*, 339–346. [CrossRef]
50. Pallesen, S.; Sivertsen, B.; Nordhus, I.H.; Bjorvatn, B. A 10-year trend of insomnia prevalence in the adult Norwegian population. *Sleep Med.* **2014**, *15*, 173–179. [CrossRef]
51. Jakobsen, G.; Engstrom, M.; Fayers, P.; Hjermstad, M.J.; Kaasa, S.; Kloke, M.; Sabatowski, R.; Klepstad, P. Sleep quality with WHO Step III opioid use for cancer pain. *BMJ Support. Palliat. Care* **2019**, *9*, 307–315. [CrossRef]
52. Collins, K.P.; Geller, D.A.; Antoni, M.; Donnell, D.M.; Tsung, A.; Marsh, J.W.; Burke, L.; Penedo, F.; Terhorst, L.; Kamarck, T.W.; et al. Sleep duration is associated with survival in advanced cancer patients. *Sleep Med.* **2017**, *32*, 208–212. [CrossRef]
53. George, G.C.; Iwuanyanwu, E.C.; Anderson, K.O.; Yusuf, A.; Zinner, R.G.; Piha-Paul, S.A.; Tsimberidou, A.M.; Naing, A.; Fu, S.; Janku, F.; et al. Sleep quality and its association with fatigue, symptom burden, and mood in patients with advanced cancer in a clinic for early-phase oncology clinical trials. *Cancer* **2016**, *122*, 3401–3409. [CrossRef] [PubMed]
54. Mercadante, S.; Aielli, F.; Adile, C.; Ferrera, P.; Valle, A.; Cartoni, C.; Pizzuto, M.; Caruselli, A.; Parsi, R.; Cortegiani, A.; et al. Sleep Disturbances in Patients with Advanced Cancer in Different Palliative Care Settings. *J. Pain Symptom Manag.* **2015**, *50*, 786–792. [CrossRef] [PubMed]

55. Nishiura, M.; Tamura, A.; Nagai, H.; Matsushima, E. Assessment of sleep disturbance in lung cancer patients: Relationship between sleep disturbance and pain, fatigue, quality of life, and psychological distress. *Palliat. Support. Care* **2015**, *13*, 575–581. [CrossRef] [PubMed]
56. Davis, M.P.; Khoshknabi, D.; Walsh, D.; Lagman, R.; Platt, A. Insomnia in patients with advanced cancer. *Am. J. Hosp. Palliat. Care* **2014**, *31*, 365–373. [CrossRef] [PubMed]
57. Yennurajalingam, S.; Chisholm, G.; Palla, S.L.; Holmes, H.; Reuben, J.M.; Bruera, E. Self-reported sleep disturbance in patients with advanced cancer: Frequency, intensity, and factors associated with response to outpatient supportive care consultation—A preliminary report. *Palliat. Support. Care* **2013**, *13*, 135–143. [CrossRef] [PubMed]
58. Delgado-Guay, M.; Yennurajalingam, S.; Parsons, H.; Palmer, J.L.; Bruera, E. Association between self-reported sleep disturbance and other symptoms in patients with advanced cancer. *J. Pain Symptom Manag.* **2011**, *41*, 819–827. [CrossRef] [PubMed]
59. Mercadante, S.; Valle, A.; Cartoni, C.; Pizzuto, M. Insomnia in patients with advanced lung cancer admitted to palliative care services. *Int. J. Clin. Pract.* **2021**, *75*, e14521. [CrossRef] [PubMed]
60. Mystakidou, K.; Parpa, E.; Tsilika, E.; Pathiaki, M.; Patiraki, E.; Galanos, A.; Vlahos, L. Sleep quality in advanced cancer patients. *J. Psychosom. Res.* **2007**, *62*, 527–533. [CrossRef] [PubMed]
61. Graci, G. Pathogenesis and management of cancer-related insomnia. *J. Support. Oncol.* **2005**, *3*, 349–359. [PubMed]
62. Fleming, L.; Davidson, J.R. Sleep and Medical Disorders. In *The Oxford Handbook of Sleep and Sleep Disorders*; Morin, C.M., Espie, C., Eds.; Oxford University Press: New York, NY, USA, 2012; pp. 505–519.
63. Matthews, E.E.; Wang, S.Y. Cancer-Related Sleep Wake Disturbances. *Semin. Oncol. Nurs.* **2022**, *38*, 151253. [CrossRef] [PubMed]
64. Mogavero, M.P.; DelRosso, L.M.; Fanfulla, F.; Bruni, O.; Ferri, R. Sleep disorders and cancer: State of the art and future perspectives. *Sleep Med. Rev.* **2021**, *56*, 101409. [CrossRef] [PubMed]
65. Schwartz, W.J.; Klerman, E.B. Circadian Neurobiology and the Physiologic Regulation of Sleep and Wakefulness. *Neurol. Clin.* **2019**, *37*, 475–486. [CrossRef]
66. Young, J.S.; Bourgeois, J.A.; Hilty, D.M.; Hardin, K.A. Sleep in hospitalized medical patients, part 1: Factors affecting sleep. *J. Hosp. Med.* **2008**, *3*, 473–482. [CrossRef] [PubMed]
67. Yennurajalingam, S.; Frisbee-Hume, S.; Palmer, J.L.; Delgado-Guay, M.O.; Bull, J.; Phan, A.T.; Tannir, N.M.; Litton, J.K.; Reddy, A.; Hui, D.; et al. Reduction of cancer-related fatigue with dexamethasone: A double-blind, randomized, placebo-controlled trial in patients with advanced cancer. *J. Clin. Oncol.* **2013**, *31*, 3076–3082. [CrossRef] [PubMed]
68. Aaronson, N.K.; Ahmedzai, S.; Bergman, B.; Bullinger, M.; Cull, A.; Duez, N.J.; Filiberti, A.; Flechtner, H.; Fleishman, S.B.; de Haes, J.C.; et al. The European Organization for Research and Treatment of Cancer QLQ-C30: A quality-of-life instrument for use in international clinical trials in oncology. *J. Natl. Cancer Inst.* **1993**, *85*, 365–376. [CrossRef] [PubMed]
69. Ancoli-Israel, S.; Liu, L.; Marler, M.R.; Parker, B.A.; Jones, V.; Sadler, G.R.; Dimsdale, J.; Cohen-Zion, M.; Fiorentino, L. Fatigue, sleep, and circadian rhythms prior to chemotherapy for breast cancer. *Support. Care Cancer* **2006**, *14*, 201–209. [CrossRef] [PubMed]
70. Liu, L.; Rissling, M.; Natarajan, L.; Fiorentino, L.; Mills, P.J.; Dimsdale, J.E.; Sadler, G.R.; Parker, B.A.; Ancoli-Israel, S. The longitudinal relationship between fatigue and sleep in breast cancer patients undergoing chemotherapy. *Sleep* **2012**, *35*, 237–245. [CrossRef] [PubMed]
71. Laugsand, E.A.; Kaasa, S.; de Conno, F.; Hanks, G.; Klepstad, P. Intensity and treatment of symptoms in 3,030 palliative care patients: A cross-sectional survey of the EAPC Research Network. *J. Opioid Manag.* **2009**, *5*, 11–21. [CrossRef] [PubMed]
72. Meuser, T.; Pietruck, C.; Radbruch, L.; Stute, P.; Lehmann, K.A.; Grond, S. Symptoms during cancer pain treatment following WHO-guidelines: A longitudinal follow-up study of symptom prevalence, severity and etiology. *Pain* **2001**, *93*, 247–257. [CrossRef] [PubMed]
73. Mystakidou, K.; Parpa, E.; Tsilika, E.; Gennatas, C.; Galanos, A.; Vlahos, L. How is sleep quality affected by the psychological and symptom distress of advanced cancer patients? *Palliat. Med.* **2009**, *23*, 46–53. [CrossRef] [PubMed]
74. Renom-Guiteras, A.; Planas, J.; Farriols, C.; Mojal, S.; Miralles, R.; Silvent, M.A.; Ruiz-Ripoll, A.I. Insomnia among patients with advanced disease during admission in a Palliative Care Unit: A prospective observational study on its frequency and association with psychological, physical and environmental factors. *BMC Palliat. Care* **2014**, *13*, 1–12. [CrossRef] [PubMed]
75. Akechi, T.; Okuyama, T.; Akizuki, N.; Shimizu, K.; Inagaki, M.; Fujimori, M.; Shima, Y.; Furukawa, T.A.; Uchitomi, Y. Associated and predictive factors of sleep disturbance in advanced cancer patients. *Psychooncology* **2007**, *16*, 888–894. [CrossRef] [PubMed]
76. Palesh, O.G.; Collie, K.; Batiuchok, D.; Tilston, J.; Koopman, C.; Perlis, M.L.; Butler, L.D.; Carlson, R.; Spiegel, D. A longitudinal study of depression, pain, and stress as predictors of sleep disturbance among women with metastatic breast cancer. *Biol. Psychol.* **2007**, *75*, 37–44. [CrossRef] [PubMed]
77. Lou, V.W.; Chen, E.J.; Jian, H.; Zhou, Z.; Zhu, J.; Li, G.; He, Y. Respiratory Symptoms, Sleep, and Quality of Life in Patients With Advanced Lung Cancer. *J. Pain Symptom Manag.* **2017**, *53*, 250–256. [CrossRef] [PubMed]
78. Strøm, L.; Danielsen, J.T.; Amidi, A.; Cardenas Egusquiza, A.L.; Wu, L.M.; Zachariae, R. Sleep During Oncological Treatment—A Systematic Review and Meta-Analysis of Associations With Treatment Response, Time to Progression and Survival. *Front. Neurosci.* **2022**, *16*, 199. [CrossRef] [PubMed]
79. Tembo, A.C.; Parker, V. Factors that impact on sleep in intensive care patients. *Intensive Crit. Care Nurs.* **2009**, *25*, 314–322. [CrossRef] [PubMed]
80. Finan, P.H.; Goodin, B.R.; Smith, M.T. The association of sleep and pain: An update and a path forward. *J. Pain* **2013**, *14*, 1539–1552. [CrossRef] [PubMed]

81. Koyama, N.; Matsumura, C.; Tahara, Y.; Sako, M.; Kurosawa, H.; Nomura, T.; Eguchi, Y.; Ohba, K.; Yano, Y. Symptom clusters and their influence on prognosis using EORTC QLQ-C15-PAL scores in terminally ill patients with cancer. *Support. Care Cancer* **2022**, *30*, 135–143. [CrossRef] [PubMed]
82. Jiménez, A.; Madero, R.; Alonso, A.; Martínez-Marín, V.; Vilches, Y.; Martínez, B.; Feliu, M.; Díaz, L.; Espinosa, E.; Feliu, J. Symptom clusters in advanced cancer. *J. Pain Symptom Manag.* **2011**, *42*, 24–31. [CrossRef] [PubMed]
83. Bernatchez, M.S.; Savard, J.; Aubin, M.; Ivers, H. Correlates of disrupted sleep-wake variables in patients with advanced cancer. *BMJ Support. Palliat. Care* **2018**. Epub ahead of print. [CrossRef]
84. Khater, W.; Masha'al, D.; Al-Sayaheen, A. Sleep assessment and interventions for patients living with cancer from the patients' and nurses' perspective. *Int. J. Palliat. Nurs.* **2019**, *25*, 316–324. [CrossRef] [PubMed]
85. Siefert, M.L.; Hong, F.; Valcarce, B.; Berry, D.L. Patient and clinician communication of self-reported insomnia during ambulatory cancer care clinic visits. *Cancer Nurs.* **2014**, *37*, E51–E59. [CrossRef] [PubMed]
86. Nesbitt, L.; Goode, D. Nurses perceptions of sleep in the intensive care unit environment: A literature review. *Intensive Crit. Care Nurs.* **2014**, *30*, 231–235. [CrossRef] [PubMed]
87. Ye, L.; Keane, K.; Hutton Johnson, S.; Dykes, P.C. How do clinicians assess, communicate about, and manage patient sleep in the hospital? *J. Nurs. Adm.* **2013**, *43*, 342–347. [CrossRef]
88. Gellerstedt, L.; Medin, J.; Kumlin, M.; Rydell Karlsson, M. Nurses' experiences of hospitalised patients' sleep in Sweden: A qualitative study. *J. Clin. Nurs.* **2015**, *24*, 3664–3673. [CrossRef] [PubMed]
89. Gellerstedt, L.; Medin, J.; Kumlin, M.; Rydell Karlsson, M. Sleep as a topic in nursing education programs? A mixed method study of syllabuses and nursing students' perceptions. *Nurse Educ. Today* **2019**, *79*, 168–174. [CrossRef] [PubMed]
90. Ritmala-Castren, M.; Virtanen, I.; Vahlberg, T.; Leivo, S.; Kaukonen, K.M.; Leino-Kilpi, H. Evaluation of patients' sleep by nurses in an ICU. *J. Clin. Nurs.* **2016**, *25*, 1606–1613. [CrossRef] [PubMed]
91. Gellerstedt, L.; Medin, J.; Kumlin, M.; Rydell Karlsson, M. Nursing care and management of patients' sleep during hospitalisation: A cross-sectional study. *J. Clin. Nurs.* **2019**, *28*, 3400–3407. [CrossRef] [PubMed]
92. Induru, R.R.; Walsh, D. Cancer-related insomnia. *Am. J. Hosp. Palliat. Care* **2014**, *31*, 777–785. [CrossRef] [PubMed]
93. Flynn, K.E.; Shelby, R.A.; Mitchell, S.A.; Fawzy, M.R.; Hardy, N.C.; Husain, A.M.; Keefe, F.J.; Krystal, A.D.; Porter, L.S.; Reeve, B.B.; et al. Sleep-wake functioning along the cancer continuum: Focus group results from the Patient-Reported Outcomes Measurement Information System (PROMIS(®)). *Psychooncology* **2010**, *19*, 1086–1093. [CrossRef]
94. Davidson, J.R.; Feldman-Stewart, D.; Brennenstuhl, S.; Ram, S. How to provide insomnia interventions to people with cancer: Insights from patients. *Psychooncology* **2007**, *16*, 1028–1038. [CrossRef] [PubMed]
95. Cheng, K.K.; Yeung, R.M. Impact of mood disturbance, sleep disturbance, fatigue and pain among patients receiving cancer therapy. *Eur. J. Cancer Care* **2013**, *22*, 70–78. [CrossRef] [PubMed]
96. Gullvåg, M.; Gjeilo, K.H.; Fålun, N.; Norekvål, T.M.; Mo, R.; Broström, A. Sleepless nights and sleepy days: A qualitative study exploring the experiences of patients with chronic heart failure and newly verified sleep-disordered breathing. *Scand. J. Caring Sci.* **2019**, *33*, 750–759. [CrossRef] [PubMed]
97. Lundeby, T.; Hjermstad, M.J.; Aass, N.; Kaasa, S. Integration of palliative care in oncology-the intersection of cultures and perspectives of oncology and palliative care. *Ecancermedicalscience* **2022**, *16*, 1376. [CrossRef] [PubMed]
98. Grutsch, J.F.; Wood, P.A.; Du-Quiton, J.; Reynolds, J.L.; Lis, C.G.; Levin, R.D.; Ann Daehler, M.; Gupta, D.; Quiton, D.F.; Hrushesky, W.J. Validation of actigraphy to assess circadian organization and sleep quality in patients with advanced lung cancer. *J. Circadian. Rhythm.* **2011**, *9*, 1–12. [CrossRef] [PubMed]
99. Araújo, T.; Jarrin, D.C.; Leanza, Y.; Vallières, A.; Morin, C.M. Qualitative studies of insomnia: Current state of knowledge in the field. *Sleep Med. Rev.* **2017**, *31*, 58–69. [CrossRef]
100. Absolon, N.A.; Balneaves, L.; Truant, T.L.; Cashman, R.L.; Wong, M.; Hamm, J.; Witmans, M. A Self-Administered Sleep Intervention for Patients With Cancer Experiencing Insomnia. *Clin. J. Oncol. Nurs.* **2016**, *20*, 289–297. [CrossRef]
101. Hugel, H.; Ellershaw, J.E.; Cook, L.; Skinner, J.; Irvine, C. The prevalence, key causes and management of insomnia in palliative care patients. *J. Pain Symptome Manag.* **2004**, *27*, 316–321. [CrossRef]
102. Garland, S.N.; Johnson, J.A.; Savard, J.; Gehrman, P.; Perlis, M.; Carlson, L.; Campbell, T. Sleeping well with cancer: A systematic review of cognitive behavioral therapy for insomnia in cancer patients. *Neuropsychiatr. Dis. Treat.* **2014**, *10*, 1113–1124. [CrossRef]
103. Elliott, J.E.; McBride, A.A.; Balba, N.M.; Thomas, S.V.; Pattinson, C.L.; Morasco, B.J.; Wilkerson, A.; Gill, J.M.; Lim, M.M. Feasibility and preliminary efficacy for morning bright light therapy to improve sleep and plasma biomarkers in US Veterans with TBI. A prospective, open-label, single-arm trial. *PLoS ONE* **2022**, *17*, e0262955. [CrossRef]
104. Palliative Care Network of Wisconsin. Fast Facts #105. Insomnia: Drug Therapies. Available online: https://www.mypcnow.org/fast-fact/insomnia-drug-therapies/ (accessed on 8 August 2022).